Handbook of Analytical Science and Instrumentation

Volume I

Handbook of Analytical Science and Instrumentation

Volume I

Edited by **Sherley Benson**

NY RESEARCH PRESS

New York

Published by NY Research Press,
23 West, 55th Street, Suite 816,
New York, NY 10019, USA
www.nyresearchpress.com

Handbook of Analytical Science and Instrumentation: Volume I
Edited by Sherley Benson

International Standard Book Number: 978-1-63238-222-1 (Hardback)

Printed in China.

Contents

Preface

While our race is advancing, more and more people are being attracted by the pull of more cerebral lines of education and career. One might say that the analytical part of a part of the human race's mind in expanding by leaps and bounds. No wonder then that research and methodologies based solely on a foundation of being analytical are becoming popular. The study of ascertaining the chemical composition of any artificial or natural matter by using classical and modern scientific techniques is called Analytical Science. Analytical science is instrumental in helping scientists perform forensic analysis, processing and manufacturing quality control as well as clinical, environmental and material analysis. Branching off from this is the Analytical Method, a highly generic procedure which combines scientific analysis with the use of any kind of formal process to solve any given problem. Its steps include a detailed need to understand the problem, forming hypotheses, performing experiments and so on. Using the analytical method might turn out to be crucial to solve many problems if it looks like the current processes are unsatisfactory and inadequate. Analytical Instrumentation also derives from the analytical sciences and the scientific method. It refers to the analytical and detailed process of using a large class of instruments to examine materials and their composition, like mass spectrometers and ion analyzers.

This book is an effort to compile and collate data on such sciences, methods and instrumentation on a regular basis to keep the scientific community updated. I am thankful to those who supported us in this endeavor.

Editor

An Ion-Pair HPLC Method for Simultaneous Determination of Exogenous Phosphocreatine and Its Metabolite Creatine and Related ATP in Rabbit Plasma and RBC: Application to a Pharmacokinetic Study

Li Lv, Heng Xi, Guozhu Han*

Department of Pharmacology, College of Pharmacy, Dalian Medical University, Dalian, China.

ABSTRACT

A specific, precise and accurate ion-pair HPLC-UV method has been developed and validated for simultaneous determination of phosphocreatine (PCr), and its metabolite creatine (Cr) as well as related ATP in plasma and red blood cell (RBC) of rabbits. After addition of TMP as IS, the samples were deproteinized with 6% PCA. The analytes were separated on a Kromasil C18 column using a tertiary gradient mobile phase composed of buffer A (0.2% KH_2PO_4 + 0.08% tetrabutyl ammonium hydrogen sulphate, pH 3.0), buffer B (buffer A adjusted to pH 7.5 with 1 mol/L NaOH) and MeOH. Detection wavelengths were set at 210 nm for PCr and Cr and 260 nm for ATP and TMP. Some blank samples were initially run for baseline subtraction. The linear detection responses were obtained for PCr concentration over a range of 10 - 7500 μg/ml (plasma) and 5 - 2500 μg/ml (RBC) and for both Cr and ATP concentrations of 10 - 1500 μg/ml (plasma) and 5 - 750 μg/ml (RBC) ($r > 0.99$). The QC samples of 3 analytes showed intra-day and inter-day precisions (RSD) of <10%, accuracy of 97% - 107%. The method was successfully used to simultaneously determine plasma and RBC concentrations of the 3 analytes and to study pharmacokinetics after iv administration of PCr to rabbits.

Keywords: Ion-Pair HPLC; Phosphocreatine; Creatine; ATP; Plasma; RBC; Pharmacokinetics; Rabbit

1. Introduction

Phosphocreatine (PCr), namely creatine phosphate, is one of important high-energy phosphates in the body and plays a vital role in energy metabolism of body by acting as a storage form of energy used for re-synthesis of ATP. Exogenous PCr has long been developed as a cardio-protective drug and extensively used for adjunct therapy of heart failure, myocardial infarction, and as an additive ingredient in cardioplegia solution during heart surgical operation [1-4], and thereby recorded in The Extra Pharmacopoeia [5].

Although there are numerous preclinical and clinical studies on pharmacodynamics of exogenous PCr [2-4, 6-10], the considerable shortages of studies on pharmacokinetics are evident. In only 3 reports available currently on plasma pharmacokinetics of PCr [3,6,11], non-specific assays without chromatographic separation, including spectro-photometry, biolumilescence, were used to determine plasma PCr concentration, so that the reliability of the obtained pharmacokinetic results is limited. A specific bio-analytical method has become prerequisite in order to elucidate pharmacokinetic behaviors of PCr; but, to date, there is no report available in literature on pharmacokinetics of PCr investigated by means of specific HPLC assay. Besides, recent studies have shown that creatine, a metabolic product of PCr in body, has extensive physiological and pharmacological activities [12-14]. However, the kinetic process of bio-transformation from PCr to creatine has not been investigated. In addition, it was previously found in our lab that after iv administration of PCr to rats the hemorheology was significantly improved, theoretically, this was related to elevated ATP level in RBC. In view of these situations, this study was designed to develop a specific ion-pair

*Corresponding author.

HPLC-UV assay for simultaneous determination of PCr, its metabolite Cr and PCr-related ATP in plasma and RBC of rabbits, so as to study the pharmacokinetics of exogenous PCr and its metabolic disposition as well as its influence on ATP level.

2. Materials and Methods

2.1. Chemicals and Reagents

Test drug PCr (>97% purity), as creatine phosphate disodium tetrahydrate formulated as sterile powder for injection, was kindly provided by Harbin Bolai Pharmaceutical Co., Ltd (Harbin, China). Reference standards of PCr, Cr and ATP (>99% purity, respectively) were purchased from Merck (Germany and USA), trimethoprim (TMP) as an internal standard (IS) (99% purity) was obtained from National Institute for the Control of Pharmaceutical and Biological Products (Beijing, China). Other chemicals including tetrabutyl ammonium hydrogen sulphate (TBA) used as ion-pair reagent (JK Co., Ltd, USA), KH_2PO_4, NaOH, Perchloric acid (PCA) and K_2CO_3 were all of analytical grade. MeOH (Tedia, USA) was of HPLC grade. The bidistilled water was used for preparation of mobile phase and solutions.

2.2. Chromatographic Conditions

The liquid chromatographic system consisted of an Agilent 1100 HPLC Chromatograph equipped with a quaternary pump, an on-line degasser, a UV-detector and a refrigerated autosampler. The system control and data processing were performed on a Chemstation 32 software (Agilent, USA). The chromatographic separation was achieved on a Kromasil C_{18} analytical column (250 × 4.6 mm, 5 μm) (Ilite, Dalian, China) protected by a pre-column (20 × 4.6 mm, 10 μm) using a gradient mobile phase, composed of buffer A (0.2% KH_2PO_4 + 0.08% TBA, pH 3.0), buffer B (buffer A adjusted to pH 7.5 with 1 mol/L NaOH) and MeOH, coupled with a changed detection wavelength of 210 nm for PCr and Cr and 260 nm for ATP and TMP as well as a changed flow rate from 1.0 to 1.3 ml/min, which was detailed in **Table 1**.

2.3. Preparation of Standards

The separate stock solutions of PCr (75 mg/ml), Cr (15 mg/ml) and ATP (15 mg/ml) were prepared in water and then serially diluted to obtain respective serial working standard solutions of 100 - 75000 μg/ml for PCr, 100 - 15000 μg/ml for both Cr and ATP. The stock solution (5 mg/ml) of TMP as IS was prepared in MeOH.

The above stock solutions were all stored at −20°C, the serial working standard solutions were freshly prepared from respective stock solutions and stored at 4°C.

2.4. Collection of Plasma and RBC Samples

A 400 μl aliquot of rabbit blood sample was collected and centrifuged at 4°C (3000 r/min) for 10 min, the plasma at upper layer was withdrawn out and stored immediately at −20°C for plasma pharmacokinetic study. The remainder at lower layer containing RBC was washed with 500 μl of physiological saline and centrifuged at 4°C (3000 r/min) for 10 min, the same procedure was repeated twice to remove co-existing plasma and other substances, the washed RBC at lower layer was further treated with addition of equal volume of distilled water to yield RBC sample.

2.5. Pre-Treatment of Samples

An aliquot of 100 μl of plasma or RBC sample was transferred into an Eppendoff tube containing 10 μl of TMP solution and 30 μl of water, the resultant mixture was deproteinized with 360 μl or 100 μl of pre-cooled 6% PCA. After centrifuging (10000 r/min × 2 min) at 4°C to remove precipitate a 300-μl or 150-μl aliquot of the supernatant was transferred into another tube and immediately neutralized with pre-cooled 2 mol/L K_2CO_3 to pH 7.0. A volume of 20 μl of neutralized supernatant following centrifugation was injected into chromatograph for HPLC analysis. The above procedures were all conducted in ice-bath and all utilized solutions were pre-

Table 1. The detailed gradient mobile phase and detection.

Time (min)	Composition of mobile phase (%)			Flow rate (ml/min)	Curve	Detection wavelength (nm)
	Buffer A	*Buffer B*	*MeOH*			
0	100	0	0	1.0	6	210
8	100	0	0	1.0	6	210
10	0	85	15	1.2	6	260
18	0	85	15	1.2	6	260
20	0	80	20	1.3	6	260
29	0	80	20	1.3	6	260
30	100	0	0	1.0	6	210

cooled in ice bath.

2.6. Preparation of Calibration Curves

Seven portions of 100 µl blank plasma or blank RBC samples were spiked with 10 µl each of appropriate standard working solutions of the above 3 analytes and 10 µl of stock solution of IS, respectively, and vortex-mixed well to make plasma standard samples containing 7500, 2500, 500, 100, 50, 20, 10 µg/ml of PCr, and 1500, 750, 300, 100, 50, 20, 10 µg/ml of each of Cr and ATP, respectively, or RBC standard samples containing 2500, 500, 100, 50, 20, 10, 5 µg/ml of PCr, and 750, 300, 100, 50, 20, 10, 5 µg/ml of each of Cr and ATP, respectively. The above plasma or RBC standard samples were further treated with 6% PCA followed by neutralization as described under the section *Pre-treatment of samples*, the chromatogram and peak area were recorded. The above procedures were repeated 3 times. Meanwhile a blank plasma or RBC sample was run for baseline subtraction. Then, the peak areas of exogenous PCr, Cr and PCr-related ATP were calculated as peak area of plasma or RBC standard samples minus respective peak area of blank samples and were used to construct the calibration curves of the 3 analytes of exogenous origin based on peak area internal standard method. The weighed least square linear regression was used for calibration curve fitting with $1/C^2$ as weighing factor.

2.7. Validation of Methodology

This HPLC assay was validated using quality control (QC) samples containing PCr + Cr + ATP, which were prepared as described under section *preparation of calibration curve*, at high (2500 + 750 + 750 µg/ml), middle (100 + 100 + 100 µg/ml) and low (20 + 20 + 20 µg/ml) concentrations for 6 repeated analysis, separately.

The validation was performed by determining the intra-day and inter-day precisions expressed as relative standard deviation (R, S, D) for 6 determinations on a single assay day and 6 consecutive assay days, assay accuracy assessed as a percent of measured concentrations based on the regression equation of the calibration curve *vs* the spiked concentration, and recovery assessed by direct comparison of respective peak area obtained from PCA-extracted QC samples with those from unextracted standard solutions having concentration identical to the QC samples.

The stability of 3 analytes in plasma and RBC was evaluated by analyzing replicates (n = 3) of the QC samples at high and low levels as described above, which were exposed to different storage conditions (time and temperature), including −20˚C for up to 1 month and room temperature for up to 8 h for QC samples, 4˚C in auto-sampler for up to 24 h for PCA-extracted QC samples and −20˚C for up to 2 months for standard stock solutions. The results were compared with those obtained from freshly prepared samples. The analytes were considered to be stable in biological matrix when 90% - 110% of the initial concentrations were found.

2.8. Application of the Methodology

The present method was used to measure concentration-time profiles of PCr, its metabolite Cr and PCr-related ATP in plasma and RBC after intravenous administration of PCr at a dose of 500 mg/kg to 5 male rabbits. The blood samples were taken prior to dosage and at 2, 4, 6, 10, 15, 20, 30, 40, 60, 90, 120, 150 min post dose and then the plasma and RBC samples were prepared in the same way as described under section *collection of plasma and RBC samples*. Pharmacokinetic parameters were estimated by compartment and non-compartment analysis using 3p97 software package (version 1.1 PK software, China Mathematical Pharmacological Society).

3. Results and Discussion

3.1. Method Development

3.1.1. Chromatography Conditions

In spite of the reports on ion-pair HPLC-UV assay for determination of myocardial PCr, Cr, ATP and other adenine nucleotides having been found in the literature [15,16], the simultaneous determination of the above 3 compounds in the plasma and RBC is not available in literature to date. The establishment of chromatographic conditions is confronted with many difficulties: 1) PCr and Cr are short of chromophoric group, end absorption wavelength 210 nm has to be used for detection, this must increase the matrix interferences; 2) PCr and Cr are polar compounds and therefore have short retention time (t_r) compared with ATP having a long t_r of >2 h on the reversed phase C18 column; 3) The pH of mobile phase has great influence on chromatographic behaviors of these compounds, low pH favors detection of PCr and Cr but not ATP, which was eluted over 3 h when pH 3.0 mobile phase was used, however, high pH favors ATP but not PCr and Cr, both of which showed serious peak tailing and extended t_r, leading to a compromised peak resolution when pH 7.5 mobile phase was used. In view of the above, the isocratic mobile phase is not practicable, the complicated gradient mobile phase involving changes in composition, pH, wavelength and flow rate has to be used. As shown in **Table 1**, tertiary mobile phase was utilized where buffer A with pH 3.0 was initially run from 0 - 8.0 min for PCr and Cr, thereafter changed to buffer B with pH 7.5 for ATP and IS, meanwhile detection wavelength changed from 210 nm to 260 nm, correspondingly. In order to shorten t_r of IS and ATP, MeOH as organic modifier coupled with increased flow rate from 1.0 to 1.3 ml/min was used after Cr and PCr were eluted.

3.1.2. Choice of Internal Standard

For reliable quantification IS selection is very important, especially in this study, majority of studies on detection of myocardial high energy phosphates employed external standard method [15] except a study by Dr A. Ally, *et al.* who used AP_4 as IS [16]. However, AP_4 is not only expensive but also is eluted after ATP and has very long t_r, leading to an unacceptable long single run time. We chosen TMP as IS from many compounds due to not only its easy to be purchased and being cheap but also its desired t_r between PCr and ATP.

3.1.3. Sample Preparation

The analytes in this study have high polarity and are difficult to be extracted with organic solvents, accordingly, PCA-deproteinization method was used to prepare plasma and RBC samples, but PCr is unstable in acid, thus all procedures must be performed under low temperature at ice-salt bath. Also, the acid-extract must be neutralized immediately to pH 7.0. The pre-treatment results in a dilution effect, which in combination with the low molar extinction co-efficiency leads to a higher μg-level of limit of quantification (LOQ), fortunately, PCr is very low in toxicity and large dose has been clinically used in human, thus such a μg-level LOQ also suffices requirements for pharmacokinetic study.

3.1.4. Quantification of Exogenous PCr, Cr and PCr-Related ATP

The 3 analytes are all endogenous physiological substances, whereas the purpose of this study is to determine exogenous PCr, Cr and PCr-related ATP. At present time, pharmacokinetic study of exogenous physiological substances is still a difficult subject, the problem lies in how to differentiate them from corresponding endogenous substances. The best way is isotope-label of exogenous substances, however, this is not desired. A simple and easy to be performed method called baseline subtraction is currently extensively used [17,18]. In this study, a PCr, Cr and ATP-free blank plasma and RBC cannot be found, therefore the baseline subtraction method is used to construct calibration curve and to study pharmacokinetics of exogenous PCr and its metabolic disposition.

3.2. Specificity

Figure 1 showed representative HPLC profiles of a blank rabbit plasma sample and a blank plasma sample spiked with standard solution and IS and a plasma sample collected from a rabbit at 20 min after a single 500 mg/kg iv dose of PCr, indicating that no interference from endogenous substances at t_r of PCr, Cr, TMP and ATP, *i.e.*, 5.8, 3.4, 17.8 and 27.2 min, respectively. The single run time was 30 min (The HPLC chromatograms for RBC samples were similar to those for plasma, and omitted).

3.3. Linearity and Limit of Quantification

Table 2 presented related parameters of calibration curves, constructed by plotting peak area ratio of each analyte to IS vs analyte concentration, good linearity for the 3 analytes over the low and high concentration range were observed with r^2 being >0.995 or >0.999 and the LOQ of 10 μg/ml plasma and 5 μg/ml RBC for the 3 analytes, respectively.

3.4. Precision, Accuracy and Extraction Recovery

Analytical precision and accuracy data were shown in **Tables 3-5**. The intra- and inter-day precisions at low to high concentrations were better than 10% for the 3 analytes in plasma, and 7% in RBC. The intra- and inter-day accuracies were between 97% - 106% for the 3 analytes in plasma and between 97% - 107% in RBC. The extraction recovery in both plasma and RBC were all between 90% - 100% for 3 analytes and IS.

3.5. Stability of Plasma and RBC Samples

It was shown that the QC samples of plasma and RBC stored at −20°C were stable for at least 1 month, but unstable at room temperature, as reflected by only 87% of initial concentration being found at 2 h-storage for PCr and <90% at 4 h for ATP and Cr. The PCA-extracted plasma and RBC samples were stable for at least 24 h when stored in auto-sampler at 0 - 4°C. The stock solutions of 3 standards were found to be stable at least 2 months during storage at −20°C.

3.6. Assay Application

Figure 2 depicted the time courses of PCr and Cr in plasma and Cr and ATP in RBC of rabbits receiving a single iv dose of PCr 500 mg/kg. It could be seen that PCr disappeared from plasma in a bi-exponential fashion with a very short elimination half life ($t_{1/2\beta}$ 24.7 min), and the metabolite Cr was detected as early as 2 min after iv dosing PCr, and rapidly reached peak concentration at 20 min followed by a slow decay in concentration with a longer half life ($t_{1/2}$ 40.6 min). It was also evident that the PCr-related ATP was not detectable perhaps due to

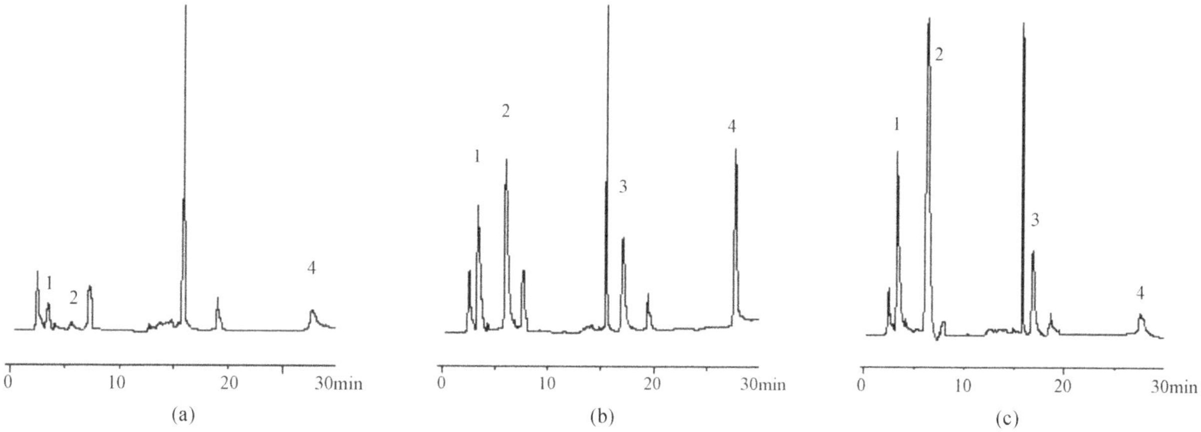

1. Cr (t_r 3.4 min); 2. PCr (t_r 5.8 min); 3. TMP (t_r 17.8 min); 4. ATP (t_r 27.2 min).

Figure 1. Representative chromatograms of rabbit blank plasma (a), a blank plasma sample spiked with standard solution and TMP (b) and a plasma sample collected from a rabbit at 20 min after a single 500 mg/kg iv dose of PCr and spiked with TMP (c).

Table 2. Linearity, LOQ and regression equations of 3 analytes (n = 3).

Biological sample	Analyte	Linearity range μg/ml	Regression equation	r^2	LOQ μg/ml
	PCr	10 - 7500	Y = 0.00236X − 0.00696	0.9987	10
Plasma	Cr	10 - 1500	Y = 0.00305X − 0.00492	0.9969	10
	ATP	10 - 1500	Y = 0.00335X + 0.00404	0.9981	10
	PCr	10 - 2500	Y = 0.00488X − 0.00049	0.9997	5
RBC	Cr	10 - 750	Y = 0.00487X − 0.00027	0.9998	5
	ATP	10 - 750	Y = 0.00394X + 0.00130	0.9989	5

Table 3. The precision and accuracy of 3 analytes in plasma ($\bar{x}$ ± SD, n = 6).

Analyte	Spiked conc (μg/ml)	Intra-day			Inter-day		
		Measured conc (μg/mL)	*RSD* (%)	*Accuracy* (%)	*Detected conc* (μg/ml)	*RSD* (%)	*Accuracy* (%)
PCr	20	19.87	3.76	99.35	19.74	4.20	98.70
	100	100.50	5.95	100.50	100.39	5.13	100.39
	2500	2549.82	3.94	101.99	2555.96	3.65	102.24
Cr	20	19.60	5.41	98.02	19.80	2.03	99.01
	100	99.19	4.51	99.19	97.86	5.72	97.86
	750	771.98	6.16	102.93	768.37	4.49	102.45
ATP	20	20.42	8.42	102.12	20.52	5.87	102.62
	100	105.32	4.61	105.32	104.68	4.11	104.68
	750	748.83	6.93	99.84	749.76	8.98	99.97

Table 4. The precision and accuracy of 3 anlytes in RBC ($\bar{x}$ ± SD, n = 6).

Analyte	Nominal conc (μg/ml)	Intra-day			Inter-day		
		Detected conc (μg/ml)	*RSD* (%)	*Accuracy* (%)	*Detected conc* (μg/ml)	*RSD* (%)	*Accuracy* (%)
PCr	10	10.07	2.87	100.66	10.09	6.02	100.88
	50	50.02	4.75	100.04	50.04	4.29	100.09
	500	531.17	3.95	106.23	521.65	4.53	104.33
Cr	10	10.15	3.66	101.53	10.07	4.74	100.72
	50	50.62	3.69	101.24	50.55	4.36	101.10
	300	297.04	5.61	99.01	296.60	2.61	98.87
ATP	10	10.35	3.63	103.45	10.50	3.00	105.04
	50	49.13	5.50	98.26	50.07	8.02	100.15
	300	293.63	6.41	97.88	293.18	4.76	97.73

Table 5. The extraction recovery of 3 analytes in plasma and RBC.

Analyte	Plasma			RBC		
	Spiked conc (μg/ml)	*Mean extraction recovery* (%)	*RSD* (%)	*Spiked conc* (μg/ml)	*Mean extraction recovery* (%)	*RSD* (%)
PCr	20	93.19	3.83	10	92.59	3.98
	100	93.11	2.45	50	93.31	2.89
	2500	93.65	2.89	500	93.86	2.37
Cr	20	92.71	3.06	10	92.15	1.98
	100	93.52	2.62	50	95.12	2.59
	750	93.70	3.53	300	94.29	2.21
ATP	20	93.36	4.1	10	93.15	5.21
	100	94.26	1.47	50	94.25	2.37
	750	92.49	3.96	300	93.51	2.18
TMP	500	95.04	1.53	500	98.50	0.39

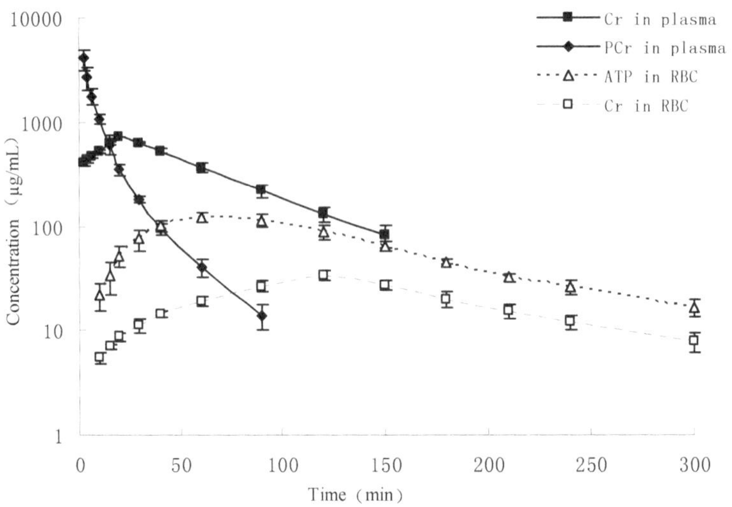

Figure 2. Mean plasma and RBC concentration-time profiles after intravenous administration of 500 mg/kg of PCr to rabbits (n = 5).

plasma being not an energy-consuming tissue.

It was shown that in the RBC no exogenous PCr could be detected, opposed to that metabolite Cr and PCr-related ATP were detectable as early as 2 min after iv dosing PCr and had t_{max} of 120 min and 67 min as well as C_{max} of 18 and 33.8 μg/ml and $t_{1/2}$ of 70.1 and 49.1 min, respectively. This indicated that plasma PCr was difficult to transport into RBC due to its high polarity, however, iv administration of PCr resulted in a significantly increased Cr and ATP in RBC. This significant increase in ATP provides a supportive evidence for improved hemorrheology after iv administration of PCr in rabbits.

4. Conclusion

An ion-pair reversed phase HPLC-UV assay coupled with a gradient mobile phase and changed detection wavelength, flow rate and buffer pH was developed for simultaneous determination of exogenous PCr, metabolite Cr and PCr-related ATP in plasma and RBC, and validated to be specific, precise and accurate. From the results obtained from rabbits receiving a 500 mg/kg iv dose of PCr, this new method has been demonstrated to allow correct quantification of the above 3 analytes by means of internal standard method and baseline subtraction method and thus to be able completely to meet the requirements for studies of pharmacokinetics and metabolic disposition of exogenous PCr in rabbits.

5. Acknowledgements

Thanks Harbin Bolai Pharmaceutical Co., Ltd (Harbin, China) for kindly providing PCr.

REFERENCES

[1] P. Kôrge, M. L. Silber and P. D. Gollnick, "Effect of Creatine Phosphate on the Contractile Activity in Acutely Failing Rat Heart," *Cardiologia*, Vol. 43, No. 12, 1998, pp. 1345-1354.

[2] V. A. Saks, I. V. Dzhaliashvili, E. A. Konorev, E. Strumia, "Molecular and Cellular Aspects of the Cardioprotective Mechanism of Phosphocreatine," *Biokhimiia*, Vol. 57, No. 12, 1992, pp. 1763-1784.

[3] V. G. Sharov, N. I. Afonskaya, M. Y. Ruda, N. M. Cherpachenko, EYa Pozin, R. A. Markosyan, II Shepeleva, M. B. Samarenko and V. A. Saks, "Protection of Ischemic Myocardium by Exogenous Phosphocreatine (Neoton): Pharmacokinetics of Phosphocreatine, Reduction of Infarct Size, Stabilization of Sarcolemma of Ischemic Cardiomyocytes, and Antithrombotic Action," *Biochemical Medicine and Metabolic Biology*, Vol. 35, No. 1, 1986, pp. 101-114.

[4] E. Strumia, F. Pelliccia and G. D'Ambrosio, "Creatine Phosphate: Pharmacological and Clinical Perspectives," *Advances in Therapy*, Vol. 29, No. 2, 2012, pp. 99-123.

[5] W. Martindale, "Martindale: The Extra Pharmacopocia," 31st Edition, The Royal Pharmaceutical Society, 1996.

[6] P. Kôrge, M. L. Silber and P. D. Gollnick, "Effect of Creatine Phosphate on the Contractile Activity in Acutely Failing Rat Heart," *Cardiologia*, Vol. 43, No. 12, 1998, pp. 1345-1354.

[7] P. Mastroroberto, L. Di Tommaso and M. Chello, "Creatine Phosphate Protection of the Ischemic Myocardium during Cardiac Surgery," *Current Therapeutic Research*, Vol. 51, No. 1, 1992, pp. 37-45.

[8] G. Prabhakar, L. Vona-Davis, D. Murray, P. Lakhani and G. Murray, "Phosphocreatine Restores High-Energy Phosphates in Ischemic Myocardium: Implication for Off-Pump Cardiacre Vascularization," *Journal of the American College of Surgeons*, Vol. 197, No. 5, 2003, pp. 786-791.

[9] MYa Ruda, M. B. Samarenko, N. I. Afonskaya and V. A. Saks, "Redtion of Ventricular Arrhythmias by Phosphocreatine (Neoton) in Patients with Acute Myocardial Infarction," *American Heart Journal*, Vol. 116, No. 2, 1988, pp. 393-397.

[10] M. L. Semenovsky, V. I. Shumakov, V. G. Sharov, G. M. Mogilevsky, A. V. Asmolovsky, L. A. Makhotina and V. A. Saks, "Protection of Ischemic Myocardium by Exogenous Phosphocreatine. II. Clinical, Ultrastructural, and Biochemicalevaluations," *Journal of Thoracic and Cardiovascular Surgery*, Vol. 94, No. 5, 1987, pp. 762-769.

[11] E. Lorenzi, G. Piacenza, E. Strumia and R. Borgoglio, "Pharmacokinetics of Phosphocreatine Following Intravenous Administration in Humans and Effect on Blood Levels of ATP," *Cardiologia*, Vol. 32, No. 9, 1987, pp. 1031-1034.

[12] M. Balestrino, M. Lensman, M. Parodi, L. Perasso, R. Rebaudo, R. Melani, S. Polenov and A. Cupello, "Role of Creatine and Phosphocreatine in Neuronal Protection from Anoxic and Ischemic Damage," *Amino Acids*, Vol. 23, No. 1-3, 2002, pp. 221-229.

[13] G. J. Brewer and T. W. Wallimann, "Protective Effect of the Energy Precursor Creatine against Toxicity of Glutamate and Beta-Amyloid in Rat Hippocampal Neurons," *Journal of Neurochemistry*, Vol. 74, No. 5, 2000, pp. 1968-1978.

[14] M. A. Tarnopolsky and M. F. Beal, "Potential for Creatine and Other Therapies Targeting Cellular Energy Dysfunction in Neurological Disorders," *Annals of Neurology*, Vol. 49, No. 5, 2001, pp. 561-574.

[15] T. Teerlink, M. Hennekes, J. Bussemaker and J. Groeneveld, "Simultanous Determination of Creatine Compoumds and Adenine Nucleotides in Myocardial Tissue by High-Performance Liquid Chromatography," *Analytical Biochemistry*, Vol. 214, No. 1, 1993, pp. 278-283.

[16] A. Ally and G. Park, "Rapid Determination of Creatine, Phosphocreatine, Purine Bases and Nucleotides (ATP, ADP, AMP, GTP, GDP) Inheart Biopsies by Gradiention-Pair Reversed-Phase Liquid," *Journal of Chromatography*, Vol. 575, No. 1, 1992, pp. 19-27.

[17] A. M. Persky, M. Müller, H. Derendorf, M. Grant, G. A. Brazeau and G. Hochhaus, "Single- and Multiple-Dose Pharmacokinetics of Oral Creatine," *Journal of Clinical Pharmacology*, Vol. 43, No. 1, 2003, pp. 29-37.

[18] M. Levine, C. Conry-Cantilena, Y. Wang, R. W. Welch, P. W. Washko, K. R. Dhariwal, J. B. Park, A. Lazarev, J. F. Graumlich, J. King and L. R. Cantilena, "Vitamin C Pharmacokinetics in Healthy Volunteers: Evidence for a Recommended Dietary Allowance," *Proceedings of the National Academy of Sciences USA*, Vol. 93, No. 8, 1996, pp. 3704-3709.

2

Application of Optical Methods in Blood Studies upon Evaluation of Severity Rate of Diffuse Liver Pathology

Margarita Kruchinina[1*], Mikhail Voevoda[1], Sergey Peltek[2], Svetlana Kurilovich[1], Abdrey Gromov[1], Vladimir Kruchinin[3], Sergey Rykhlitsky[3], Vladimir Volodin[3], Vladimir Generalov[4]

[1]Institute of Internal Medicine, SB RAMS, Novosibirsk, Russia; [2]Institute of Cytology and Genetics, SB RAS, Novosibirsk, Russia; [3]Rzhanov Institute of Semiconductor Physics, SB RAS, Novosibirsk, Russia; [4]State Research Center of Virology and Biotechnology, Ministry of Public Health and Social Development of the Russian Federation, Koltsovo, Russia.

ABSTRACT

Possible early diagnostic application of optical methods (dielectrophoresis, spectral and imaging ellipsometry, Fourier-transform infrared spectroscopy, Raman spectroscopy) in studies of red blood cells and serum of patients with diffuse liver disease with varying degrees of fibrosis has been evaluated. Application of combined optical methods was confirmed to significantly improve the performance of sensitivity, specificity, and accuracy index as well as to achieve the reliable results in diagnosis of both severe fibrosis and slight ulterior liver fibrosis. Identified diagnostic potential of optical methods can be effectively utilized in noninvasive screening evaluation of stages of diffuse liver disease of various geneses.

Keywords: Optical Methods; Blood Serum Studies; Erythrocytes; Diffuse Liver Disease; Degree of Fibrosis; Dielectrophoresis; Ellipsometry; Raman Spectroscopy; IR Fourier Spectroscopy

1. Introduction

Evaluation of the stage of the disease and the severity rate of liver fibrosis remains to be one of the key issues in verification of the diagnosis in patients with diffuse liver pathology as this evaluation predetermines prognosis and administration of adequate therapy. The issue of diagnosis of early stages of the liver fibrosis remains to be critical as they are known to be reversible [1]. Numerous methods of determining stages of liver fibrosis have been proposed, *i.e.* biochemical, macro-morphological, semiquantitative ones etc. [1,2]. Along with obvious advantages they also have some significant defects, *i.e.* a complicated sample preparation procedure, high cost of chemical agents and equipment, their specificity and insufficient sensitivity rate. Liver biopsy remains to be the "gold standard" in resolution of this issue. Even liver biopsy raises some issues of a subjective nature of the results obtained, the necessity to study several biopsy materials and frequent objections on the part of patients under study.

The optical methods applied in evaluation of the formed elements and blood serum are known to have some advantages that are significant for studies of biological objects. First of all, their high sensitivity rate when measuring low concentrations of bioorganic compounds in solutions or when applying these solutions to hard substrates is very crucial. Second of all, they do not destroy biological objects, numerous of which being of a complicated structure. Third of all, these methods enjoy high operational efficiency of measurements which do not require any special conditions (*i.e.* high vacuum, heating or cooling of a sample under study, special ionizing radiation, and the use of fluorescent tags).

The findings of previous studies of electrical and visco-elastic parameters of erythrocytes, ellipsometric indices of blood serum in patients with diffuse liver disease enabled to obtain the encouraging pilot results concerning successful diagnosis of the stage of the disease [3, 4].

This research aims to evaluate the potential value of application of optical methods in studies of erythrocytes and blood serum in early diagnosis of patients with diffuse liver disease with various degrees of fibrosis.

*Corresponding author.

2. Material and Methods

2.1. Study Groups

Total 62 males aged 38 - 69 with diffuse liver pathology who gave their written consent for involvement in the clinical trials were included into the study. Control comprised 21 patients aged 35 - 60 in whom biochemical and instrumental studies revealed no clinically manifested chronical diseases of inner organs and whose biochemical and hematological blood findings were normal.

The severity rate of fibrosis was confirmed by liver biopsy performed in 21 patients with diffuse liver diseases. In 41 patients this index was determined by visual macromorphologic studies (ultrasonography, computed tomography, nuclear magnetic resonance tomography). For ethical principles patients in control were not performed liver biopsy. All the patients under study were performed biochemical and instrumental examinations, ultrasonography of liver, spleen, and portal vessels. The viral aetiology of the disease was determined based on the presence/absence of serological markers by method of enzyme multiplied immunoassay (EME) and/or DNA and RNA of viruses by method of polymerase chain reaction (PCR); and the alcoholic one was diagnosed based on negative results of EME of viral hepatitis serological markers and confirmed by extensive alcohol consumption.

Liver biopsy was used to separately define the histological index of activity of chronic viral hepatitis and the degree of severity of liver fibrosis. The majority of the patients (14 subjects) were diagnosed to have a moderate degree of histological activity (from 9 to 12 points); in 5 patients histological activity was light (from 4 to 8 points) and in 2 males it was severe (from 13 to 18 points).

Based on the results of cumulative analysis three groups have been formed depending upon the degree of severity of liver fibrosis. Group 1 comprised control subjects (21) with no traces of liver fibrosis (F0). Group 2 included 39 patients with the first or second degree of fibrosis (slightly pronounced fibrosis; F1-F2). Group 3 comprised 23 patients with the third or fourth degree of fibrosis (severe fibrosis; F3-F4).

The patients with F1-F2 fibrosis were diagnosed to have chronic viral hepatitis (hepatitis C in 9 patients, hepatitis B in 15 patients, hepatitis of B + C genesis in 4 patients) with minimal (16 subjects) and moderate (12 subjects) degree of biochemical activity. 12 patients of this group would consume alcohol on the regular basis, *i.e.* they had a combined viral and alcohol-related etiology. The liver pathology genesis of 11 patients was determined as the alcoholic one; no markers of viral infection have been revealed in them. The histological activity of the process in this group ranged from 4 to 8 points (A1) that would speak in favor of fibrosis.

9 patients with an F3-F4 degree of fibrosis were diagnosed to have chronic viral C hepatitis; chronic viral B hepatitis was confirmed in 14 subjects. In 10 of the above 23 patients the chronical viral infection was associated with a long-term alcohol abuse. In the majority of patients with liver cirrhosis (*i.e.* hepatitis of the fourth degree) the compensation factor by Child-Pugh corresponded to 8 - 10 points (class B), and in 3 individuals it ranged from 12 to 14 points (class C). Biopsy studies of 11 patients revealed a moderate degree of histological activity that ranged from 9 to 12 points (A2) and in 3 patients it was strongly pronounced and ranged from 13 to 18 points (A3).

The trial was performed in compliance with the approval granted by the Committee of Biomedical Ethics with the Institute of Internal Medicine of the Siberian Branch of the Russian Academy of Medical Sciences. All the subjects under study would fill out a standard form, *i.e.* the Informed Written Patient's Consent for participation in the trial and the standard form about the history of alcohol consumption.

2.2. Dielectrophoresis Researches

The patients with verified diagnosis and males of control were examined their electric as well as viscoelastic parameters of erythrocytes by method of dielectrophoresis (DEPh) in nonhomogeneous alternating electric field (UAEF) with the help of an automated specialized plant applying an electro-optical system of cells detection [5]. The following parameters were evaluated: conductivity of membranes, indices of aggregation and destruction of erythrocytes, cellular membrane capacity, velocity of erythrocyte motion to electrodes, position of synchronous frequency, amplitude of deformation of erythrocytes, polarizability of cells, summarized rigidity and viscosity, values of induced dipole moment and charge.

The *CELLFIND* original software package was applied in computer processing of the data obtained.

2.3. Processing of Thin Films from Blood Serum

The centrifugation method was applied in this study to obtain thin films from blood serum [6,7]. Techniques of films processing and evaluation of their quality have been thoroughly described before [3]. The samples of fresh blood serum (~0.1 ml) were applied onto the central part of silicon wafers Si(111) KDB-10 of 76 mm in diameter preliminary washed in ammonia-peroxide etching solution at room temperature. The rotation velocity was 2500 r/min; the spin time equalled to 10 seconds. Prior to the ellipsometric characterization the films had been dried in the open air at room temperature within 2 hours. All the patients under study underwent the ellipsometric characterization of blood serum by methods of spectral and scanning ellipsometry.

2.4. Ellipsometric Study of Blood Serum Films

Ellipsometric spectra $\Psi(\lambda)$ and $\Delta(\lambda)$ were measured by means of "*ELLIPS*-1771" spectral ellipsometric complex (*SB RAS Institute of Semiconductor Physics*) [8] within the range of wave length of $\lambda = 250 - 1000$ nm, the spectral resolution of apparatus being 2 nm, the recording time of one spectrum did not exceed 20 seconds; the angle of incidence of a light beam onto the sample was 70°. A four-zone measuring was applied, with further averaging for all the four zones. Computation of dispersion response of retraction index $n(\lambda)$ and the values of film thicknesses have been achieved by solving the inverse problem of ellipsometry applying the optical model, *i.e.* substrate-film-medium. The below experimental parameters of films have been estimated: the value of net thickness (d) and dispersion of $n(\lambda)$ retraction index [3].

To study the homogeneity degree of coating the silicon wafers with serum films on the surface have been measured with the "*Microscan*" high resolution scanning ellipsometer (*ISP SB RAS*). Reading of ellipsometric Ψ and Δ angles was done in the central zones of the wafers (Ø76 mm) of 15×15 mm^2 in size, the pitch on "x" and "y" axes being 0.1 mm. The angle of incidence of light beam on the sample equaled to 60°.

2.5. IR Spectroscopy

Registration of the IR spectra was performed using the *Bruker Vector* 22 IR-Fourier-spectrometer at the wave length ranging from 400 to 7000 cm^{-1} with resolution of 2 cm^{-1}. The *OPUS* 6.0 software was used to remove the oscillations connected to multiple-beam interference and the lines of water vapor and carbon dioxide absorption.

2.6. Raman Spectroscopy

Raman scattering of light of patients blood liquid serum was registered at room temperature by means of the *T*64000 spectrometer (*Horiba Jobin Yvon*) with the triple monochromator. The Ar^+ line of a laser with the wave length of 514.5 nm was used for excitation, the spectral resolution being 1.5 cm^{-1}. A silicon photosensor array cooled by liquid nitrogen was used as a detector. An additional attachment was used for microscopic studies of Raman scattering of light. The power of laser beam reaching the sample comprised 2 - 3 mW, the spot size equalling 5 - 6 μm.

2.7. Statistical Treatement of Data Obtained

The statistical treatment of data obtained was performed using the *SPSS*, *ver.*10 software. Significance of indices difference was computed using the Student and Pearson criteria (in case of normal distribution); in cases of deviation of distribution from the normal values the non-parametric criteria were used (U-criterion of Mann-Whitney, Kholmogorova-Smirnova).

3. Results and Discussion

Evaluation of parameters of erythrocytes by means of electrooptic systems revealed that at high frequencies (5×10^5 Hz and 5×10^6 Hz) erythrocytes of healthy subjects were moving to the closest electrode with high forward velocity (positive dielectrophoresis) and displayed a clearly marked ability to deform under the influence of non-uniform alternating electric field (NUAEF). No deformation of cells was registered at low frequencies (5×10^4 Hz, 5×10^5 Hz). Erythrocytes would push off the electrodes (negative dielectrophoresis), some single cells being destroyed by the field.

Interaction of erythrocytes with NUAEF in patients with diffuse liver pathology (DLP) tended to differ greatly, *i.e.* at high frequencies the amplitude of deformation (at the background of summarized indices of viscosity and rigidity), the value of dipole moment, velocity of the forward motion of cells against the electrodes tended to be significantly lower as compared to control ($p < 0.001 - 0.03$) (**Figure 1**).

Excessive hemolysis of cells was observed at all frequencies of the range under study ($5 \times 10^4 - 5 \times 10^6$ Hz). It

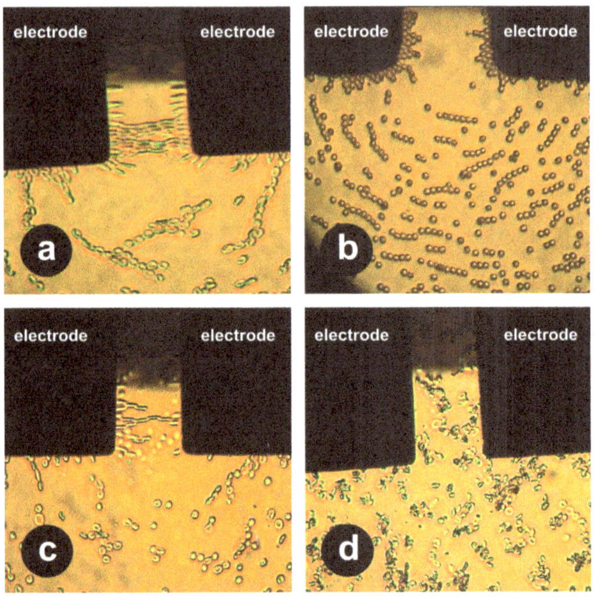

Figure 1. Division and deformation of erythrocytes influenced by nonhomogeneous alternating electric field at frequency of 1 MHz: (a) High deformation amplitude in control; (b) Low deformation amplitude in patients with severe liver fibrosis F3-F4; (c) Increased hemolysis of erythrocytes in patients with severe liver fibrosis F3-F4; (d) Increased development of aggregation in patients with severe liver fibrosis F3-F4.

matched the enhanced capability of erythrocytes to form large and medium size aggregates after transferring of erythrocytes into the 0.3 M sucrose solution ($p < 0.001$) (**Figure 2**).

Along with aggravation of fibrosis (from F1-F2 to F3-F4) erythrocytes of the patients with DLP tended to become more rigid with lower capacity of deformation, tending to form aggregates. These erythrocytes had lower resistance to stress factors of various intensity, and therefore they would easily get destroyed by NUAEF both at low and high frequencies. Reduced biological activity of red blood cells correlated with the reduced levels of polarization of cells at all frequencies ($r = 0.57$; $p < 0.02$). The membranes of erythrocytes would thicken (that displayed significant decrease of the level of electric capacity; $p < 0.03$. They conducted electrical current well (*i.e.* they possessed high electrical conductivity), although the walls of red blood cells would normally be dielectric. The reduced superficial negative charge of erythrocytes in cases of severe F3-F4 fibrosis demonstrated significantly lower indices of velocity of cells motion to the electrodes and the ones of the dipole moment ($p < 0.01 - 0.05$) (**Table 1**).

Thus, it has been determined that the presence of diffuse liver pathology causes depolarization of erythrocytes. In this case the amplitude-and-frequency-related characteristics of cells in NUAEF serve as a good marker of physical and chemical properties of its membrane and cytoplasm as well as display the condition of its biological activity.

Summarized evaluation of the findings of analysis of viscoelastic and electrical parameters of erythrocytes by method of dielectrophoresis demonstrated a 71.4% sensitivity rate and a 6% specificity of the method as compared to the biopsy findings in diagnosis of various stages of liver cirrhosis.

Special attention was paid to the significance of a positive ("+") prognostic value of the result of the method, *i.e.* F3-F4 fibrosis was successfully diagnosed in 83.3% of cases, and the accuracy index equaled to 68.4%. Nevertheless, the negative ("–") prognostic value of the result of the method aimed to diagnose the absence of

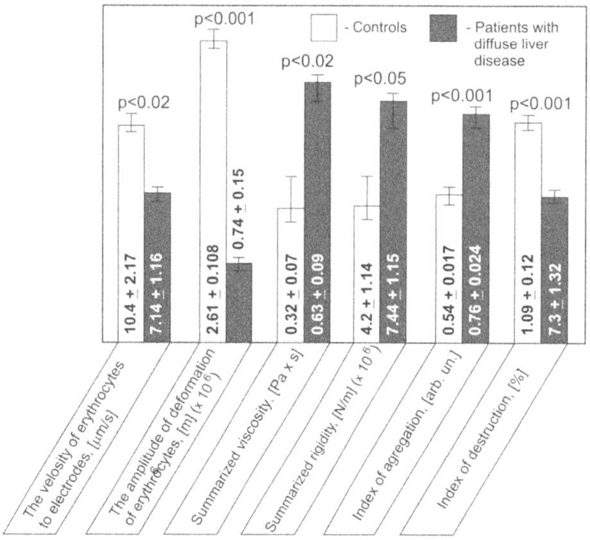

Figure 2. Electric and visco-elastic parameters of erythrocytes in patients with diffuse liver disease and control.

Table 1. Viscoelastic and electrical properties of red blood cells (Er) in patients with diffuse liver disease with various degree of fibrosis (M ± m).

Er indices	Group 1 fibrosis state (F0) n = 21	Group 2 fibrosis state (F1-F2) n = 39	Group 3 fibrosis state (F3-F4) n = 23
The amplitude deformation of Er, [m]	$(2.75 \pm 0.53) \times 10^{-6}$	$(0.92 \pm 0.28) \times 10^{-6**}$	$(0.61 \pm 0.05) \times 10^{-6***\wedge}$
Total rigidity, [N/m]	$(4.05 \pm 0.79) \times 10^{-6}$	$(7.4 \pm 1.5) \times 10^{-6*}$	$(12.1 \pm 1.4) \times 10^{-6***\wedge\wedge}$
Total viscosity, [Pa·sec]	0.34 ± 0.07	$0.67 \pm 0.14^{*}$	$0.91 \pm 0.13^{***\wedge}$
Index of Er aggregation [con.un.]	0.54 ± 0.05	$0.72 \pm 0.04^{**}$	$0.82 \pm 0.04^{***\wedge}$
Index of Er destruction [%]	1.12 ± 0.13	$6.1 \pm 1.5^{**}$	$8.4 \pm 2.3^{**}$
Polarization [m^3] (at 10^6 Hz)	$(9.78 \pm 0.27) \times 10^{-15}$	$(6.43 \pm 0.97) \times 10^{-15***}$	$(4.90 \pm 0.87) \times 10^{-15***\wedge}$
Electric conductivity [Cm/m]	$(2.89 \pm 0.42) \times 10^{-5}$	$(6.33 \pm 0.55) \times 10^{-5**}$	$(8.53 \pm 0.49) \times 10^{-5***\wedge\wedge}$
Cellular membrane capacity [F]	$(6.97 \pm 0.46) \times 10^{-14}$	$(5.07 \pm 0.37) \times 10^{-14**}$	$(3.2 \pm 0.41) \times 10^{-14***\wedge\wedge}$
Velocity of Er motion, [mcm/sec]	10.4 ± 2.5	6.9 ± 2.1	$4.9 \pm 2.0^{*}$
Position of crossover frequency Hz]	$(4.96 \pm 0.47) \times 10^{5}$	$(6.94 \pm 0.49) \times 10^{5**}$	$(9.5 \pm 0.7) \times 10^{5***\wedge\wedge}$
Dipole moment [Kl m]	$(14.2 \pm 3.7) \times 10^{-22}$	$(6.4 \pm 1.7) \times 10^{-22*}$	$(3.8 \pm 1.0) \times 10^{-22**}$

Note 1: M stands for mean values, m-stands for mean error, *statistical significance of differences in Group I ($^{*}p < 0.05$, $^{**}p < 0.02$, $^{***}p < 0.0001$); $^{\wedge}$statistical significance of differences in Group II ($^{\wedge}p < 005$, $^{\wedge\wedge}p < 0.02$, $^{\wedge\wedge\wedge}p < 00001$); Note 2. Dipole moment was calculated at electric field tension of 8.85×10^{-12} F/m.

fibrosis, and potentially its early stages turned out to be rather low, *i.e.* 42.9%.

The need to increase the levels of specificity and sensitivity tfor diagnosis of the degree of liver fibrosis predetermined the necessity to apply such optical methods as ellipsometry, IR-spectroscopy and Raman spectroscopy of blood serum of the patients under study.

Along with aggravation of the liver disease more pronounced changes in blood serum occur. They are due to dysfunction of hepatic protein synthesizing function, aggravation of syndromes of cytolysis, cholestasis, and the immune inflammatory one. Similar disturbances cannot but affect the optical characteristics of blood serum that was demonstrated in our previous pilot studies [3].

Evaluation of ellipsometric indices of thin film findings obtained from blood serum of patients with various degrees of severity of liver fibrosis enabled to reveal significant increase of retraction index in combination with decrease of film thickness along with fibrosis increase ($p < 0.001$ - 0.002) (**Figure 3**).

In cases of severe fibrosis similar changes can be attributed to reduction of synthesis of the number of compounds in the liver (*i.e.* blood serum protein, coagulation factors, cholesterol, triglycerides and others) and they become crucial for the film thickness (correlation for the levels of total cholesterol $r = 0.56$, $p < 0.001$; triglycerides $r = 0.44$, $p < 0.01$; total protein $r = 0.57$, $p < 0.02$). Prevailance of nonhomogeneous films in cases of severe fibrosis can be obviously attributed to the degree of biochemical disbalance in blood serum induced by liver cirrhosis. The index of refraction is to a great degree determined by the components of cytlolytic syndromes (for ALT $r = 0.46$, $p < 0.03$, for AST $r = 0.41$, $p < 0.05$) and the ones of cholestasis (for total bilirubin $r = 0.52$, $p < 0.01$, for GGT $r = 0.48$, $p < 0.033$); the degree of intensity of the components tends to increase in severe fibrosis [2].

Evaluation of homogeneity of the surface of blood serum films by means of the "Microscan" high resolution scanning elipsometer revealed that the films obtained from the patients with F3-F4 liver fibrosis tended to be extremely nonhomogeneous in their thickness and had large breaks as compared to the ones of patients with F1-F2 fibrosis ($p < 0.05$) (**Figure 4**).

Such changes were determined based on significant dislocations on the IR-spectra, where the peak areas and intensity determining the secondary structure of proteins are registered. Special attention was paid to evaluation of the Amide I, Amide II, Amide III amide groups which are connected with absorption of IR-radiation by the O=C-N-H fragments, the latter being sensible to the secondary structure of protein molecules (α-helix, β-sheet, random coil). Changes of the secondary structure of protein molecules, in their turn, are judged by the presence of pathological process in the organism, including the diffuse liver pathology with various degree of fibrosis.

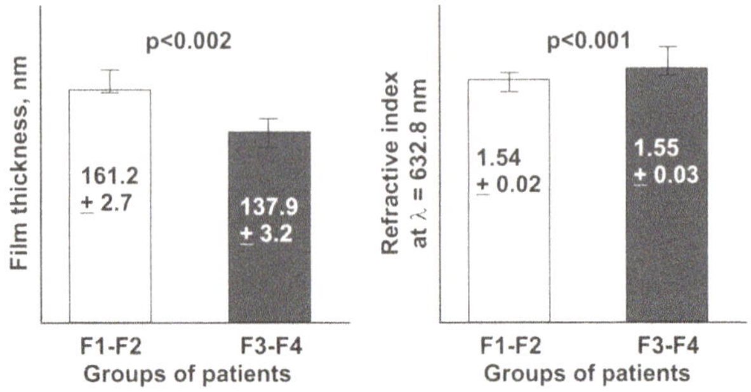

Figure 3. Ellipsometric indices in patients with different degree of liver fibrosis.

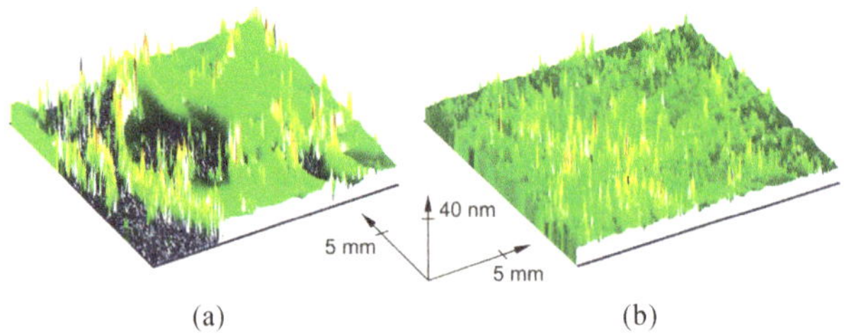

Figure 4. Thickness distribution of three-dimensional thin films on the surface of plates processed from blood serum of patients with severe F3-F4 liver fibrosis (a) and light F1-F2 liver fibrosis (b).

IR-spectra of patients with severe F3-F4 liver fibrosis demonstrated significantly higher peaks of amide III at 1280 cm^{-1} in spectrum range of essential amide absorption bands and the peak at 3190 cm^{-1} reflecting the N-H structures as compared to the ones in patients with F1-F2 fibrosis. Also, the above spectra revealed a significantly more frequent presence of protein molecules with secondary spiral-shaped structure (α-helix) ($p < 0.001$ - 0.05) (**Figure 5**).

The patients with F3-F4 fibrosis in contrast to the ones with F1-F2 fibrosis demonstrated significantly lower peaks at 776, 818, 889 cm^{-1} and 966 cm^{-1} characteristic for the C-C, C-O bond stretches of deoxyribose and the PO_4^- fragment of molecules of nucleic acids [9] ($p <$ 0.02 - 0.05). Significant decrease in cases of pronounced fibrosis was demonstrated by some peaks at 1399 cm^{-1} reflecting symmetrical valent vibrations of CH_3 groups of skeletal fragments of protein molecules; peaks at 1450 cm^{-1} reflected deformed oscillations of CH_3 groups of lipids and proteins; peaks at 1570 cm^{-1} (amide II); 1581, 1620 cm^{-1} (carbonile structures) ($p < 0.01$ - 0.05). A significant decrease of the component at 1635 cm^{-1} revealed upon breakdown of complex amide peaks in patients with F3-F4 liver fibrosis is attributed to the reduction of the amount of protein molecules possessing a structure in the form of a folded sheet (β-sheet) ($p < 0.05$).

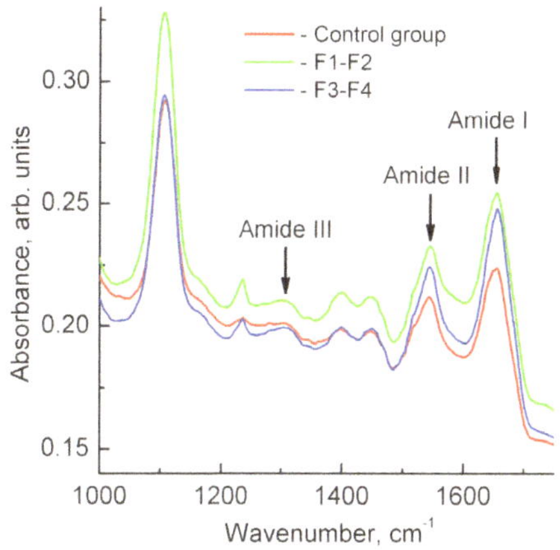

Figure 5. IR-spectra of patients blood serums with different degree of liver fibrosis and of experimental groups (fragments of IR-spectra with various degree of liver fibrosis.

In spectra of the combined light dispersion belonging to samples of liquid blood serum in the range of 750 - 1750 cm^{-1} one could observe the three typical absorption peaks, *i.e.* at 1005, 1157 and 1520 cm^{-1}. Their intensity rate tended to be significantly lower in cases of F3-F4 fibrosis as compared to the ones of F1-F2 ($p < 0.0058$ - 0.02) (**Table 2**, **Figure 6**).

No shared opinion has been expressed yet in modern literature on the origin of these peaks; however the majority of authors attribute them to carotins. For example, the authors of [10,11] reported that in oncological patients reduction of content of carotin in blood serum was registered that results in rapid fall of signal intensity at 1005, 1157 and 1520 cm^{-1} in Raman-spectrum. In case of diffuse liver pathology evaluated in this we can assume that reduction of areas of the above specified peaks in cases of pronounced finbrosis is connected with significant changes in carotin metabolism, their transition into the A vitamin, that is closely connected with the liver function [1,2].

An interesting experimental fact, observed in this study, was associated with the effect of the laser beam on blood serum of patients with various degree of fibrosis when conducting the spectroscopy of combined light dispersion. The patients with light form of fibrosis or without it demonstrated a nonhomogenous surface of their serum. Optic miscroscopic study revealed separate clearly defined areas of less than 0.5 μm.

Laser irradiation caused the interflowing and consolidation of areas up to 10 μm and more (**Figures 7(a)** and **(b)**). In case of severe fibrosis the areas under study remained practically unchanged (**Figures 7(c)** and **(d)**).

Local heating of blood serum by a laser beam can cause some changes associated with conformational changes. A more decisive influence associated with denaturation of blood serum protein molecules and fusion

Table 2. Peak areas of the Raman spectra in patients with diffuse liver disease of various degree of fibrosis (M ± m), p-statistical significance.

Peak position, cm^{-1}	Peak areas, arb. units		p <
	Patients group with fibroses stage F1-F2	Patients group with fibrosis stage F3-F4	
1005	840 ± 133	357 ± 90	0.02
1157	2257 ± 411	917 ± 114	0.015
1520	2750 ± 174	712 ± 110	0.0058

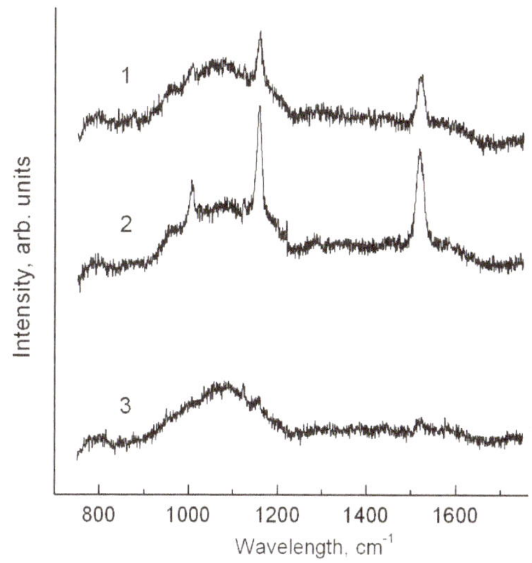

Figure 6. Raman spectra of the samples of liquid blood serum in patients with light F1-F2 fibrosis (1), severe F3-F4 fibrosis (3) and in patients of control F0 (2).

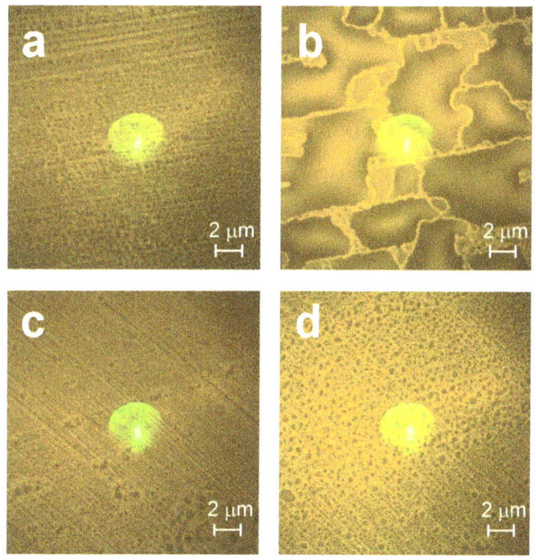

Figure 7. Raman microphotos of surfaces of liquid blood serum of patients with light (F1-F2) fibrosis (a), (b) and severe (F3-F4) fibrosis (c), (d) before (a, c) and after (b), (d) effect of Ar^+ 2 - 3 mW laser beam with wave length of 514.5 nm within 3 min.

of breakdown products into larger areas can result in temperature-influenced interflowing of lipoproteins into the larger clusters. Such changes are obviously of adaptational nature and occur due to the influence of some severe pathologic factors upon the human organism.

In this respect the role of the thermotropic phase transfers presents the greatest interest, *i.e.* stepwise changes in the degree of membrane proteins and lipids ordering under the influence of such factors as temperature, bioactive substances, ultraviolet irradiation and others [12]. It is quite possible that in case of light (medium) fibrosis the state of thermotropic phase transfers in the molecules of proteins and lipids (which composition does not undergo any significant changes) is close to the normal physiological one, that predetermines the possibility of changes within the adaptional limits. As it has been specified in our previous publications [13] a progressing liver pathology is associated with the appearance of abnormal thermotropic phase transfers in proteins and lipids, where the high temperature phase transfer is recognized to be the most stable one and to reflect the degree of the liver damage. Upon the presence of external pathological factor this circumstance tends to significantly limit the interaction of blood serum components. Probably such changes result in a "frozen" image of serum in patients with severe fibrosis following laser radiation affect.

The attempt of further combined application of optical differences in parameters of erythrocytes and blood serum used to determine the degree of fibrosis is based on the well known facts about strong interrelation between red blood cells and blood serum components. There is a constant interchange of lipid components between them. Viscosity level and erythrocytes aggregation at diffuse liver disease (DLD) are known to be influenced to a definite degree by thrombocyte aggregation, hypercholesterolemia, hypertriglyceridemia, fatty acids, and ionic composition. Increase in blood viscosity upon hyperlipidemia is predetermined by the capacity of high density lipoproteids to affect the functional state of regular blood elements, to reduce their electro-kinetic potential and as the result to increase the aggregation capacity and rigidity of membranes.

For example, one can observe inhibition of anticoagulating activity with strong depression of fibrinolysis and activation of blood coagulation service, degradation of rheological blood indexes [9,14] under the influence of alimentary hyperlipidemia. T. Y. Leonova and co-authors [15] have determined that erythrocytes can absorb on their surface lipoproteins of blood plasma charged by cholesterol (in the form of VLDL, HDL and LDL). This capability is attributed to the presence of lipoproteins in blood serum and the need of erythrocytes in cholesterol. Constant cholesterol metabolism is known to happen between erythrocytes and lipoproteins adsorbed on their surface. The most intensive metabolism of cholesterol is observed between red blood cells and HDL. In this case a blood serum enzyme, *i.e.* lecithin-cholesterol-acyltransferase (LCAT) is known to play a significant role. Its synthesis takes place in the liver and is to a great extent predetermined by its state. One can conclude that diffuse liver pathology is accompanied by decrease of LCAT level with further increase of free cholesterol in blood serum and later on in erythrocytes. This is secured by cholesterol reducing capacity of erythrocytes that, on the one hand, influences upon the intensity rate of cholesterol metabolism between the lipoproteins of blood serum and blood cells, and, on the other hand, it supports a physiological level of cholesterol in erythrocytes. Disturbances in this biologically important mechanism of cholesterol metabolism can result in hypercholeterolemia, the latter being an important risk factor in development of atherosclerosis, IHD and a potential cause of pathologic condition of optical parameters of blood serum [16].

It has been established that large-molecule proteins of blood serum tend to settle on erythrocyte membranes. This results in "screening" of the charge of erythrocytes as well as reduction of its dipole moment and the surface negative charge. Streiff *et al.* established [17] that zeta potential of erythrocytes correlated with changes in concentration of protein fractions. Stoltz *et al.*, in their turn, revealed that the globulin fraction favored rapid decrease of electrophoretic mobility of erythrocytes while the albumin one would practically not influence upon their mobility in the electrical field. Other researchers [9,18] proved that changes in the electrical charge would mainly depend upon the ratio of protein fractions, *i.e.* the greater the shift from the normal physsiological state is, the lower the electrical charge becomes. The decrease of electrical charge of erythrocytes and increase of blood serum globulin level are attributed to the absorption of globulin molecules on the cells surface, the latter traditionally possessing the lower electrical charge. The idea that globulins can be adsorbed on the surface of erythrocytes irrespective of their antigenic structure has been confirmed by numerous researchers, including the results of Pirofsky B., Cordowa M., Imel T. (1962). Dysproteinemia associated with the increased γ-globulins has been observed in cases of DLD upon the pronounced immuno-inflammatory syndrome [2]. It can be assumed that decrease of electrical charge of erythrocytes in patients with DLD can be predetermined by changes in ratio of protein fractions.

On the other hand, it has been established that in cases of diffuse liver damages increase of high-molecular weight protein (fibrinogen) in blood plasma is observed, which reduces the negative charge of erythrocyte and hampers their approach [19]. According to V. A. Levtova and the co-authors [20], the fibrogen molecules serve as bridges connecting the cells. Adsorption on the surface of erythrocytes of plasma proteins (fibrin, fibrinogen) tends to favor aggregation and adhesion of erythrocytes. It can serve as one of the reasons of the increase of erythrocytes aggregation index in patients with DLD revealed in this study upon examination by method of dielectrophoresis.

In our previous study [3] we have established the correlation of the optical indices of blood serum, intensity of IR-spectra resonances with lipid and protein components. Consequently, the relations discovered between electrical and visco-elastic parameters of erythrocytes and the optical indices of blood serum seem to be logic. We have also determined the direct correlation between thickness of films with the amplitude of erythrocytes deformation ($r = 0.49$, $p < 0.02$), between polarization at high frequencies ($r = 0.57$, $p < 0.033$ for 1 MHz; $r = 0.41$, $p < 0.05$ for 0.5 MHz) as well as the reverse one related to the total indices of viscosity ($r = -0.44$, $p < 0.048$), rigidity ($r = 0.40$, $p < 0.05$), electroconductivity ($r = -0.60$, $p < 0.015$). The index of refraction would correlate directly with the electrical conductivity ($r = 0.53$, $p < 0.05$), the indices of destruction ($r = 0.36$, $p < 0.03$), aggregation ($r = 0.39$, $p < 0.05$) and would be reversely related to the capacity of erythrocyte membranes ($r = 0.46$, $p < 0.05$), the value of dipole moment ($r = 0.39$, $p < 0.04$), amplitude of cells deformation ($r = 0.41$, $p < 0.05$).

Areas of the peaks of the amide III zone at 1280 cm^{-1}, 3190 cm^{-1} on IR-spectra of blood serum tended to correlate directly and significantly with amplitude of deformation of erythrocytes ($r = 0.62$, $p < 0.03$ for 1280 cm^{-1}; $r = 0.58$, $p < 0.014$ for 3190 cm^{-1}) and polarization at all frequencies (at frequency of 1 MHz – $r = 0.57$, $p < 0.02$ for 1280 cm^{-1}; $r = 0.48$, $p < 0.01$ for 3190 cm^{-1}). The same peaks of the IR-spectra showed the feedbacks with the total indices of viscosity ($r = -0.42$, $p < 0.05$ for 1280 cm^{-1}; $r = -0.49$, $p < 0.048$ for 3190 cm^{-1}), rigidity ($r = -0.35$, $p < 0.033$ for 1280 cm^{-1}; $r = -0.50$, $p < 0.05$ for 3190 cm^{-1}), the index of destruction ($r = -0.60$, $p < 0.027$ for 1280 cm^{-1}; $r = -0.45$, $p < 0.018$ for 3190 cm^{-1}) and electroconductivity of erythrocytes ($r = -0.40$, $p < 0.033$ for 1280 cm^{-1}; $r = -0.37$, $p < 0.05$ for 3190 cm^{-1}).

The results of comperative evaluation of the set of optical methods for assessment of erythrocytes and blood serum in patients with diffuse liver pathology aimed to diagnose the degree of fibrosis and the results of liver biopsy ("gold standard" for estimation of the degree of fibrosis) are specified in **Table 3**.

These findings demonstrate sufficiently high values of sensitivity (78.6%) and specificity (87.7%), prognostic value of positive (91.7%) and negative result (66.7%) as well as the index of accuracy (81%) [21]. The increased levels of these values as compared to the isolated study of erythrocytes enable to assume the necessity of application of the "set of optical methods" for blood studies in cases of non-invasive diagnosis of the degree of liver fibrosis. The most important are the perspectives of diagnosis of the early stages of liver fibrosis (according to the data of prognostic value of the negative result).

Table 3. Results of evaluation of the set of optical methods aimed to determine the degree of liver fibrosis vs the ones of liver biopsies.

Results of application of optical methods	Liver biopsy results	
	F3-F4 group (n = 14)	F1-F2 group (n = 7)
F3-F4 group (n = 12)	True-positive result, F3-F4 group (n = 11)	False-positive result, F1-F2 group (n = 1)
F1-F2 group (n = 9)	False-negative result, F3-F4 group (n = 3)	Truly slight fibrosis F1-F2 group (n = 6)

Sensitivity 11: (11 + 3) × 100% = 78.6%; Specificity 6: (1 + 6) × 100% = 87.7%; The estimated value of positive result of the method 11: (11 + 1) × 100% = 91.7%; The estimated value of negative result of the method 6: (3 + 6) × 100% = 66.7%; Accuracy index (11 + 6)/(11 + 1 + 3 + 6) × 100% = 81%.

4. Conclusions

Thus, application of the combination of the "set of optical methods" for studies of erythrocytes and blood serum of patients with diffuse liver pathology enables to significantly increase the indices of sensitivity, specificity, the index of precision as well as to achieve some real results in diagnosis of not only severe cases of fibrosis but to also gain the possibility to diagnose slightly pronounced forms of liver fibrosis.

The discovered diagnostic potential of optical methods can become promising in noninvasive screening diagnosis of stages of diffuse liver pathology of various geneses.

5. Acknowledgements

The study was supported by The Ministry of Education and Science of the Russian Federation, Project # 8041 of 24.08.2012 (Application # 2012-1.1-12-000-1001-004 filed in the Information Computerized System).

REFERENCES

[1] V. T. Ivashkin and T. L. Lapina, "Gastroenterologiya. Natsionalnoyerukovodstvo (Gastroenterology. National Guide)," GEOTAR-Media, Moscow, 2008.

[2] S. D. Podymova, "Liver Diseases," Meditsina, Moscow, 1993.

[3] M. I. Voevoda, S. E. Peltek, M. V. Kruchinina, S. A. Kurilovich, V. N. Kruchinin, K. P. Mogilnikov and S. V. Rykhlitskii, "Studying Thin Films Obtained through Cen-

trifugation of the Human Blood Serum by Methods of Spectral Ellipsometry and Infrared Spectroscopy," *Opto-electronics, Instrumentation and Data Processing*, Vol. 46, No. 4, 2010, pp. 382-393.

[4] M. V. Kruchinina, S. A. Kurilovich, M. V. Parulikova, A. A. Gromov, T. S. Bakirov, V. M. Generalov and A. V. Pak, "Viscoelastic and Electrical Properties of Red Blood Cells in Patients with Various Degrees of Liver Fibrosis," *Vestnik NGU*, Vol. 4, 2005, pp. 43-52.

[5] V. M. Generalov, M. V. Kruchinina, A. G. Durymanov, A. A. Medvedev, A. S. Safatov, A. N. Sergeev, G. A. Buryk, S. A. Kurilovich and A. A. Gromov, "Dielectrophoresis in Diagnosis of Infectious and Noninfectious Diseases," ZERIS, Novosibirsk, 2011.

[6] Y. Dimitriev, Y. Ivanova and R. Iordanova, "History of Sol-Gel Science and Technology (Review)," *Journal of the University of Chemical technology and Metallurgy*, Vol. 43, No. 2, 2008, pp. 181-192.

[7] K. Vorotilov, V. Petrovsky and V. Vasiljev, "Spin Coating Process of Sol-Gel Silicate Films Deposition: Affect of Spin Speed and Processing Temperature," *Journal of Sol-Gel Science and Technology*, Vol. 5, No. 3, 1995, pp. 173-183.

[8] S. V. Ryhlitsky, E. V. Spesivtsev, V. A. Shvets and V. Yu. Prokopiev, "Spectral Ellipsometric Complex Ellipse, 1771," *Pribory I Tekhnika Eksperimenta*, Vol. 2, 2007, pp. 160-161.

[9] G. I. Kozinets and V. A. Makarov, "Issledovaniyesistemykrovi v Klinicheskoypraktike (Studies of Blood System in Clinical Practice)," Triada-X, Moscow, 1997.

[10] R. Manoharan, Y. Wang and M. S. Feld, "Histochemical Analysis of Biological Tissues Using Raman Spectroscopy," *Spectrochimica Acta Part A*, Vol. 52, No. 2, 1996, pp. 215-249.

[11] T. C. B. Schut, G. J. Puppels, Y. M. Kraan, J. Greve, L. van der Maas and C. G. Figdor, "Intracellular Carotenoid Levels Measured by Raman Microspectroscopy: Comparison of Lymphocytes from Lung Cancer Patients and Healthy Individuals," *International Journal of Cancer, Predictive Oncology*, Vol. 74, No. 1, 1997, pp. 20-25.

[12] E. A. Chernitsky and A. V. Vorobey, "Structure and Function of Erythrocyte Membranes," Nauka I Takhnika, Minsk, 1981.

[13] M. V. Kruchinina, "Structural and Functional Properties of Red Blood Cells in Somatic Pathology of Alcoholic Genesis," Avtoref. dis. kand. Med. Nauk, Novosibirsk, 1994.

[14] V. V. Novitsky, N. V. Ryazantseva and E. A. Stepovaya, "Physiology and Pathophysiology of the Erythrocyte," Izdatelstvo TGU, Tomsk, 2004.

[15] T. J. Leonova, "On the Issue of Erythrocytic Mechanism of Control of Cholesteremia in Cases of Experimental Hypercholesteremia and Ischemic Heart Disease," Avtoref.diss.kand. Med. Nauk, Novosibirsk, 1982.

[16] F. J. Schiffman, "Pathophysiology of Blood," Edition Group "BINOM", Moscow and "Nevsky Dialogue", Saint-Petersburg, 2000.

[17] L. Strayer, "Biochemistry," Mir, Moscow, 1984.

[18] A. I. Miroshnikov, V. M. Fomchenkov and A. Yu. Ivanov, "Electrophysical Analysis and Separation of Cells," Nauka, Moscow, 1986.

[19] A. L. Chizhevsky, "Biophysical Mechanisms of Erythrocyte Sedimentation Rate," Nauka, Novosibirsk, 1980.

[20] V. A. Levtov, R. S. A. Egirer and N. H. Shadrina, "Blood Rheology," Meditsina, Moscow, 1982.

[21] T. Grinhald, "Evidence-Based Medicine," GEO-TARG-Media, Moscow, 2004.

3

Quantitative Registration of the Degree of the Voluntariness and Involuntariness (of the Chaos) in Biomedical Systems

Valery M. Eskov, Alexander A. Khadartsev, Valery V. Eskov, Olga E. Filatova

Surgut State University, Surgut, Russia.

ABSTRACT

What does it mean: certainty and uncertainty, chaos and order? The understanding of such definition and real interpretation of chaotic behavior the third type of systems in nature was presented. The practical realization of different medical problems was presented too.

Keywords: Chaos; Attractor; Phase Space; Tremor

1. Introduction

At present we know of at least a dozen hypotheses about the mechanisms of the brain activity related to the notion of dynamic chaos. However, there are no quantitative methods in all these hypotheses, which could describe the characteristics of these chaotic regimes, *i.e.* parameter of chaos. The problem of the chaos control stays still more uncertain. Thus, the problem of complex registration of chaos parameters (and not the fact of its emergence) and the problem of dynamic chaos control in the framework of traditional science, that is, if it is based on a deterministic approach or stochastic approach (all these we now define as deterministic-stochastic paradigm-DSP) currently remain open. Problem of such measurements in real biomedical research is still uncertain.

However, now in Russia, a new third approach is being developed (differs from DSP) [1,2], which attempts to combine the notions of chaos and order in a unit in the new theory of chaos-selforganization (TCS)). The necessity to create TCS arose from attempts to resolve the contradictions between H. Haken's synergetics and I. R. Prigogine's complexity. In fact, we had to create original compartment-cluster theory of neural brain networks, and then all of biosystems-(CCTB) for the mathematical interpretation of the first H. Haken's postulate [3]. Established in 70 - 80 years CCTB led to a new understanding of the basic principles of the mammalian brain and then the new principles of artificial neural networks and neuroemulators-neurocomputers. At the same time the impossibility of using neurocomputers was proved when a single iteration for the purpose of identification of the order parameter (OP) in the various purposes of medical diagnosis (neurocomputer is not applicable in the diagnosis of the main diagnostic features in medicine at a single iteration). Since each iteration (for example, the binary classification problem) determines its own special set of OP [4,5].

The result of all these efforts is the formation of the five basic principles of the functioning of any complex biological systems (including the neural brain networks) and 13 differences of special systems of the third type (STT), which are in constant and continuous chaos [1,2, 6-11]. This chaos is emerged in continuous, chaotic motion, the system state vector (SSV) in the phase states space (PSS). Number of outstanding scientists paid attention to it but quantitative identification methods [3,4,11] of these special properties and a special dynamic behavior of STT had not been established. The current report is aimed to the practical application of these methods in neurosciences. In this case, we show the new restrictions neurocomputers and new opportunities of TCS, its identification methods and chaos control in TTS [1,2,11,13, 14].

2. The New Methods of Measuring of Neuro-Muscles System's Chaotic Dynamics

According to the theory of chaos-selforganization we have uninterrupted movements of SSV $x = x(t) = (x_1, x_2, \cdots, x_m)^T$ in phase space of state (PSS). Let us introduce a series of postulates (assumptions) according to TCS which are presented the formal description of TTS:

1) Any biological dynamical system as the third type of systems can be described by a state vector $x = (x_1, x_2, \cdots, x_m)^T$ in an m-dimensional phase space of states. The components of this vector can be any variables (parameters) of the cardiorespiratory system, biochemical parameters of the blood, parameters of various physiological and psychophysiological human functions, etc. In this case, the main point is that methods and devices exist for their objective measurement. This has already been discussed in earlier papers [1,6], when for example the parameters were constructed and measured for models of respiratory neural networks (or other neural networks) within the compartmental/cluster approach, where as x_i we took the bioelectric activity of the efferent nerves or muscles.

2) When the maximum set x_i exists, we can always minimize the dimensionality of the phase space of states, *i.e.*, we can go from m to k where $k < m$ (or $k \ll m$). These procedures allow us to minimize the dimensionality of the phase space and to obtain some compact model of the system under study. For these purposes, we use three approaches: 1) minimization of the dimensionality of the phase space of states within the compartmental/cluster theory of biological dynamical systems; 2) according to neural net technologies we construct the algorithm of older parameters identification, and the method for analysis of the parameters of quasi-attractors for the motion of the human state vector in the phase space of states. The latter is based on the third postulate of TCS, but gives a realistic assessment of the meaning of x_i [1,2,11].

3) The variability (the degree of spread in the biosystem state vector in the phase space of states, as a certain analog to fluctuations in physics) of the parameters of the biological dynamical system (the values of x_i) is the information for evaluating the state of biosystems and predicting their transition to other states (other regions of phase space, change in the order parameters, transition to other phase spaces of states). It is important that the level of variability in the biological dynamical system be a measurable quantity and that it can be monitored long-term, and compared for different groups of biosystems or for the same biosystem found in different functional states [2,11,13,14].

Based on these three assumptions (postulates), we can measure the parameters of the quasi-attractors (QA) for regions of phase spaces within which motion of the biosystem state vector occurs. Here we do not deny the existence of real attractors for the motion of the human state vector, but in principle they are not accessible in measurements. A certain analog to the Bernoulli theorem (the law of large numbers) is postulated in measuring the parameters of these real attractors, *i.e.*, if the number of experiments (repetitions of the measurements) is made infinite, then the parameters of the quasi-attractors will approximate the parameters of an ideal attractor (as in statistics, $P^*(A) \to P(A)$ as the number of measurements $n \to \infty$). But the five properties of biological dynamical systems indicated early [1,2] impose fundamental limits on the inaccessibility of measurements of the parameters of an ideal attractor, since a biological dynamical system cannot be kept in approximately the same state for as long as desired (evolution of the biological dynamical system occurs, its death is prescribed teleologically, it is constantly changing dramatically—a human gets sick, ages, changes lifestyle, etc.). Since the number of repetitions n is finite and small, we have to make do with the parameters of quasi-attractors. The algorithm for calculating them contains several operations which are executed each time during processing of the data obtained in measurements of x_i from a group of biological objects found in approximately the same functional state (arbitrarily identical disease, identical environmental conditions, identical physiotherapeutic treatments, etc.).

The algorithm for calculating the parameters of quasi-attractors for the human state vector based on chaotic analysis of SSV of patients (in medicine). The data obtained from one or a group of test subjects by repetitions of the measurements in the form of a set m of data blocks (compartments) where m is the number of measured biological indices, are transferred in the form of points to an m-dimensional phase space of states, in which, fixing the extreme left-hand and right-hand values of the parameters for the human state vector along each coordinate x_i, they generate a quasi-attractor in the form of an m-dimensional parallelepiped, for which we determine the volume V_g, the center, and the asymmetry value.

Using these three quantities (absolute or relative), we decide if the recovery is effective by comparing these parameters before and after treatment. Essentially, each group of test subjects forms a certain "cloud" in the phase space of states, which has its own limits Δx_i along each of the coordinates x_i ($i = 1, 2, \cdots, m$). These limits Δx_i are the faces of the m-dimensional parallelepiped in the m-dimensional phase space, and each face represents the level of variability of the ith parameter (diagnostic index), a component of the vector x in the phase space of states. The total volume of this parallelepiped is the volume V_g of some quasi-attractor, which for different groups of test subjects has its own parameters before and after treatment. These parameters include: the volume of the quasi-attractor V_g, the coordinates of its center C in the phase space of states, the asymmetry r. We determine either the absolute values ($\Delta V_g = V_{g2} - V_{g1}$, where V_{g2} are the values of V_g after treatment, and V_{g1} are the values before treatment) or the relative values in percent (($\Delta V_g/V_{g1}) \times 100$) of these parameters. If the relative changes in the parameters of the quasi-attractor exceed the uncertainty in the measurement of the diagnostic indices, then they are already considered significant (in many cases, they are 5% or 10% or more). The larger the changes in these three overall (integrated) indices, the more effective is recovery.

One simple example of the algorithm realization includes the investigation of the role of voluntary or involuntary control (by the human brain) of postural tremor. It is evident according to classic physiological theories that postural tremor isn't voluntary movement because we have not any possibility for achievement of stationary mode ($dx/dt \neq 0$ and $x_i \neq const$) for every time t [15,16]. But the brain (its control) can provides the regulation of postural tremor quasi-attractor volume. Such volumes changing present the proofs of the role of voluntary control from the brain in postural tremor organization. But the control has chaotic form (see below, chapter 2).

According to tremorgrams (I_1) we obtain amplitude-frequency diagram (AFD), see the **Figure 1** at (II_1). The main point of the experiment is static keeping of the forefinger under the plate securing the hand position for two coordinates-micro-movement of fingers (x_1, m) and the speed of micro-movement ($x_2 = dx_1/dt$, m/s). Each subject takes part in the experiment being in different organism state three times: before sighting, during sighting and after sighting. It means every subject has a task to keep goal. It is obvious that on the second stage of the experiment mind activity process has been involved, that is important to keep a hand in one position (statically), control its fixity consciously. We must remark that for anytime t we never have the condition: $dx/dt = 0$, where $x = (x_1, x_2)^T$ is a state vector of biomechanical system (tremor). And for every little period of time—we cannot obtain two equivalent phase trajectory or full equivalent simple dynamic $x_i(t)$. So we have nonrepeatable processes (for AFD (II_2) it is evident) and full uncertainty for future prognoses of $x_i(t)$, value. It is chaotic process for ever t and AFD we can't use for investigation of tremor. It is the first reason of our article presentation.

In addition to the amplitude-frequency diagram, we have considered a two-dimensional or six-dimensional phase space, which describes the dynamics of the human organism state vector (HOSV) at two coordinates x_1 and x_2 or six coordinates (x, y, z, $\cdots$ respectively). We have determined the area of the obtained quasi-attractors—QA. Having identified the characteristics that correspond to a chaotic dynamics (quasi-attractors in the parameters and the values of entropy for comparison with DSP).

Experimental findings have been treated in the phase coordinates x_1 (the absolute movement of the finger) and $x_2 = dx_1/dt$. According to two-dimensional vector space $x = (x_1, x_2)^T$ it is easy to represent the tremor in the phase plane. In general, we have made the monitoring of the human body motion in three-dimensional space (three coordinates x, y, z of space) and then the dimension of the phase space of states is equal to six ($m = 6$), where three are real physical coordinates (x, y, z) and three others are their speed, *i.e.* $x = (x_1, \cdots, x_6)^T$.

Such an approach is made in the framework of the classical interpretation of biophysics as science (see Hill "Why biophysics" [17]), using methods of physics if we study biological systems. In one of our articles [2,3,5-10] coordinates x_i of SSV will have no physical meaning, but it will be also the coordinates of SSV in the phase space of states (PSS), they also characterize the dynamics of movement of SSV in PSS, and this dynamic has a pronounced chaotic character like our the first example-tremor.

Now we present a typical example for the subject of the experiment No. 1—D. D. (**Figure 2**) and general number of all subjects is 120 people (information is similar, but the numerical results are individual). For all such systems (third type with chaotic dynamics of SSV in PSS) we can calculate the volume V_G of PSS (this volume we define as quasi-attractor), and the parameters of the quasi-attractor have a significant informational value as opposed to a specific AFD and others value of SSV behavior in PSS, which has no any information about chaos.

As we take two instrument readings-speed ($x_2 = dx_1/dt$, m/s) and motion of fingers of the subject (x_1, m), the latter is measured in conventional amplitude values-in volts, then to construct the phase portrait (**Figures 1** and **2**) representing the dependence of the rate on volt movement are converted to meters. So, we consider the results of QA calculations and some special changing of its volume V_G for such three type of complexity system (effector system —the second cluster of two cluster's hierarchical system). The **Figures 1** and **2** presents the result of quasi-attractor calculation to the two stages of mister D. D. It is easy to see that voluntary brain control changes the volume of quasi-attractor and we transform the quasi-attractor III_1 (see **Figure 1**) into quasi-attractor III_2 (see **Figure 2**). The changes of quasi-attractor volume V_G (under such transformation) present the role of voluntarism in involuntary process-postural tremor (see **Table 1**). So we can use the volume of quasi-attractor V_G for registration of human

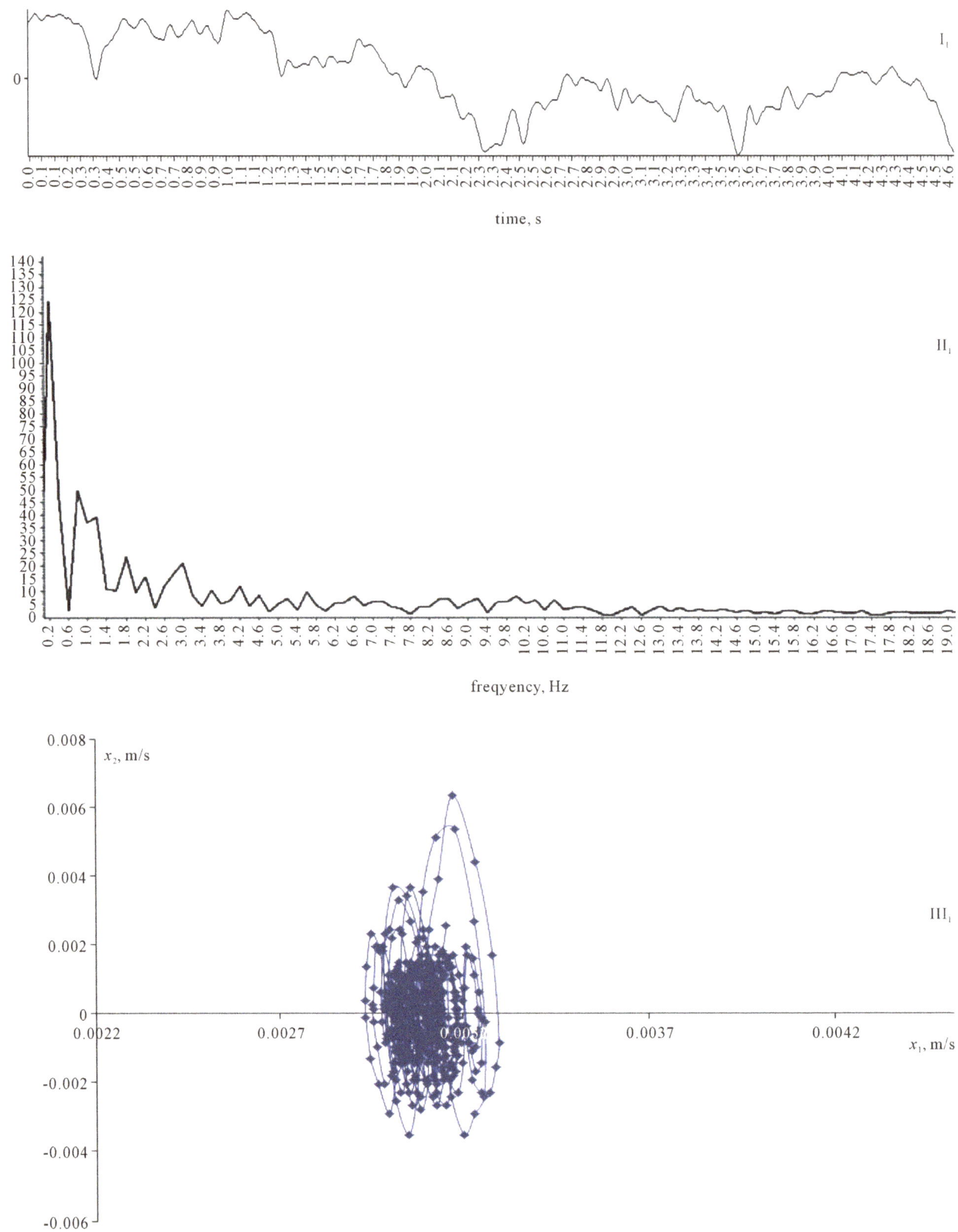

Figure 1. Real tremorgrams (I_1), their amplitude-frequency diagram-AFD (II_1) and phase portraits (III_1), taken from the fingers of the subject D.D. at the pre-sighting stage. The tremorograms and AFD changes every second and only volume of quasi-attractor is constant for constant physiological state of human organism.

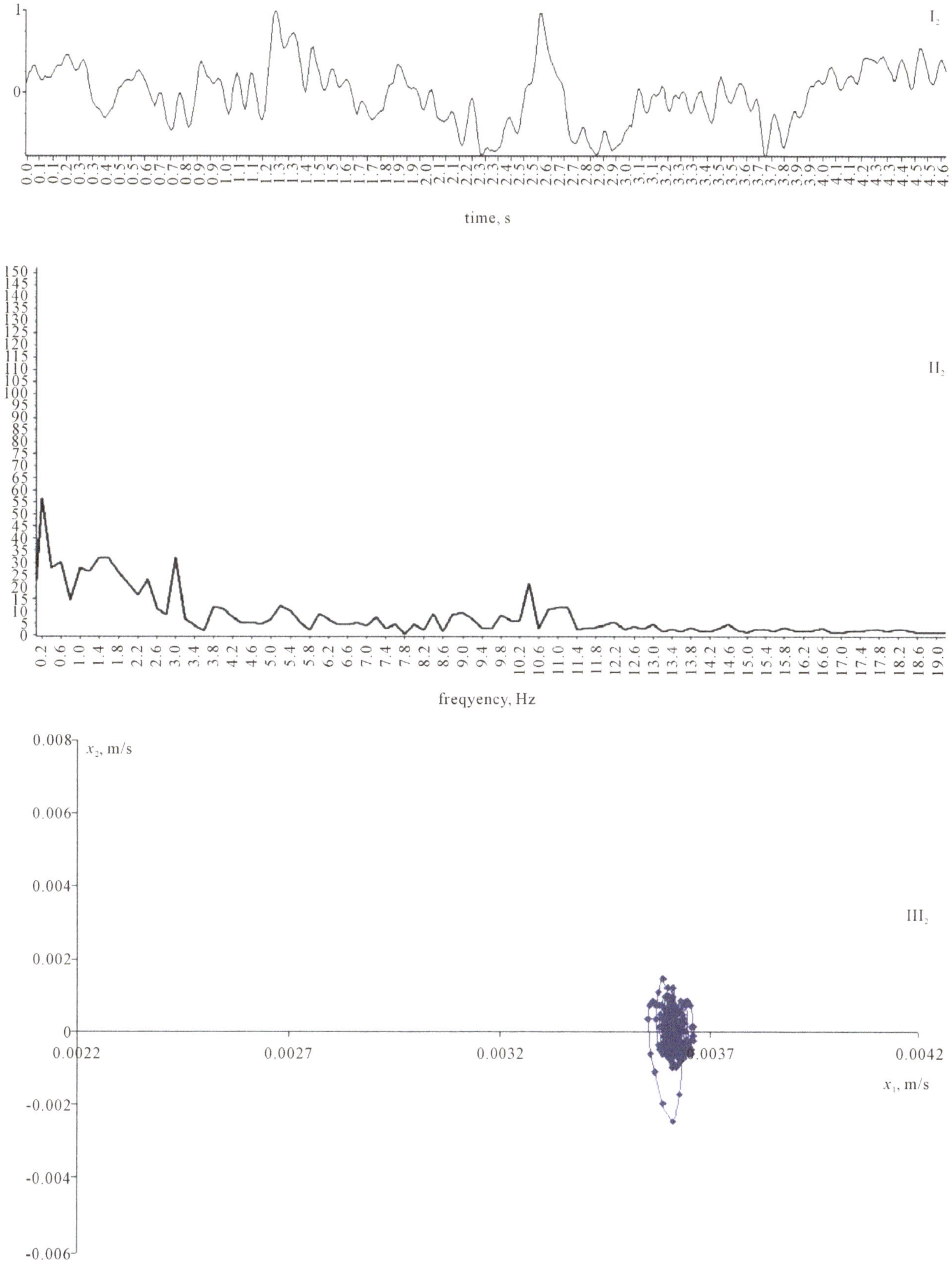

Figure 2. Real tremorgrams (I_2), their amplitude-frequency diagram (II_2) and phase portraits (III_2), taken from the fingers of the subject DD at the while-sighting stage.

organism stage (normal, pathological …) under different changing (different treatment in medicine, ecological factors in ecology, different training in sport …).

So it is easy to see that **Figure 2** presents the role of voluntary (purposeful)-control (from brain) but chaos is uninterrupted process (see **Figure 3** below). We have global chaos but the volume of quasi-attractors presents the value of voluntarism (the role of brain control). The volumes of represented quasi-attractors are shown in **Table 1**, where it is easy to see that the smallest QA size we have at while-sighting stage (**Figure 2**), by order of magnitude different from the state before and after sighting (**Figure 1**). Such result we use for selection of sportsmen. If sportsmen almost rate the great changing of V_G (before and after sighting stage) he has a good future of his sport career. The V_G present now for us the result of therapy in medicine, in psychology, etc.

So we use such approach for identification of effectiveness of different medical or other treatment. The volume V_G before treatment and after it presents the effectiveness of our voluntary treatment for man with pathological state and its transformation into normal state. It is new method in medicine, sport physiology, ecology, etc. We have a great number of results of such V_G calculations (50 books in Russsia).

3. How Does the Brain Realize the Control of Effector Systems?

Some general approach is presented in this message in evaluation of the work of brain neuronets and artificial neuronets from positions of the new and actively developing theory of chaos-self-organization (as a new consciousness and new understanding of nature (life system)). The principles of biological objects work as special third type systems (TTS) are shown in a continuous chaotic movement. As a whole, the work of a brain can be considered as one chaotic quasi-attractor influencing the work of various regulatory systems by means of which it keeps in a mode of a homeostasis and keeps all functional systems of an organism (FSO) of the person in a homeostasis. But all of this FSO are chaotic biosystems

Table 1. Quasi-attractor volume of the subject Mr. D. D. under special voluntary control and like quasi-voluntary control from the brain.

	Pre-sighting stage V_{G1}	While-sighting stage V_{G2}
Quasi-attractor volume (V_G)	3.6×10^{-6}	8.3×10^{-7}

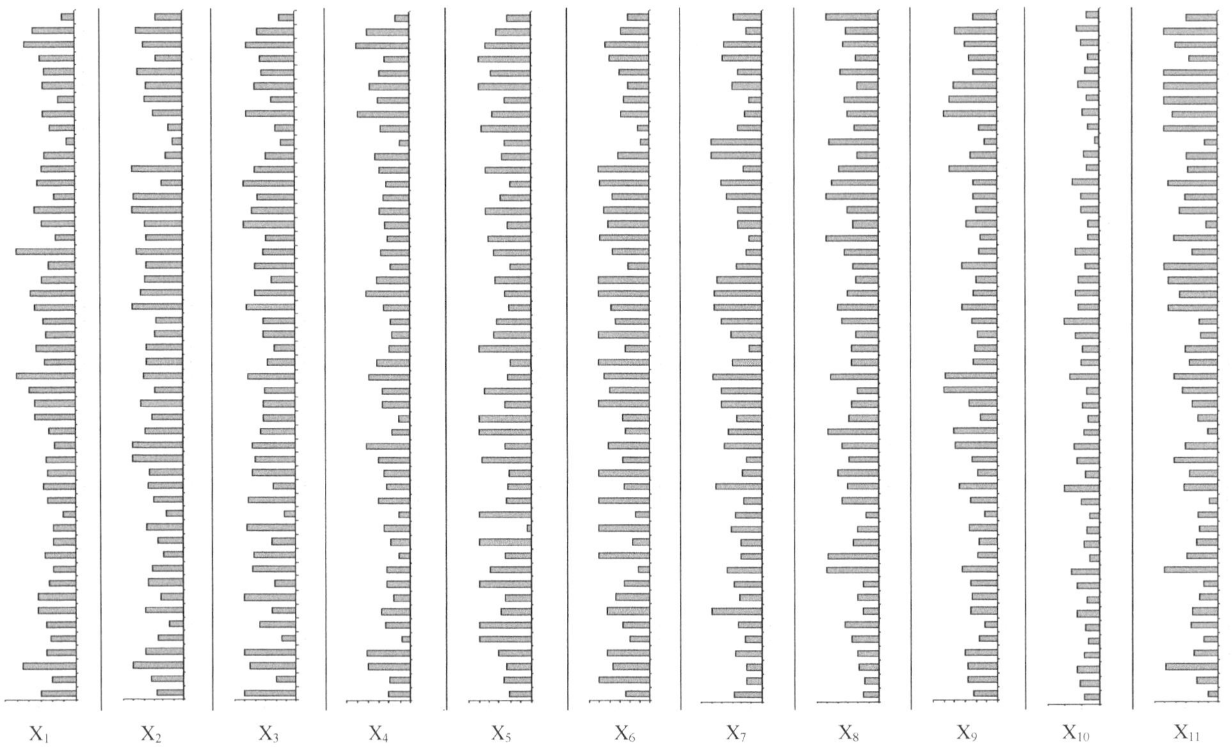

Figure 3. Diagram of distribution of weights for each parameter (x_i) for each of the j-th repetition of equal artificial neural network training (gradient descent method, $j = 1, \cdots, 50$).

for which the stationary regime take the other interpretation (opposite to traditional science) [1,2,5-14].

So if we use the artificial neuron network in medicine we must answer to the question: why can't we use the neuroemulator in medicine for diagnostic procedure? The answer to the question is connected with the second question: why the brain (as the first cluster) has chaotic behavior? Very often in medicine we have the task of binary classification. The objectives of detection of distinctions between two (or more) groups of objects having some general signs quite often arise in medicine and biology. These can be two groups of patients having an identical disease, but being, for example, in various conditions (ecological, physiological, mental) or exposed, for example, to treatment by two different methods. In these cases the problem of detection of the importance of influence on the state of an organism of these external influencing factors or conditions (medicines, methods of treatment, action of adverse ecological factors, etc.) is solved at the same time. For these two groups of people (two selections) diagnostic signs of x_i which form the m-dimensional phase space of states (PSS), that is the vector of a state $x = (x_1, x_2, \cdots, x_m)^T$ is set. It is obvious that thus in PSS these two groups occupy some areas and the neuroemulator can reveal if these areas differ significantly or insignificantly.

Meanwhile the neuroemulator (NE) is set an objective: to reveal whether these two selections (groups) significantly differ or don't differ. It is obvious, that similar tasks are also in the field of recognition of images (for example, do two similar photos represent different people or one and the same person?). Let's admit that such a task can be solved and the neurocomputer reaches the goal—it distinguishes these two groups, *i.e.* the problem of diagnostics has its solution. At once we note that irrespective of the methods of neuronet control the task is either solved or not, *i.e.* the objective is either reached or not. However, if we still solve the problem and ask the neuronet how it solved it, for example, we will request the weight of communications between neurons or we will demand to present the weight figures of used diagnostic signs, it will appear that the more we force a neuronet to solve the same problem (at the same initial selections) the more variants of weight figures for x_i we will receive. Million times we will repeat an identical task of binary classification and one million times we will receive different weight figures. The internal architecture of a neuronet, character and the weight of communications, the importance of the diagnostic signs forming m-dimensional phase space in which these two groups of patients are set will be various each time [1,2,11,13,14]. So the one cycle of NE presents the value of x_i (the order parameter-OP) and this OP isn't valid. The NE is not usefulness in medicine in such regime.

As it was stated in our early publications [1,2,11] the distinctive feature in the work of special third type of systems is the "glimmering property", *i.e.* their continuous, chaotic movement of its system state vector (SSV) of such TTS in phase space of states. This movement of SSV in PSS shouldn't be perceived as real (mechanic) movement since for neuronets it can be shown in continuous change of a network architecture (any neurons can join, be switched off, can change the weight of synoptic communications, etc.). Thus the general architecture (morphology) won't undergo any changes.

Externally the neuronet (and the neuroemulator) will show a steady result. For example, the problem of binary classification will repeat (the exit will be identical) and the number of elements, their arrangement in a network won't change. In rough consideration all this seems permanent. However, the internal state of a network (neuroemulator) will be thus in continuous chaotic change. Thus any of such instant internal states won't repeat. For example, muscles can't identically repeat this or that movement (it repeats within quasi-attractors), we can't execute two identical respiration, two identical knee-bends. Moreover, hand suspension (or a finger) in space doesn't lead to a stationary mode, *i.e.* $dx/dt \neq 0$ and we observe the postural tremor. This tremor is caused by chaos in the main controller—the brain, its neural networks and in the work of an effector—a group of muscles which are carrying out any movement (finger suspension in space), but it's done involuntarily, *i.e.* chaotically, within a quasi-attractor. On the **Figure 1** we are presented diagram of distribution of weights for each parameter (x_i) for each of the j-th artifical neural network training (method of linked gradients, $j = 1, \cdots, 50$).

The 11-th dimensions of phase space of state consists some special biological parameters for characterization of state of cardio-vascular system (CVS) of group of patients. Here x_i are parameters of CVS: x_1—NN, mc; x_2—SpO_2, %; x_3—SIM, x_4—PAR, x_5—SDNN, x_6—HRV, x_7—IB, x_8—VLF, x_9—LF, x_{10}—HF, x_{11}—LF/HF (the biological means of x_i we presented early [1,2,10]).

All underlying structures also start working chaotically (though they themselves also undergo continuous chaos within their own quasi-attractors). However, the change of the QA parameters of the hierarch (the brain) can lead to an essential change of the QA parameters of the underlying clusters. For example, changing a brain state (within QA at the change of the basic purpose *i.e.* any movement) can lead to the change of the QA parameters in the organization and effector system in the form of a tremor, chaotic work of heart, biochemical parameters of blood, and many other effector systems connected with the brainwork. We will present this fact and in other messages and now we will show an example of internal chaotic dynamics in the work of quasi-attractors as the real models of work of living neuronets (*i.e.* brainwork) [1,2,10]. It is evident: if we have the brain (or artificial neuron network) chaotic behavior we can't obtain the order parameter in medicine and the behavior or effector systems of human organism will be chaotic too. But we can use the parameters of quasi-attractor as a diagnostic method in medicine and biology. Now we calculate the matrix of quasi-attractor distances for analyses of effectiveness of therapy [1,2,10,13,14].

4. Conclusions

1) Every complex biological dynamic systems as the third type of systems—TTS (which is describing by system's state vector (SSV) $x = x(t) = (x_1, x_2, \cdots, x_m)^T$ in phase space of state—PSS) has uninterrupted movements of SSV in PSS and $dx/dt \neq 0$ at all.

2) The traditional science (DSP) hasn't possibility for the TTS description and we present the TCS and its quasi-attractors for such description.

3) The quasi-attractor presents the real chaotic behavior of brain, neuroemulator and many other TTS with chaotic dynamics. In medicine we must do the monitoring of quasi-attractors parameters. For such chaotic systems we can't use the neuro emulator for identification of order parameters (main diagnostic parameters) in medicine. Now we create the special procedure with neuroemulator for such purposes (see the next our presentation).

REFERENCES

[1] V. M. Eskov, V. V. Eskov and O. E. Filatova, "Characteristic Features of Measurements and Modeling for Biosystems in Phase Spaces of States," *Measurement Techniques* (*Medical and Biological Measurements*), Vol. 53, No. 12, 2011 pp. 1404-1410.

[2] V. M. Eskov, V. V. Eskov, O. E. Filatova and M. A. Filatov, "Two Types of Systems and Three Types of Paradigms in Systems Philosophy and System Science," *Journal of Biomedical Science and Engineering*, Vol. 5. N. 10. 2012. pp. 602-607.

[3] H. Haken, "Principles of Brain Functioning: A Synergetic Approach to Brain Activity, Behavior and Cognition (Springer Series in Synergetics)," Springer, Berlin, 1995.

[4] M. M. Churchland, J. P. Cunningham, M. T. Kaufman, J. D. Foster, P. Nuyujukian, S. I. Ryu and K. V. Shenoy, "Neural Population Dynamics during Reaching," *Nature*, Vol. 487, No. 7405, 2012, pp. 51-58.

[5] V. M. Eskov and O. E. Filatova, "Respiratory Rhythm Generation in Rats: The Importance of Inhibition," *Neurophysiology*, Vol. 25, No. 6, 1993, pp. 420-426.

[6] V. M. Eskov, "Direct Control by Dissipation Factor in Respiratory Neuron Networks," *Neural Network World*, Vol. 6, 1994, pp.663-670.

[7] V. M. Eskov, "Models of Hierarchical Respiratory Neuron Networks," *Neurocomputing*, Vol. 11. 1996, pp. 203-226.

[8] V. M. Eskov and K S. V. ulaev, "Role of Connectedness Configuration at Neuron Network Behaviour. The Compartmental Approach," *Advances in Modeling. Automatic Control* (*Theory* and *Applications*), Vol. 59. No. 1, 2004, pp. 9-24.

[9] V. M. Eskov, "Hierarchical Respiratory Neuron Networks," *Modelling, Measurement & Control*, Vol. 48, No. 2, 1995, pp. 47-63.

[10] V. M. Eskov, "The Dependence of Activity of Cyclic Respiratory Neuron Network with Subcycles on Damping Coefficient," *Neural Network World*, Vol. 1, 1996, pp. 57-67.

[11] V. M. Eskov, "Third Paradigm. Samara. (Russia)," Ofort, Samara, 2011.

[12] V. M. Eskov, "Indirect Control by Chemoreceptor Drive in Respiratory Neuron Networks," *Modelling, Measurement & Control*, Vol. 48, No. 3, 1995, pp. 1-12.

[13] M. A. Filatov, D. Y. Filatova, O. I. Himikova, J. V. Romanova, "Matrixes of Quasiattractor Distances at Identification of Human Psychophysiology Function," *Complexity Mind Postnonclassic*, Vol. 1, 2012, pp. 19-24.

[14] T. V. Gavrilenko, A. A. Baltikova, D. A. Degtyarev and A. S. Pashnin, "The Comparison of the Efficiency of Classic Stochastic Theory and Theory of Chaos-Selforganization (TCS)," *Complexity Mind Postnonclassic*, Vol. 1. 2012, pp. 81-91

[15] W. Weaver, "Science and Complexity. Rockefeller Foundation," American Scientist, New York, 1948.

[16] I. Prigogine, "The Die Is Not Cast. Futures," *Bulletin of the Word Futures Studies Federation*, Vol. 25. No. 4 2000, pp. 17-19.

[17] A. V. Hill, "Why Biophysics?" *Science*, Vol. 124. No. 3234, 1956, pp. 1233-1237.

4

Ultrafast Electron Microscopy for Chemistry, Biology and Material Science

Sergei A. Aseyev[1], Peter M. Weber[2], Anatoli A. Ischenko[3*]

[1]Institute of Spectroscopy of the Russian Academy of Sciences, Troitsk, Russia; [2]Department of Chemistry, Brown University, Providence, USA; [3]Moscow M. V. Lomonosov State University of Fine Chemical Technologies, Moscow, Russia.

ABSTRACT

For the past thirty years, intense efforts have been made to record atomic scale movies that reveal the movement of atoms in molecules, the fast dynamical processes in biological tissues and cells, and the changes in the structure of a solid confined to nano-scale volumes. A combination of sub-nanometer spatial resolution with picosecond or even femtosecond temporal resolution is required for such atomic movies. Additional important information can be obtained when the energy of the electron beam transmitted through the sample is measured. The four dimensional (4D) spatially and temporally resolved ultrafast electron microscopy method is made possible by the extremely high detection efficiency that is reached in 4D electron microscopy. Using ultra-short electron bunches for the visualization of biological tissue can also improve the spatial resolution compared to conventional electron microscopes, thereby enabling the study of complex biological samples of relevance to the life sciences. Of particular interest to a broad audience is the possibility to create a video, and in the future, a real atomic movie, using 4D electron tomography.

Keywords: Ultrafast Electron Microscopy; Dynamic Processes; Structural Dynamics; Atomic Movie; Femtosecond Temporal Resolution; Atomic Spatial Resolution; Electron Tomography; Spectral-Spatial-Temporal Resolution

1. Introduction

Since the 1980s, intense efforts have been made to record an atomic movie showing the movements of atoms within molecules, the fast dynamical processes in biological tissues and cells, and the changes in the structure of a solid in a nano-scale volume [1-3]. Such an atomic movie requires a combination of sub-nanometer spatial resolution with picosecond or even femtosecond temporal resolution.

The aim of any microscopy is to study the structure, composition, and a variety of physical and chemical characteristics of samples (usually solids) on a very small length scale, with linear dimensions of irregularities that are on the order of micrometers or nanometers. While in most microscopies, photon beams (in optical microscopy) or corpusclar beams (for example in electron microscopy) are used as the probes, tunneling microscopies use a very sharp tip that is scanned across surfaces. Electron microscopy and optical microscopy use light and a focused electron beam for imaging, respectively. Beyond those, other microscopic methods may use ions, protons, positrons, or neutrons, or acoustic or microwave radiation in addition to other, less common, methods [3-6]. In each of them, the specificity of the interaction of a beam of the particles (or photons) and the molecules or atoms of a sample yield unique and rather useful information about the structure, composition and microscopic inhomogeneities of the sample, and the nature of their intermolecular interactions [3,4,7]. For example, the slow neutrons in neutron microscopy and spectroscopy barely interact with the electrons, but interact strongly with the nuclei of the atoms.

There are two main approaches to achieve imaging in classical microscopy: 1) In transmission mode, a large area of the sample is illuminated by a beam of light or particles. Specifically designed lenses project the transmitted beam onto a screen or detector to form an image of the structure of the sample. 2) In reflection mode, the beam reflected from the sample, or secondary particles or photons generated by the incident beam in the sample, form an image of the sample surface or a thin region underneath the surface [8].

Because of absorption of the optical radiation by matter, transmission optical microscopy is limited to thin

*Corresponding author.

samples, typically less than 1 mm thick. In electron microscopy, the sample has to be much thinner, because the electrons interact with matter more strongly than light. In transmission electron microscopy it is therefore possible to get images only of very thin slices, with film thicknesses of much less than 1 micrometer. Such samples have to be prepared using special ultramicrotomes that can produce film thicknesses down to a few tens of nanometers.

There are two different ways to obtain an image [9,10]: 1) "all at once", when the sample is irradiated entirely with light or electrons and the transmitted or the reflected beam is detected at once using, for example, a film, photographic plate, photosensitive recorder or digital camera; 2) in "scanning microscopy" mode, where the sample is scanned by a beam focused to a spot of rather small diameter, and the resulting image is assembled point by point.

2. Transmission and Scanning Electron Microscopy

In conventional scanning microscopy, images are formed by a beam of charged particles or photons that is focused to a very small diameter. This probe is moved across the surface of an object, and signal of any physical nature, but in particular reflected particles or photons, is detected and the image obtained "point by point". The method is quite analogous to the generation of an image using a scanning electron beam in a traditional cathode ray tube (CRT), or in modulating the brightness of each luminous point in a LCD monitor [11]. Scanning microscopy is a rather convenient tool to study the morphology and topography of various geometric objects, to investigate elemental compositions, to measure the electric and magnetic fields in micro-volumes, (the so-called "micro-fields", as introduced by Prof. GV Spivak), or to test electrical parameters of different semiconductor crystals (e.g., lifetime and diffusion length of charge carriers).

In scanning electron microscopy, the scan of the electron beam should be synchronous with a sweep of the probe on a sample. Like in a TV image, the raster image consists of individual small dots or pixels, the brightness and make-up of which, formed after signal amplification, define the image. Thus, the elements of the image appear one after the other, as if we move our fingers to spell a complicated book of life, but not all at once, as in the usual optical or transmission electron microscopes. This "point-wise" map of the object greatly facilitates the processing and interpretation of images [12].

In a transmission electron microscope, the electrons illuminate and pass through a very thin sample. An image of the sample is formed by means of electrostatic or magnetic lenses on the fluorescent screen or equivalent position sensitive detector [13]. In 1931, the first such device was demonstrated by M. Knoll and E. Ruska, who imaged a wire mesh using an incident electron beam and axially symmetric magnetic electron lenses with a narrow annular gap and a sharp maximum of the magnetic field on the axis. The first commercial transmission electron microscope was developed and released by "Siemens" in 1939 under the supervision of E. Ruska, who received the Nobel Prize for Physics in 1986 for his invention of the electron microscope.

The transmission electron microscope (TEM) is the now the most common tool, designed to study very thin samples with thicknesses of about 10 - 100 nm at electron accelerating voltages of up to 200 kV [14]. Its scheme is similar to the optical system of an optical light microscope in that it creates an image of the sample based on the transmitted radiation beam. **Figure 1** compares the main units of the transmission electron microscope to those of the optical microscope and the scanning electron microscope.

The transmission electron microscope consists of an electron gun and electromagnetic lenses. **Figure 1** shows the path of the rays in a microscope with three levels of zoom and one condenser lens, which is used to pre-focus of the electron beam. The two common modes of its operation are the imaging mode and the recording of micro-diffraction patterns. In imaging mode, the plane of a sample conjugates to the screen, where it is assumed that the sample is very thin compared to the focal length of the imaging objective lens. In the diffraction mode, the object of observation conjugates to the back focal plane of the objective lens, where the distribution of the amplitudes of the electron beam, diffracted off the sample, corresponds to the Fraunhofer's diffraction of the electrons. The resulting picture is observed on a remote screen. It is possible to determine the composition of the sample by analyzing the diffraction pattern [16].

In the phrase "micro-diffraction", the word "micro" implies that the diffraction signal is observed only from a rather small area of the sample. The area is chosen either by a special selector diaphragm, installed in the plane of the first intermediate image (see **Figure 1**), or by irradiation of the desired part of the sample with a sharply-focused electron beam with a diameter of about 1 micrometer or less.

In imaging mode, the so-called aperture diaphragm is installed in the focal plane of the object lens in order to limit the aperture of the beam, *i.e.* the opening angle of the cone of rays emerging from the sample. This helps to decrease the spherical aberration, a basic imaging error associated with the projection of a point in the sample to a point of the image.

Modern microscopes are usually equipped with a two-lens condenser, which allows for a rather high electron

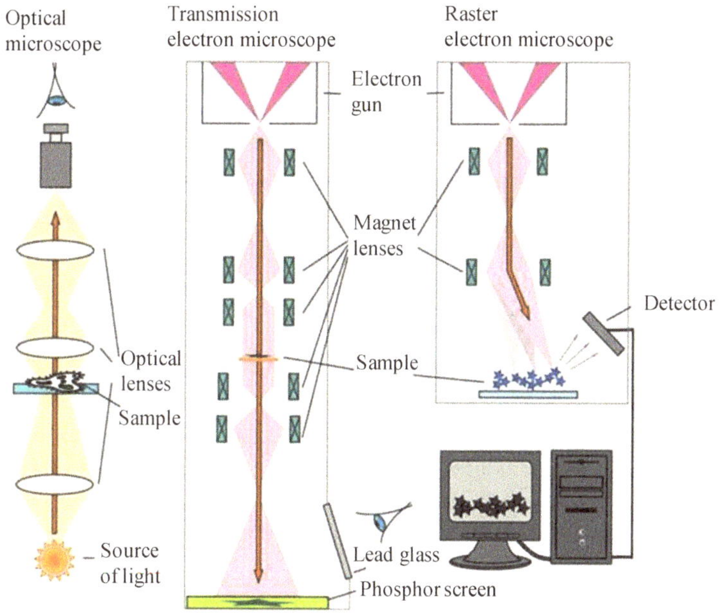

Figure 1. Schematic of the optical, transmission and scanning electron microscopes [15].

flux to a small area of the object with characteristic size of 1 - 5 microns. This eliminates the growth of the films due to hydrocarbon polymerization of oil vapors as a result of electron bombardment of the unobserved sample areas in microscope where the column is pumped by diffusion pumps.

According to statistical data, tens of thousands of transmission microscopes with accelerating voltages of 80 - 100 kV have been produced. Hundreds of devices with an accelerating voltage of between 200 kV and 500 kV were created for the study of thicker objects, with a thickness of up to 10 microns. And there are only a few dozen microscopes intended for the voltage range from 1 MV to 3.5 MV.

For a typical transmission electron microscope with an accelerating voltage of 100 kV, the de Broglie wavelength of the electron is 3.7 pm. This is much smaller than the diameter of a hydrogen atom, which equals approximately 100 pm. This would seem to make it easy to observe any atomic particle, as long as the resolution of the electron microscope is on the order of the electron wavelength as is the case in optical microscopes. But in fact, it is extremely difficult to observe individual atoms because of the poor quality of the electron lenses compared to the light-optical lenses: at the same angular size of the light and electron beams, the electron lenses create almost of two orders of magnitude more aberrations and image distortion [13].

At a kinetic energy of 100 eV, the de Broglie wavelength of the electron is equal to 0.1 nm, *i.e.* on the order of atomic dimensions. Electrons with such energies are strongly diffracted by atomic lattices. This property is used in low-energy electron diffraction to investigate the structure and composition of surfaces and near-surface layers, because the penetration depth of the electrons is very small.

3. Stroboscopic Electron Microscopy

Electron microscopy takes advantage of two of the three main properties of the electron: the small de-Broglie wavelength of the electron, which enables very high spatial resolution; and the electric charge, which allows to control its movement by electric and magnetic fields. About 60 years ago, the third property of the electron, its small mass (*i.e.*, low inertia) began to be used in so-called stroboscopic electron microscopy for the study of periodic processes [17]. The essence of the method is quite simple: a sample with periodically time-varying characteristics, with a fixed rate of change, should be irradiated by short bunches of electrons with the same frequency and fixed phase. The synchronicity freezes the image, which corresponds to the state of the sample at the time of arrival of the electron pulses.

In the early 60's, in the Soviet Union as a result of the work of Prof. G.V. Spivak, and in other countries, stroboscopic techniques were developed for all major types of electron microscopes: translucent, raster, emission and mirror types [18,19]. These developments allowed the exploration of processes in rapidly varying thin films (transmission electron microscopy) and in the surface layers of solids (raster, emission and mirror microscope): re-polarization of various ferroelectrics; distinctions of the domain structure during re-magnetization of thin magnetic films; heterogeneity of alternating fields of magnetic heads; and local defects of p-n junctions during their fast switching between the locked and open states [20]. Additionally, from a theoretical point of view, it is possible to study the microstructure of solution-gels in periodic reactions such as the Belousov-Zhabotinsky reaction using electrons with low kinetic energies, less than 100 eV, and to investigate the periodic sorption-desorption processes of catalysis in a similar fashion [20].

The initial work on stroboscopic microscopy was performed with micro- and nanosecond time resolution. Later, several laboratories started to investigate periodic processes with picosecond temporal resolution, usually using stroboscopic scanning electron microscopy. The main difficulties arose from the gating, the interruption, or the modulation of the electron beam. Generally, one uses one of two methods: first, it is possible to change of the intensity of the beam by locking or unlocking it with appropriate voltage pulses applied across the cathode or across the modulator in the electron gun of the microscope; or secondly, one can deflect a continuous electron beam by an electric or magnetic field in the vicinity of a skipping aperture of small diameter. The blanked, locked or unlocked electron pulses should be properly frequency and phase synchronized with the signals on the sample in order to avoid a loss of resolution.

The larger the duty cycle of the blanking pulses, *i.e.* the higher the ratio of the repetition period to the duration of the front of the electron pulse, the better the temporal resolution of the stroboscopic microscope, but the lower the brightness of the image. Consequently, each application requires a careful assessment of the best trade-offs. Moreover, a very short front of the pulse causes the resolution of the microscope to deteriorate because of non-ideal edges of the pulse, the appearance of chromatic aberration in transmission electron microscopy and the manifestation of the Bursch effect, which describes the transition of the longitudinal component of the electron velocity to a transverse component due to Coulomb repulsion.

4. Dynamic Transmission Electron Microscopy

Transmission electron microscopy, with its rather wide range of its tools, has long been a powerful technique in many fields of science and technology. While allowing to reach a spatial resolution on the sub-nanometer scale, conventional transmission electron microscopy does not permit an ultrashort temporal resolution, *i.e.* a time resolution in the picosecond or femtosecond regime. In contrast, optical microscopy using fluorescent probes, such as green fluorescent proteins, made it possible to visualize the processes occurring in vitro [2]. However, despite the high temporal resolution of optical methods, their spatial resolution is usually limited by the wavelength to the range of 200 - 800 nm. A review article [21] and a recently published monograph [2] describe the development of the methods of 4D, dynamic transmission electron microscopy (DTEM) and ultrafast electron microscopy (UEM), respectively.

The images and diffraction patterns in the work [22] were obtained at an accelerating voltage of 120 keV for non-organic materials (single crystals of gold, amorphous carbon and polycrystalline aluminum) and for biological cells of the intestine of rats [23]. In those studies, the gated beam contains on the average only one electron per pulse, with the total dose being limited to a few electrons per $Å^2$. Note, that these electron pulses were completely controlled in space and in time. Conceptually, this works is based on the methods of time resolved electron diffraction (TRED) or ultrafast electron diffraction (UED) [1,24-26,78,112], ultrafast electron crystallography (UEC) [23] and the other DTEM studies [21], but with one important difference-namely, the implementation of synchronized pulses of single electrons to form an image. The sketch of this 4D DTEM is shown in **Figure 2**.

This single-electron approach differs quite remarkably from the one used in the work [27], which used single pulses with -10^8 electrons and pulse durations of -20 ns.

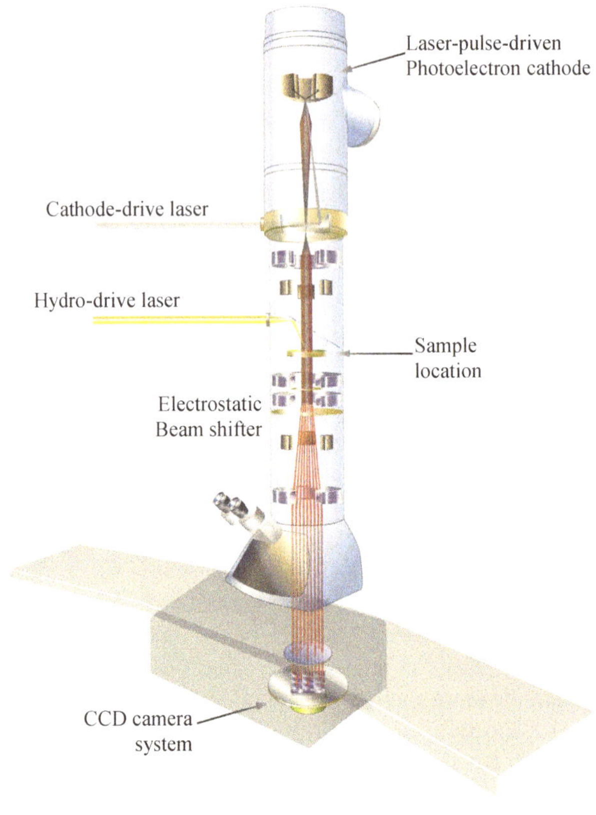

Figure 2. The schematic of a dynamic transmission electron microscope (DTEM) [21].

Those pulses, used to study laser-induced melting of different metals, contain so many electrons that high temporal resolution is not possible. Moreover, as noted in [28], the uncertainty in the spatial resolution in these experiments due to statistical interferences was of the order of micrometers, because the temporal window for imaging was on a nanosecond scale. It was found, that the use of amplified femtosecond light pulses, with much higher intensity of about $10^{12}\ \mathrm{W/cm^2}$, led, at first, to the formation of an electron beam that was insensitive to the focusing system of the microscope, and, secondly, to a significant lengthening of the pulse, which was due to electron-electron Coulomb repulsion.

The single-electron per pulse approach of 4D UEM [22,29] was used to obtain images from polycrystalline and single crystals [2]. To verify the authenticity of the experiment, some images were collected when the femtosecond pump laser pulses were blocked. In this case, the pattern was not observed, confirming that the time resolved signal generated in the electron microscope was indeed induced optically, and that the contribution of thermal electrons that are continuously generated by the cathode can be neglected [29]. It is useful to remember that electron microscopes can operate in UEM and DTEM mode. To reach atomic scale spatial resolution, the UEM diffraction regime should be used by introducing an intermediate lens to select the back focal plane of the lens as the object.

5. Controlling the Motions of Free Electrons by Femtosecond Light Pulses

Femtosecond lasers play a key role in 4D electron microscopy. Ultrashort pulses are necessary for the optical excitation of ultrafast processes in the sample and to form the ultrashort probe electron pulses. Additionally, pulsed laser radiation makes it possible to measure the temporal characteristics of the photoelectron bunches, to implement additional compression of electron pulses, and to generate electron bunches with durations that may potentially reach into the attosecond regime.

The last statements need to be amplified on, because they imply that is possible to control free electrons with a laser beam. At first glance, this statement conflicts with the well-known fact that free electrons cannot absorb the electromagnetic (EM) radiation because the laws of conservation of energy and momentum cannot simultaneously be satisfied. Nevertheless, free electrons can scatter EM radiation. From a quantum mechanical point of view, this is the result of stimulated Compton scattering in a strong laser field.

The possibility to control the translational degrees of freedom of charged particles, in particular of free electrons, by a spatially inhomogeneous electromagnetic field was demonstrated theoretically by Gaponov and Miller in the mid 1950's [30]. This mechanism leads to the ejection of charged particles from the strong field. In the case of high-power laser radiation, the Gaponov-Miller, or ponderomotive (gradient) force (*PF*) [31], which is defined as the spatial gradient of the so-called ponderomotive potential U_p, can reach sufficiently high values to form a basis for the effective control of electron pulses in vacuum.

For non-relativistic electrons, the concept of the ponderomotive potential in a spatially inhomogeneous electromagnetic field $E = E_o(r)\sin(\omega t)$ can be introduced by averaging the Hamiltonian H over the fast oscillations with frequencies ω and 2ω:

$$\begin{aligned} H &= \left\langle \left(p + eE_o(r)\sin(\omega t)/\omega\right)^2 \right\rangle / 2m_e \\ &= p^2/2m_e + \left[eE_o(r)\right]^2/\left(4m_e\omega^2\right) \\ &= p^2/2m_e + U_p \end{aligned}$$

Here, m_e and e are the mass and the charge of the electron and $\langle\cdots\rangle$ denotes an averaging over optical cycles. As a result, the expression for the *PF* can be written as the following expression:

$$F^{(\mathrm{pond})}(r,t) = -\left[e^2\lambda^2/64\pi^2 m_e\varepsilon_0 c^3\right]\nabla I(r,t)$$

Here ε_0 is the dielectric constant, c the speed of light and λ the wavelength of the laser radiation. It follows that for a tightly focused laser pulse with an intensity of $10^{15}\ \mathrm{W/cm^2}$ in the center of a focal spot of 2 μm diameter (at the level of $1/e$) and λ=800nm, the force, $F^{(\mathrm{pond})} \approx 10^{-11}\,\mathrm{N}$, is approximately equal to the strength of the interaction between two electrons separated by a distance of 5 nm. In this example, the ponderomotive potential is $U_p \approx 60\,\mathrm{eV}$.

It is evident that ultrashort laser radiation allows to control free electrons in vacuum on a femtosecond scale. This makes possible both the creation of ultrashort photoelectron bunches and the measurement of their duration [32,33]. The determination of the temporal characteristics of a pulsed electron beam is based on scanning the time delay between the laser radiation that forms the photoelectric bunch and the tightly focused laser beam that changes the velocity distribution of the photoelectrons as a result of the *PF*. Such a method compares favorably with a standard streak camera in its final characteristics and can be used for femtosecond electron beams [33].

In a streak camera, the electrons are deflected by a high-speed and high-voltage electric field with an amplitude of about 2 - 5 kV before reaching a position-sensitive detector. The rate of change of the high electric field determines the temporal resolution of the device. The currently best time resolution is in the subpicosecond range and reaches a value of ~300 fs. In addition to this, it should be noted that the femtosecond temporal synchronization of the high-voltage electrical pulse in the streak camera and the femtosecond laser pulse that forms the ultrashort photoelectron bunch poses serious technical challenges.

The study of the behavior of the free electrons in spatially inhomogeneous electromagnetic fields began shortly after the appearance of the theory of Gaponov and Miller and the first experiments were devoted mainly to the possibility of creating traps using microwave technology [34]. Here, special attention was focused on the observation of the passage of the electrons with a certain kinetic energy through the ponderomotive potential, which allowed, for example, to determine the value of U_p [34,35]. The first demonstration of scattering of the photoelectrons by the ponderomotive potential, created by an intense sub-nanosecond laser pulse, was done in the work [36]. Here the multi-photon ionization of Xe, especially bleeding in the vacuum system, was used to prepare the pulsed photoelectron beam with kinetic energies less than 5 eV. This experiment demonstrated the control of low-energy photoelectrons by an optical ponderomotive potential with the height of about 10 eV.

Special attention should be given to the measurement of the value of the *PF* with which tightly-focused femtosecond laser radiation acts on the electrons propagating in vacuum. This may be useful for *in situ* space-time diagnostics of the laser fields of high intensity, and can be used to validate the measurement of ultrashort electron pulse durations based on the irradiation of the electron bunches with the laser pulses of high intensity. Here it should be noted that the use of laser radiation with intensities of more than $10^{14}\ \mathrm{W/cm^2}$ usually is accompanied by the photoionization of the residual gas in the vacuum system and thus by a possible deformation of the electron trajectories that pass through a cloud of charged particles.

As was noted above, electrons are ejected from their field-free paths by a strong electromagnetic field. This is illustrated in **Figure 3**, which shows two counter-propagating tightly-focused laser beams forming a standing wave from which the electron pulse is deflected. Such a scheme has a dual purpose. First, it allows to determine the duration of the initial electron bunch by scanning the timing of the laser pulses that create the standing wave, and using a position-sensitive scheme to detect the rejected, but in general, diffused, electron beam components. Secondly, the deflected electron pulses potentially have a shorter time duration compared to the incident electron pulses.

Two important remarks should be made here. The scheme in **Figure 3** requires the spatial-temporal matching of two femtosecond laser pulses from two directions and a (sub)picosecond electron bunch from another direction. The use of a single laser beam for the electron scattering would, of course, be simpler. However, in the standing wave the spatial inhomogeneity of the electromagnetic field is about $\lambda/2$, a rather small value that is technically difficult to obtain with a single amplified femtosecond laser beam. For example, a parabolic mirror can usually focus 0.8-micron amplified femtosecond laser radiation to a spot diameter of $d_{1/e} \sim 6$ microns. Therefore, for a given laser pulse energy, laser wavelength and focusing conditions, the scattering of electrons in a pulsed standing wave will be more pronounced.

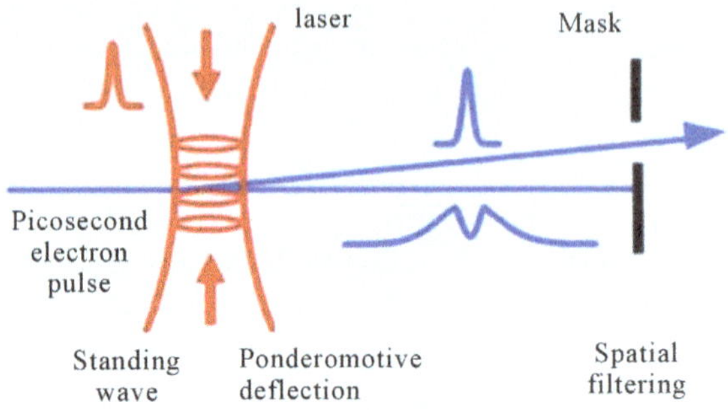

Figure 3. Deflection of an electron beam as a result of the optical ponderomotive force generated by an ultrashort laser pulse [2].

Another important consideration relates to the ultimate possibilities of this approach. This is important because it is rather interesting to know the accuracy of the electron pulse duration measurement as well as the duration of the deflected electron bunch. These characteristics are directly related to the transit time of the fast electrons through the laser focal spot,

$$\tau_{\text{fin}} \approx \tau_{tr} = d_{1/e}/v_e\ ,$$

where v_e is the speed of the electron beam. Using $d_{1/e} = 6\ \mu\text{m}$ and $v_e \approx 10^8\ \text{m/s}$ for electrons with kinetic energy of 30 keV, we find $\tau_{\text{tr}} \approx 60\ \text{fs}$. Note that the use of a standing wave formed by two laser beams (**Figure 3**) will not lead to a decrease in τ_{fin}, and hence the accuracy of measurement for this example will remain of the order of 100 fs. Therefore, to obtain a femtosecond electron bunch with $\tau \sim 10\text{-}50$ fs and its full diagnostics it is important to ensure the tight focusing of intense laser radiation, and/or to use electrons with higher kinetic energy. Of course, in order to deflect these ultrafast electrons it is necessary to implement intense laser fields.

An alternative way of creating ultrashort electron bunches also employs a standing light wave, but in a collinear geometry [2]. This approach is shown schematically in **Figure 4**.

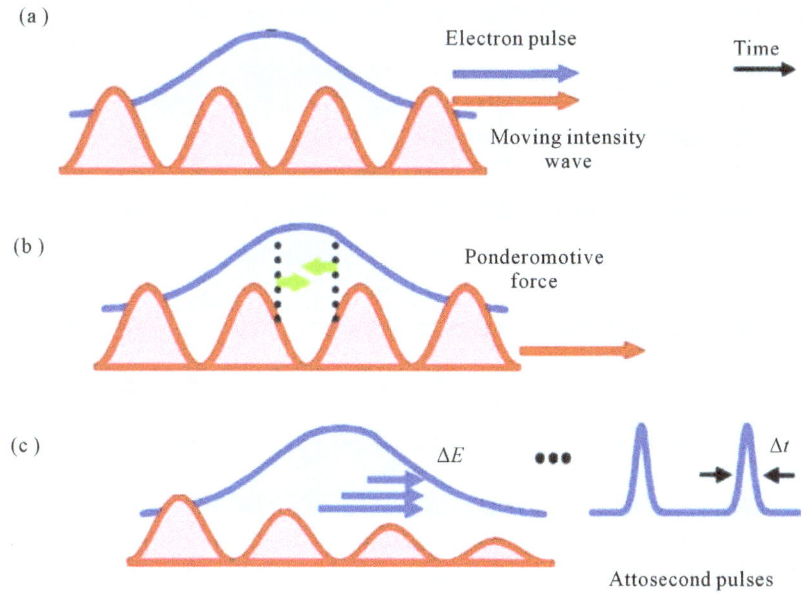

Figure 4. Schematic diagram of the creation of attosecond electron pulses. (a) An electron bunch with sub-picosecond duration is irradiated by two counter-propagating laser beams (forward and after the bunch) with different wavelengths. In the reference frame of the electrons, the wavelengths are the same as a result of the Doppler effect, and a standing light wave is formed; (b) Ponderomotive forces push electrons away from the strong laser field; (c) As a result, the compression leads to the formation of electron bunches with attosecond duration. Strictly speaking, the envelope duration of the initial electron bunch remains constant, but importantly, the attosecond spikes appear inside the envelope [2].

In this context it should be recalled that on the basis of the uncertainty principle, $\Delta E \Delta t \geq \hbar$, the existence of ultrashort, in this case attosecond, bursts needs a wide spectrum of energy ΔE_e. For example, a bunch with 100 as (10^{-16} s) duration has at least $\Delta E_e \approx 6\text{eV}$. The initial picosecond photoelectron pulse does not exhibit such spectral properties. As shown in **Figure 4**, the required broadening is due to the work of the ponderomotive forces [2]. Let us illustrate this with a simple estimate, by writing down the expression for the work in the traditional sense, where the energy gain of the electrons is equal to $\Delta E_e = F^{(\text{pond})} \tau_p v_e$ (τ_p-the laser pulse duration), and the force is approximately defined as $F^{(\text{pond})} \approx 2U_p / \lambda^*$ (λ^*-wavelength of the laser radiation in the reference frame of the fast electron). In the conditions of the numerical experiment [2] $U_p \approx 0.2\text{eV}$, $\lambda^* = 370\text{nm}$, $v_e \approx 10^8\ \text{m/s}$ and $\tau_p = 300\text{fs}$, one finds $\Delta E_e \approx 32\ \text{eV}$. The calculated value ΔE_e exceeds by 2 orders of magnitudes the spectral width of "normal" sub-picosecond electron pulse, emitted from the solid photocathode under the action of femtosecond laser radiation.

As considered in the monograph [2], this scheme is rather elegant. Even so, it is important to emphasize that the measurement of the duration of the formed attosecond electron bursts is a quite a challenging experimental problem. It is likely that it belongs to the class of experiments in which it is easier to create an ultrashort bunch than to measure it.

6. Ultrafast Electron Microscopy

An early antecedent of modern methods to investigate fast processes is the English photographer Eadweard Muybridge, who used multiple cameras to temporally resolve fast motions. For example, in 1871 Muybridge captured the moment when a galloping horse separated all four hooves from the surface of the earth (**Figure 5**).

Of interest to the scientific world today would be movies showing the movement of atoms in molecules, the fast dynamical processes in biological systems, or the changes in the geometrical structures of nanoscale devices. To estimate the magnitudes of time scales involved, we can extrapolate from the observation of macroscopic objects by using as the maximum velocity the speed of sound. Thus, a time resolution of 1 μs suffices to freeze in time motions of objects with spatial dimensions of 1 mm. In a system with a characteristic length scales of nanometers, a picosecond time resolution is appropriate. And, for molecular systems, femtosecond time resolution is necessary to capture the motions of individual atoms and functional groups.

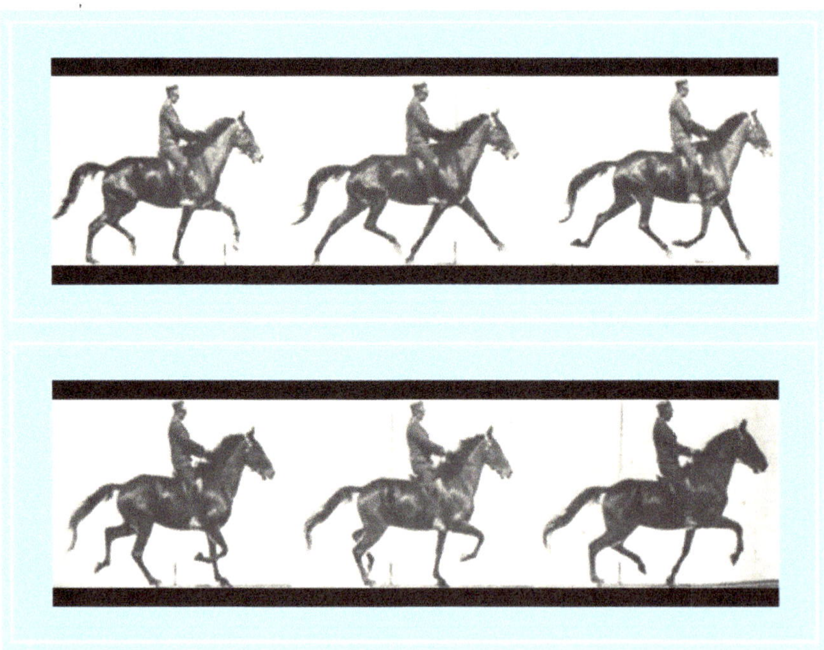

Figure 5. One of the first movies in history. These images were obtained by E. Muybridge in 1871. The time delay between the exposures was about 50 ms [37].

The ultrafast electron microscope built at the California Institute of Technology [2] is designed to explore a class of problems that occur on picosecond or sub-picosecond time scales. A schematic illustration of this instrument is shown in **Figure 6**.

It is convenient to conceptualize the setup as consisting of three main sections: 1) the formation and acceleration of the sub-picosecond pulsed electron beam; 2) the interaction unit, in which the sample is excited by femtosecond laser pulses and its spatial-temporal structure probed by ultrashort electron bunches; and 3) the position-sensitive detection of the electrons after their interaction with the sample. The use of electron beams of subpicosecond duration imposes strict requirements on the detector, because the probe pulse cannot contain a large number of electrons. As it will be illustrated below, the main cause of the electron pulse temporal spreading is the space-charge interaction between the electrons within each pulse on account of their Coulomb repulsion. Therefore, the detection efficiency for electrons in this microscope should be as high as possible. Much preparatory work was done to achieve this [2]. As a result, the detection efficiency for the time-resolved instrument is almost an order of magnitude higher than that of a standard electron microscope. This also provides an interesting opportunity to significantly reduce the exposure (the total number of electrons passing through the sample to form an image on the detector), which is particularly important for imaging biological samples that are very sensitive to radiation induced damage.

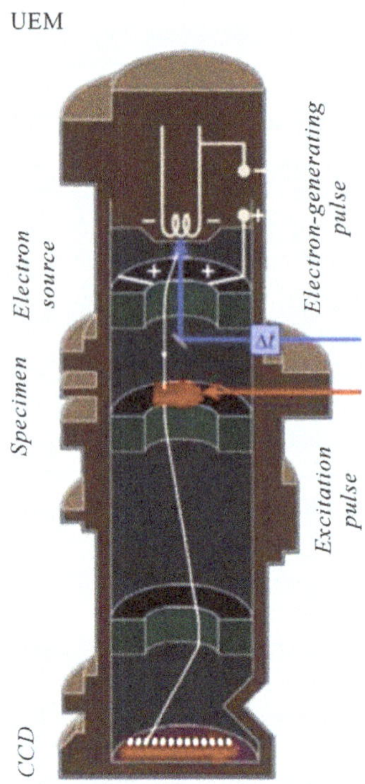

Figure 6. Schematic illustration of an ultrafast electron microscope. A commercial electron microscope, operating in a continuous mode, can be taken as a basis in order to create such a device. To adapt it to 4D microscopy requires a femtosecond laser, which provides a pulsed excitation of the sample (corresponding to a laser beam shown in red), and allowing for the preparation of the photoelectron probe pulse of ultrashort duration (ray marked in blue) [2].

The ultrashort electron bunch is the essence of 4D electron microscopy. Therefore, the question of its duration should be given special attention. The duration of the photoemission of electrons from the surface of a solid cathode and therefore the initial duration of the photoelectron bunch are determined by the duration of the laser pulse. But during the propagation from the photocathode to the target, the electron beam is stretched in time. Here we consider three major factors that lead to a smearing of the electron bunch.

First, the acceleration of photoelectrons in a static electric field near the cathode causes so-called time-of-flight chromatic aberration for the electrons (TFCA). The temporal resolution limit is given by:

$$\Delta\tau_{EF} = \sqrt{2m_e \Delta E_e} / eF \ ,$$

where ΔE_e is the initial spread of the kinetic energy of the electrons as they emerge from the cathode and F is the electric field in the acceleration gap.

The expression for TFCA can be obtained by solving a quadratic equation that describes the uniformly accelerated motion of the electrons in the acceleration gap of the length of l:

$$v_0 t + (eF/m_e)t^2/2 = l$$

Differentiating the both sides of the equation with respect to time t, we get:

$$\Delta v_0 + (eF/m_e)\Delta t \approx 0$$

from which immediately follows the expression for TFCA. Let us consider a numerical example. At $\Delta E_e = 0.3$ eV and $F \approx 10^8$ V/m, which approximately corresponds to the maximum value of the field (higher values lead to vacuum breakdown even with special electrodes that underwent thorough mechanical, electronic and ionic treatment), this contribution is $\Delta t_{EF} \approx 60$ fs.

Secondly, the movement of electrons through the field free flight tube with slightly different kinetic energies is accompanied by an elongation of the time of flight, which is expressed as

$$\Delta\tau_{TOF} = t_{TOF}(\Delta E_e / 2K)$$

Here, t_{TOF} is the time of flight of the electrons with kinetic energy K from the photocathode to the sample. At $K = 30$ keV, $\Delta E_e = 0.3$ eV and $t_{TOF} \approx 3$ ns, corresponding to the length of 30 cm between the cathode and the target, we get $\Delta t_{TOF} \approx 15$ fs.

Thirdly, the Coulomb repulsion between the electrons within a pulse causes a swelling of the photoelectron bunch. To assess the effect of this mechanism on the temporal properties of the beam, we consider a simple model, in which the electron bunch is travelling through a field-free region. (Here the field-free region is taken for the simplicity.) The total energy of the bunch, consisting of N electrons, is conserved:

$$V + W = 1/2\left(\sum e^2/4\pi\varepsilon_0 \left|\boldsymbol{r}_i - \boldsymbol{r}_j\right|\right) + 1/2\sum m_e v_i^2$$

Therefore, the characteristic value of the distribution of the velocities, δv, resulting from the electrons pushing apart each other can be qualitatively estimated by using the expression

$$m_e \delta v^2/2 \approx e^2 N/(16\pi\varepsilon_0 \delta r),$$

where δr is the initial size of the photoelectron cloud. In this model, the velocity spread of the electrons causes the elongation of the order of $\Delta\tau_C \approx t_{TOF}\,\delta v/v_e$, which can be rewritten as:

$$\Delta\tau_C = (t_{TOF}/v_e)\sqrt{e^2 N/(8\pi\varepsilon_0 \delta r m_e)}$$

At $v_e \approx 10^8$ m/s, $N \approx 10^4$, $t_{TOF} \approx 3$ ns and $\delta r \approx 10$ mm, corresponding to the size of the focal spot of the tightly focused laser beam, this contribution in the framework of the simple model is equal to $\Delta\tau_C \approx 5$ ps. This shows that the Coulomb repulsion of the electrons in the electron bunch is the main factor limiting the ability of ultrafast electron microscopy. Naturally, the use of ultra-high accelerating voltages, such as 500 kV instead of 30 kV, can reduce the swelling of the electron bunch due to Coulomb repulsion, but can also lead to destruction of the sample. Use of very fast electrons, in the MeV range, can further reduce the space-charge broadening because of the onset of relativistic effects [110,111]. It should also be noted that a system based on a TOF electro-static mirror, a so-called reflectron, can be used to re-compress the electron bunches after their original spreading [26,38]. At the present time, a radio frequency compression scheme is already in use for this purpose.

A serious challenges in the field of 4D electron microscopy remains the task of observing an image or a diffraction pattern produced by a single, ultrashort electron pulse. The solution to this problem is rather important, because it would enable the study of ultrafast irreversible processes. An early breadkthrough has been achieved by a collaboration of scientists at Brown University and at SLAC, who used ultrashort electron pulses with 5.4 MeV of energy, generated using the Gun Test Facility at SLAC, to record single-shot diffraction patterns of 160 nm thick aluminum foils [110,111]. Their work showed that MeV electrons should make it possible to achieve sub-100 fs time resolution. Subsequently, physicists in Japan were able to collect diffraction patterns of thin gold films with single subpicosecond electron bunches [39]. They applied a magnetic sector in order to compress the electron beam in the experiment [39].

Non-Thermal vs Thermal Melting of a Solid Irradiated by a Femtosecond Laser

The study of ultrafast phase transitions in condensed matter is one of the most interesting fields of 4D electron microscopy, viz. ultrafast electron crystallography. To induce an ultrafast phase transition such as the melting of a solid, it is necessary to insert energy into the lattice within a small period of time and to reach the critical temperature as quickly as possible. The most suitable tool to accomplish this is a femtosecond laser pulse. In metals and semiconductors, the photons excite the electrons (and the holes) to high-lying electronic states, resulting in a change of the charge density. Initially, the absorbed energy is stored in electronic degrees of freedom. With the decrease of electron density between the atoms, the bonds in the system become weaker, and at some degree of excitation, the lattice is no longer fixed. Such processes may lead to non-thermal melting of a solid, provided that they are fast compared to the relaxation of the electron energy into the lattice vibrations (phonons). Preliminary studies of aluminum by a femtosecond laser pump-probe have shown that in this material, the transition from a solid to a liquid phase occurs within 500 fs, *i.e.* faster than the energy transfer from the excited electrons to the phonons [40].

The direct observation of melting of Al with subpicosecond temporal resolution has been achieved by ultrafast electron diffraction [41]. A 20-nm thick film of polycrystalline aluminum was used as the sample in this experiment. The results are shown in **Figure 7**.

Figure 7 clearly demonstrates, that the lattice is heated and remains in the solid phase with a distinct diffraction pattern on a time scale of up to 1.5 - 2.5 ps after the excitation by laser pulse, which is consistent with the thermal melting. Current research continues in this direction.

7. Phase Transitions in Nanoparticles

Vanadium dioxide, VO_2, undergoes a phase transition of the first type from a low-temperature monoclinic phase (**M**) into a high-temperature tetragonal phase of rutile (**R**) at −67°C. This phase transition has been the subject of intensive research since its discovery almost half a century ago. The work [29] describes the first study of this material using 4D UEM. Because single electron probe pulses were used, Coulomb (space-charge) repulsion was absent. The experiment demonstrated the possibility of obtaining a sequence of images (*i.e.* a movie) with atomic-scale spatial resolution combined with an ultrashort temporal resolution. In particular, it was shown that it was possible to investigate the ultrafast metal-insulator phase transitions in vanadium dioxide. A structural phase transition in VO_2 nanoparticles is exhibited in the diffraction patterns (at atomic scale) and UEM images (at nanometer scale) with a characteristic hysteresis with a temporal resolution on the order of 100 fs (**Figure 8**).

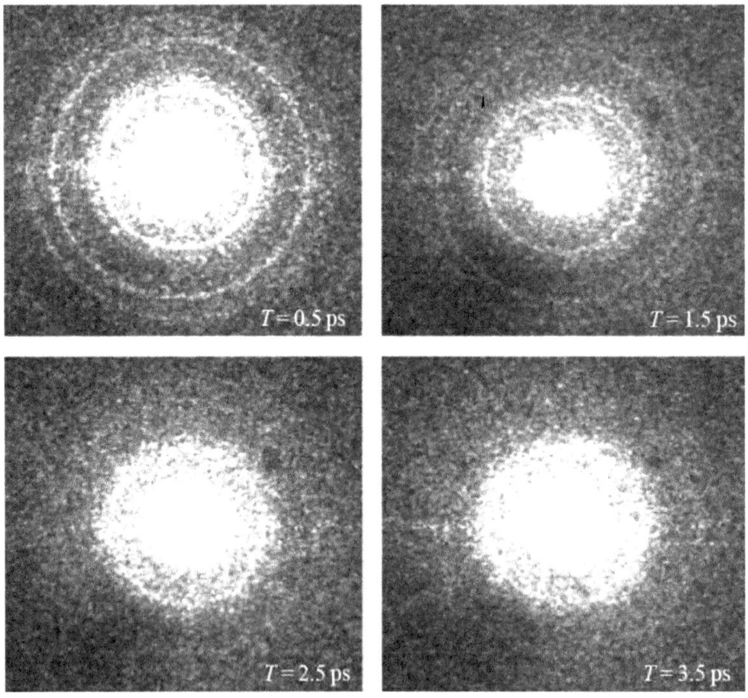

Figure 7. Electron diffraction patterns from a film of polycrystalline Al, recorded at different times after the excitation by a femtosecond laser pulse as labelled. The energy density of the laser pulse is about 70 mJ/cm^2 [41].

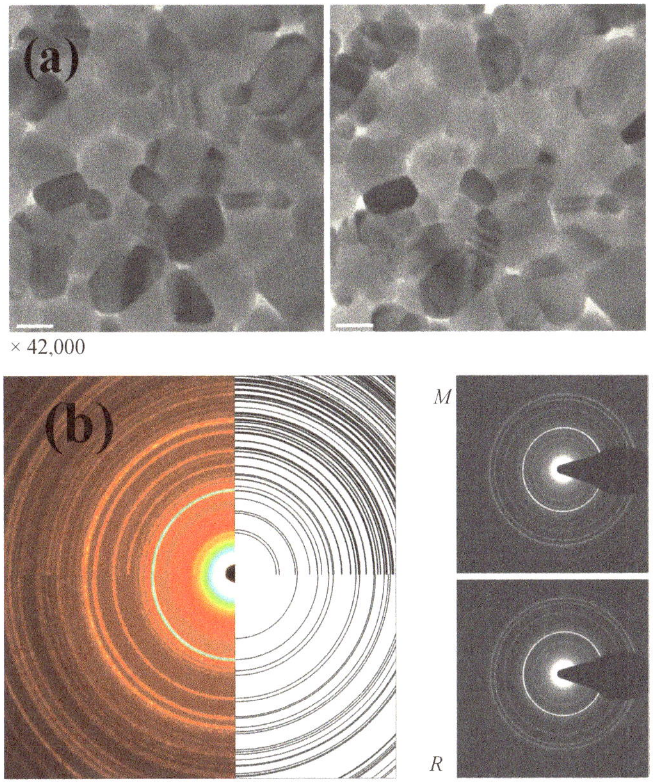

Figure 8. (a) Images corresponding to a phase transition in the films of VO_2 (left) and after the phase transition (right), were collected in UEM with the magnification of 42,000 (the bar corresponds to 100 nm). It should be noted that these images could not be observed when the generation of ultrashort photoelectrons was locked. (b) The diffraction patterns, corresponding to the phase transition in the films of VO_2 (right) and after the phase transition (left), were obtained by UEM. The diffraction patterns of two phases (the monoclinic phase M and the high-temperature phase of tetragonal rutile, R), the experimentally observed (left side of (b)) and constructed by calculations (right side of (b)). The analysis was described in the paper [29].

4D UEM with femtosecond laser irradiation of the sample in the near IR allowed to record the movement of atoms during the phase transition in all three dimensions and in time. This study showed that at first, the vanadium atoms were separated from each other for a short time period, and then began to move towards their final positions.

8. Visualization of Mechanical Displacements of a Cantilever

The development of ultrafast electron microscopy made possible the direct observation of vibrations of nanostructures in the ultrasound range, covering the range from kHz to GHz [42]. In this connection it should be recalled that the methods of optical interferometry cannot ensure complete three-dimensional spatial resolution, while the possibilities of standard electron microscopy are usually limited to a temporal resolution of about 30 ms.

The work [42] used a pseudo-one-dimensional molecular material, crystal Cu (TCNQ), which exhibits highly anisotropic electrical and optical properties, to prepare a nano- and microcantilever with a thickness of 300 nm and 2 mm, respectively. Opto-mechanical motions can be initiated in such a crystal by a charge transfer from the radical $(TCNQ)^-$ to copper Cu^+.

In the experiment, the charge transfer occurred as a result of irradiation of the crystal by nanosecond laser pulses with a wavelength of 671 nm, a pulse intensity of about $160\,mJ/cm^2$ and a pulse repetition rate of 100 Hz. This exposure induced resonant cantilever vibrations in the megahertz frequency range. The tiny displacements of the tip were visualized using the 120 keV photoelectrons (**Figure 9**).

The results so obtained allowed to determine the Young's modulus, measure the resonance frequency of the sample, and to calculate the energy stored due to the opto-mechanical expansion of the crystal Cu (TCNQ).

9. 4D Microscopy of Fatty Acids—Toward the Study of Cell Tissue Materials

The study of organic and biological structures is an extremely important field of electron microscopy. However, the investigation of such objects entails special

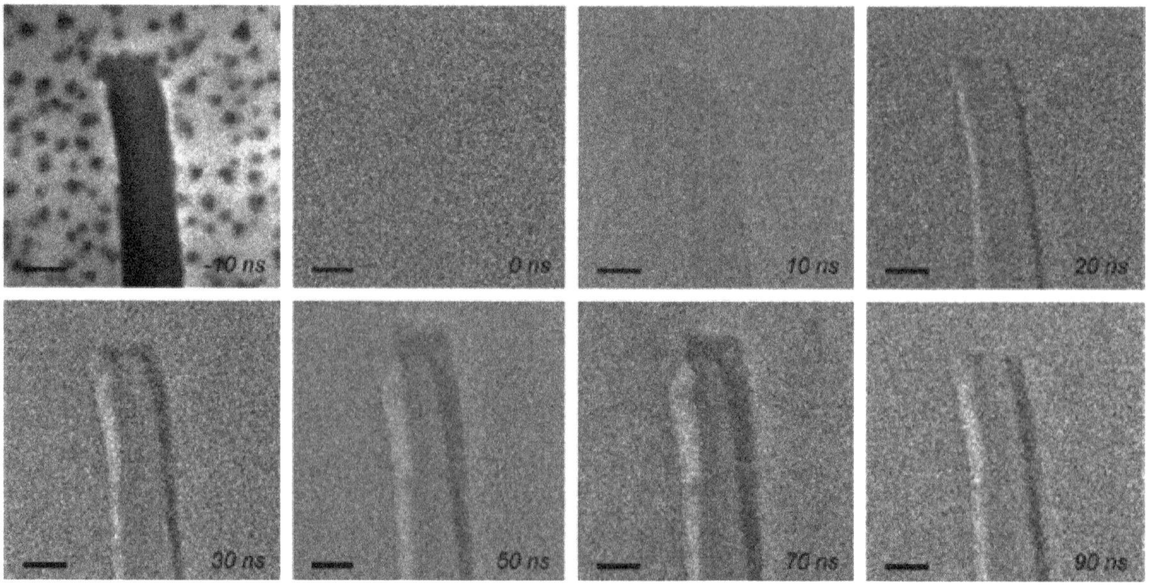

Figure 9. 4D electron micrographs of a nanoscale cantilever. The frames show difference images, *i.e.* time dependent images where a reference frame is subtracted out. The image recorded at negative delay (−10 ns) relative to the 671 nm pump laser pulse served as a reference frame. A scale bar in the lower left corner corresponds to 200 nm [42].

demands on the microscope. This is primarily due to the increased sensitivity of the bio-structures to possible irreversible deformation and damage that can be caused by a beam of fast electrons. Therefore, for successful experiments in this field it is necessary to ensure an extremely high detection efficiency of the electrons passing through the sample. Note that the efficiency of the detector is directly related to the time of exposure. In this connection it is helpful that the high detection efficiency in the 4D microscopy allows working with a smaller number of electrons in the bunch.

Amongst the most common biological structures are membrane-type layers of fatty acids. In the laboratory, two-dimensional crystals can be created from long fatty acid hydrocarbon chains. It is possible to use the Langmuir-Blodgett technique to prepare the monomolecular layers of organic material on a solid substrate. It is rather interesting to observe the transition from a two-dimensional to a three-dimensional crystal of the long molecular chains and to visualize the self-assembly processes.

$C_{20}H_{42}$ chains with a length of 95Å on a silicon substrate were considered in computer simulations [2]. In the model, an ultrafast pulsed heating of the substrate leads to further changes in the inter-atomic distances of $C_{20}H_{42}$, reflecting the coherent motion of the atoms in the molecular chain. Each step of the calculation corresponds to a time interval of 0.5 fs, while the total number of steps reaches 200,000. Model calculations have shown an increase in the length of the fragment $-CH_2-CH_2-CH_2-$ that is located in the immediate vicinity of the silicon surface, by about −0.08 Å during 5 ps.

The 4D UEM method should allow us to study more complex biological samples that are directly related to the life sciences. It should be recalled that the use of ultra-short electron bunches for the visualization of biological tissues can improve the spatial resolution of the device compared to conventional electron microscopy. In some ways the situation is similar to making a snapshot by a photographic camera with a long exposure time: the resulting image becomes quite fuzzy (**Figure 10**), but admittedly, can lead to a new artistic perception of reality.

The 4D electron microscopy succeeded in imaging stained cellular material in rat tissue (**Figure 11**), and of crystalline proteins, or cells in the vitreous bodies.

10. Musical Nanoscale Instruments: From the Drum to the Harp and the Piano

Ultrathin plates of graphite, gold, or other material, subjected to pulsed laser heating, can experience an ultrafast deformation (bending) as a result of anisotropic expansion of the atomic structure. As it turned out, the vibrations that accompany such deformations may be surprisingly interesting.

Electron bunches at 200 keV were used to visualize mechanical vibrations of graphite nano-sheets (the membranes) with thicknesses of 75 nm. Vibrations with submillisecond temporal intervals resulted from exposure to pulsed laser radiation with a wavelength of 532 nm and a repetition rate of 5 kHz [2]. The energy density of the focused laser pulses reached a value of about 7 mJ/cm^2.

Immediately following the pulsed laser heating of the

Figure 10. Photograph of a city at night, made with a long exposure time. (The figure is carried over from [43]).

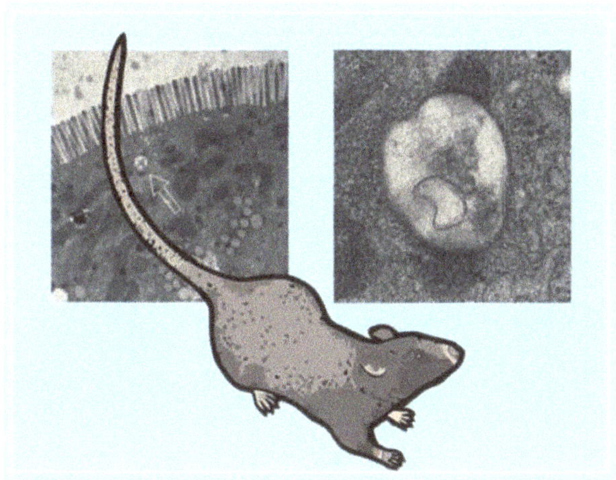

Figure 11. Images of stained cells of rat-related tract, obtained using ultrafast electron microscopy. The micrograph shows the micro-villas in the intestinal epithelium of newborn rats and numerous small vesicles (blisters) in the cytoplasm. An enlarged image of the vesicles, pointed by the arrow on the left frame, is given on the right [2].

membrane, individual carbon atoms were found to vibrate in a random order. But after a few tens of microseconds, the corresponding modes of the membrane began to be synchronized (**Figure 12**), resembling a "drumbeat" with a frequency of around 1 MHz.

The processing of the images (**Figure 12**) used a cross-correlation function defined as

$$\gamma(t';t) = \left[\Sigma_{x,y} C_{x,y}(t) C_{x,y}(t')\right] \Big/ \sqrt{\left[\Sigma_{x,y} C_{x,y}(t)^2 S_{x,y} C_{x,y}(t')^2\right]}$$

where the contrast is

$$C_{x,y}(t) = \left[N_{x,y}(t) - \langle N(t) \rangle\right] / \langle N(t) \rangle.$$

Here, $N_{x,y}(t)$ is the number of counts corresponding to the picture element (pixel) with the coordinates of (x, y), and $\langle N(t) \rangle$ is the average number of the counts. 2000 images, collected with an interval of 50 ns, were used in **Figure 12**. The micrograph of the graphite membrane is represented in the lower left corner.

The success with the graphite nano-sheets inspired the search for other musical nano-instruments. A harp and a piano, constructed at the nanoscale, were made using arrays of cantilevers (**Figure 13**). They were prepared by micro-structuring of multilayer workpieces of $Ni/Ti/Si_3N_4$ using sharply-focused ion beams [44]. The workpieces consisted of 30 nm layers of nickel and titanium, consistently applied to 15 nm film of Si_3N_4. For the piano, the cantilevers had almost the same lengths, approximately 4.6 μm, but their width varied from about 400 nm to 2.3 μm. In the harp, the lengths and widths of individual elements varied in the range of 1.2 - 9.1 μm and ~300 - 600 nm, respectively.

In the experiment, a sample (the harp or the piano) was heated by nanosecond 519 nm (or 532 nm) pulses from a Nd:YAG laser with a pulse repetition rate of 1 kHz. The energy density per pulse reached 2 mJ/cm^2. As a result of the heating of the sample, consisting of the layers with different coefficients of thermal expansion, the mechanical deformation of the cantilever triggered its vibrations. The mechanical movements were observed using the photoelectrons that were obtained by illuminating the photocathode with the fourth harmonic of a Nd:YAG laser. The results for individual cantilevers p1, p5, h1 and h5 (for convenience they are shown in **Figure 13**) are presented in **Figure 14**. The combined results of the observations are given in **Figure 15**.

In the next series of the experiments, the femtosecond laser with MHz pulse repetition rate was used "to play an instrument." Here a fiber laser generator with pulse repetition frequency of about 25 MHz was used. The energy density per pulse was in the range of 20 - 30 mJ/cm^2.

Beyond their academic interest, the musical nanosystems can have very specific applications. For example, the authors [44] note that such layered nanostructures can be used to precisely measure the temperature of a device with a spatial resolution of 10 μm in combination with a microsecond temporal resolution.

11. 4D Electron Tomography

Electron tomography began to develop as early as the 1960-ies [45]. Progress in this field was mainly limited by the development of computers, because it is important to process a large number of 2-dimensional electron diffraction patterns and combine them in a single movie. A humorous illustration to these words is given in two images of (**Figure 16**), taken from the book [45].

The use of modern laser technology has allowed to add to electron tomography the fourth dimension, time, and thus to create a completely unique device (**Figure 17**).

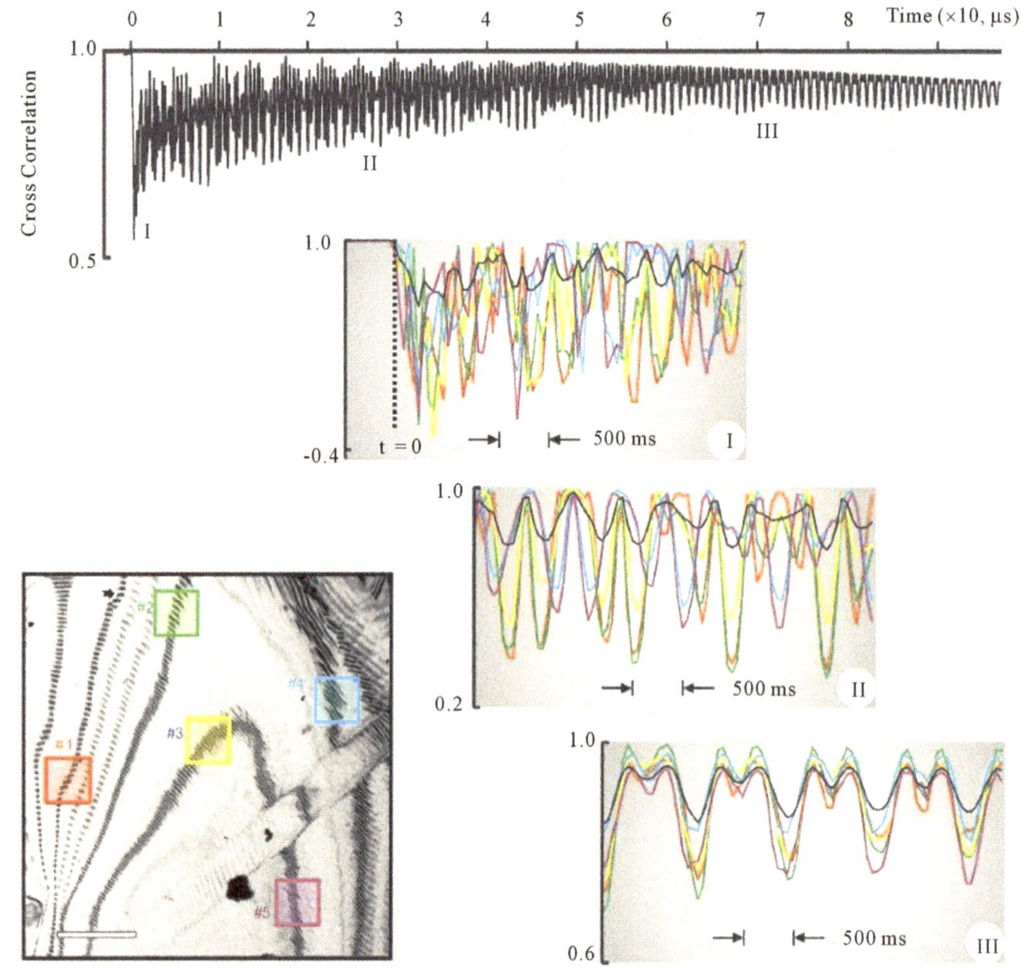

Figure 12. Time-dependence of the cross-correlation function. The scale in the lower left corner corresponds to 5 microns. Five areas with different initial dynamics are marked in red, yellow, green, blue and purple. The transition from chaotic behavior to coherent oscillations is shown in the panels I-II-III. They correspond to different temporal intervals as indicated in the figure [44].

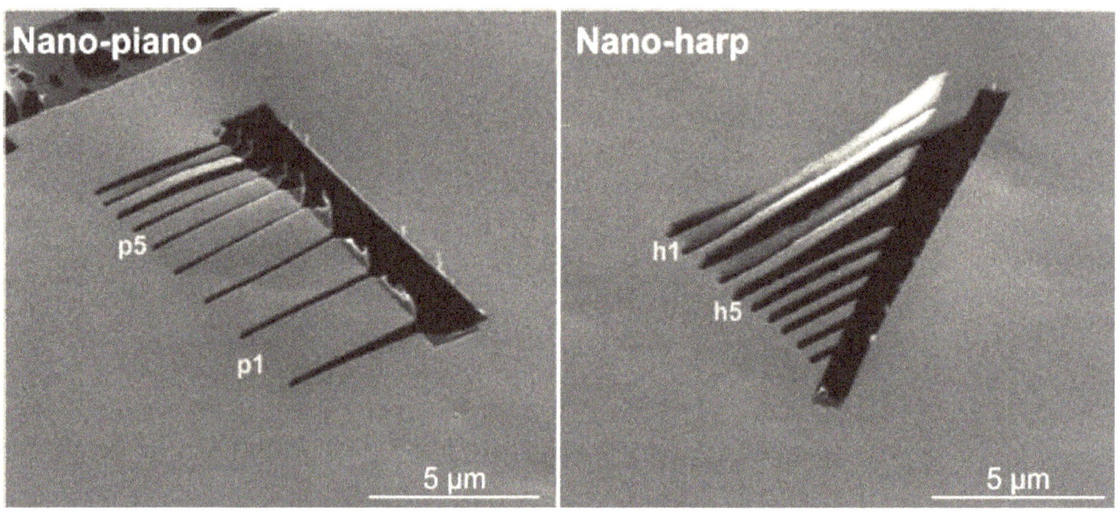

Figure 13. Images of musical nano-instruments. These frames were obtained with a scanning electron microscope. Note that two of the thinnest cantilevers, p7 and p8 on the left, and two pairs of cantilevers, h1/h2 and h3/h4 on the right, are partially melted in the central part as a result of micro-structuring by the focused ion beam [44].

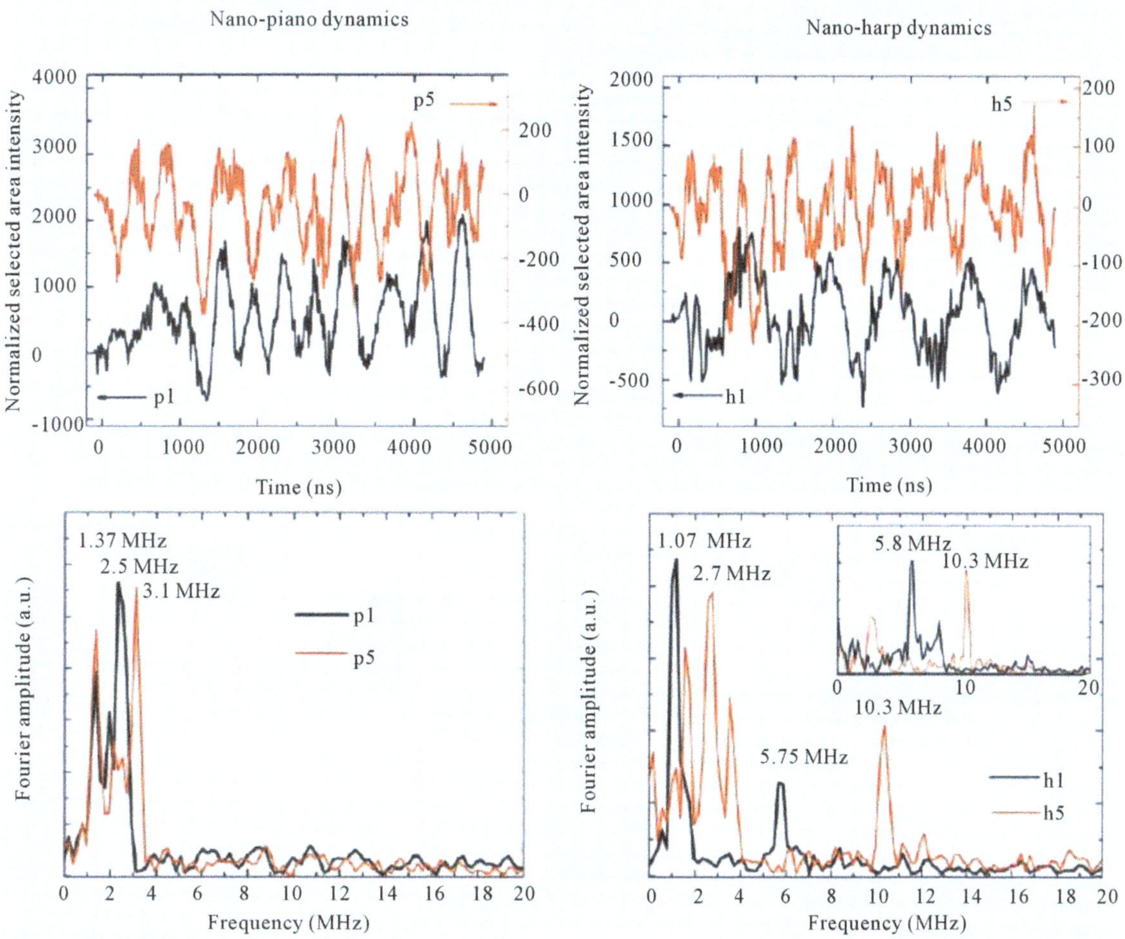

Figure 14. The oscillations of the cantilevers were caused by the pulsed heating of the sample (top). The results of the fast Fourier transform are presented below. The experimental data obtained in the tilted position of the sample with the angle of 30° are given below (insert in the right). The tilted position made it possible to record the movements of the cantilevers, performed outside of the plane of the nano-harp [44].

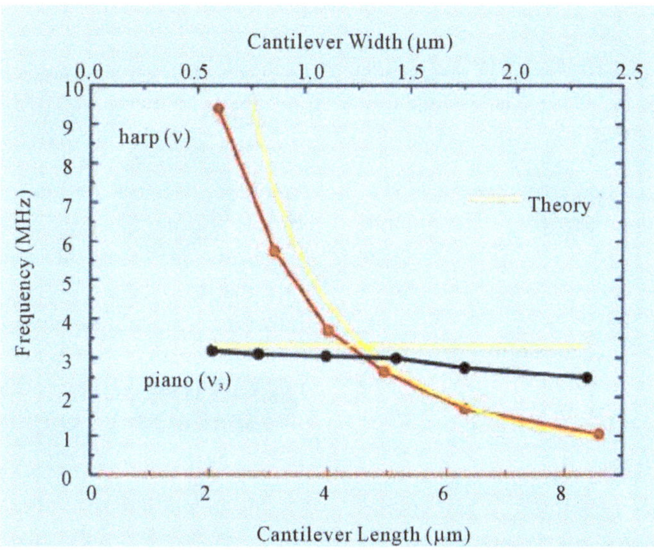

Figure 15. The dependence of the oscillation frequency on the size of the respective cantilever. Only the vibrations that occur out of the plane of musical nano-instruments are considered in this case [44].

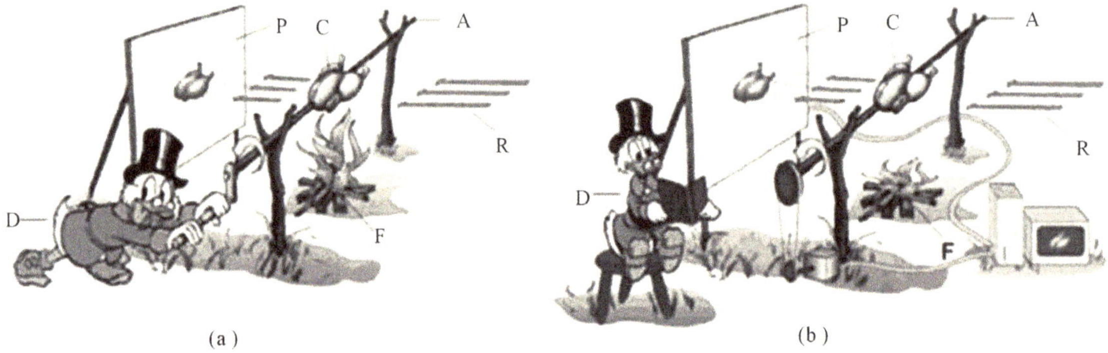

Figure 16. The humorous evolution of 3D tomography [45].

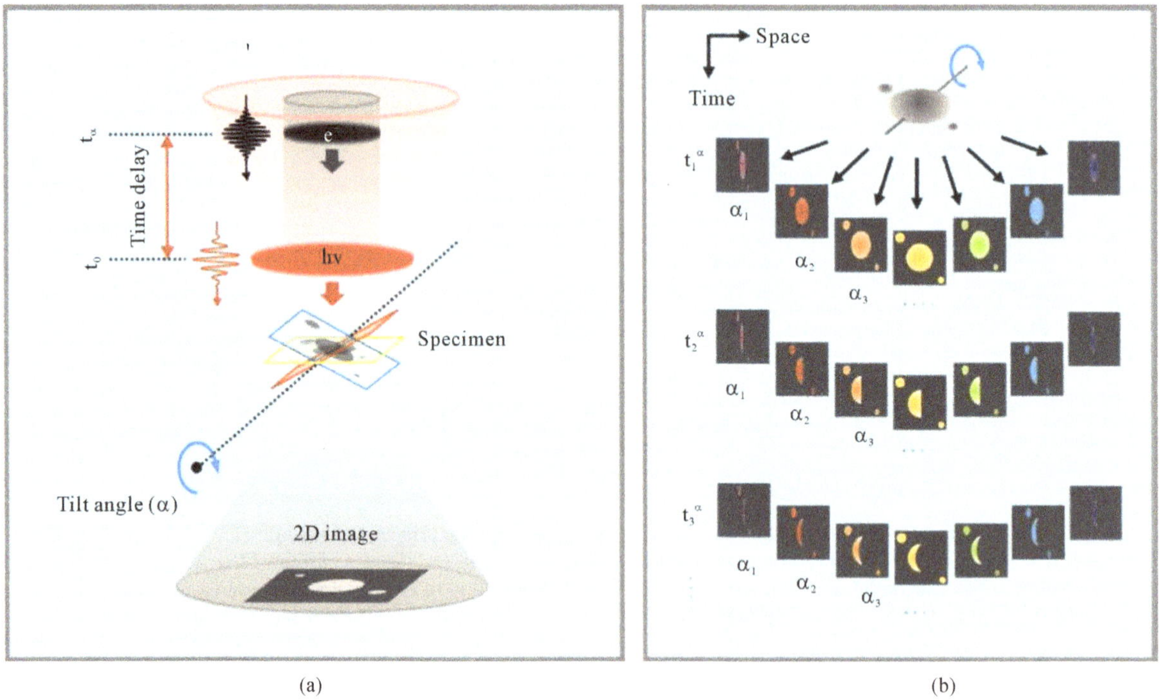

Figure 17. (a) Schematic diagram of 4D electron tomography. (b) A series of two-dimensional electronic images taken at different angles α and time delays. In [46] the angle range from −58° to +58° with increments of 1°. The total number of projections reaches 4000 [46].

The temporal behavior of carbon nanotubes was studied by 4D tomography in the work [46]. The sample was twisted in the form of a bracelet. The measured length of the nanotube was $L \approx 4.4$ microns. The images in **Figures 18** and **19** were obtained using the 200 keV electron pulses. Femtosecond laser excitation led to the heating of the nanotube and caused the subsequent structural changes that began on a picosecond temporal scale.

Of particular interest to a broad audience is that it is possible to create a video, and in the future, a real atomic movie, using the results of 4D electron tomography. It should be emphasized that the pioneering experiment [46] did not induce irreversible changes in the structure of the nanotube, because the total dose received by the carbon sample during the experiment was about 2 orders of magnitude less than the value at which the irreversible deformation occurs. This is a tribute to the extremely high detection efficiency reached in 4D electron microscopy.

12. Electron Microscopy with High Spectral, Spatial and Temporal Resolution

The combination of spatial nano-resolution and subpicosecond, eventually femtosecond temporal resolution, forms the basis of the atomic movie. Naturally, when the energy of the electron beam, transmitted through the sample, is measured in a similar geometry, additional and

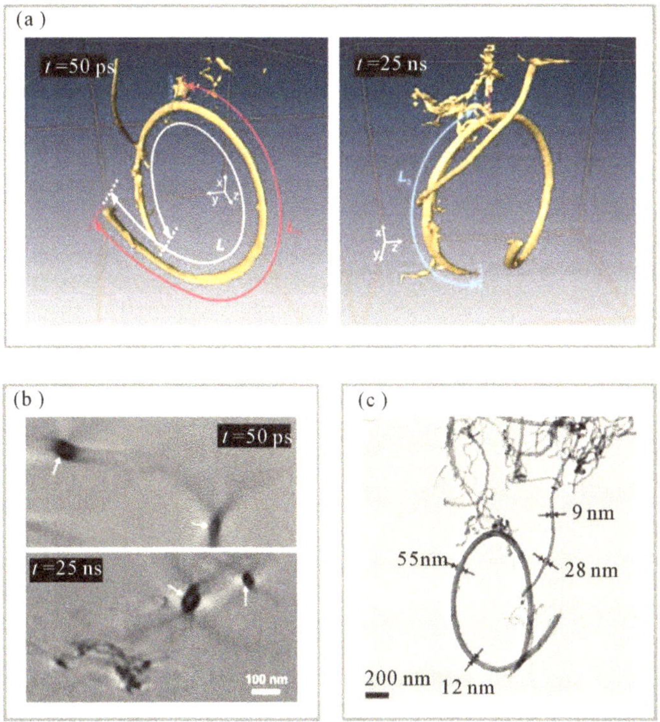

Figure 18. (a) 3D images of multi-walled carbon nanotubes at different times. The lengths of the respective segments around a fixed area are denoted as L1 and L2; (b) Cross section of the 3D images. Two-dimensional slices in the plane xy are shown. Slice thickness is 4.6 nm. Dark areas of the arrows point to the nanometer-thick walls. According to these micrographs, the spatial resolution of the method is sufficient for the correct imaging of channels with a diameter of ~10 nm; (c) An image, taken with a transmission electron microscope [46].

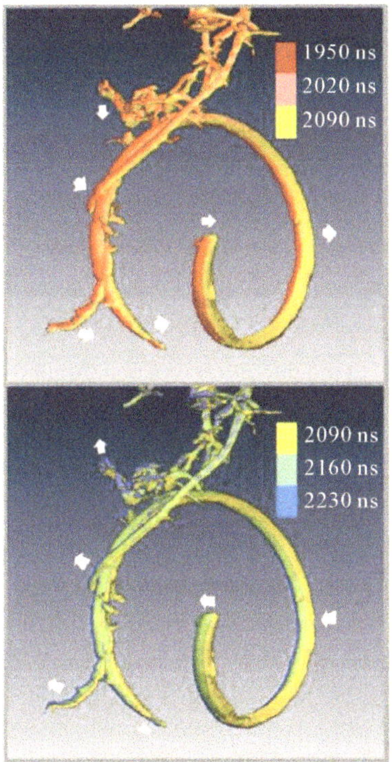

Figure 19. Mechanical vibrations of a carbon bracelet. The arrows indicate the direction of the individual sections of a carbon nanotube [46].

very important information is obtained. Note that for most modern research laboratories, such a study would pose quite a serious experimental challenge. Therefore, the precedent in this field plays an important role.

This idea was realized in the work [47] using a single metal nanostructure. In this experiment, the silver particle of triangular shape with a characteristic length of 130 nm and thickness of 20 nm, placed on a substrate of graphene (**Figure 20**), was irradiated by femtosecond laser pulses with a photon energy of 2.4 eV. The plasma oscillations in silver, excited by the optical radiation, were probed by an ultrashort electron beam with a diameter of 10 nm. The electron beam could be moved over the surface of the sample. Here the energy gain of the electrons, passed through the sample, was measured in addition to the spatial and temporal characteristics.

It is important to emphasize that due to the field localized on the surface of a metal particle, the electrons can not only lose kinetic energy, as would be the case with a standard transmission electron microscope, but also to acquire energy (**Figure 21**). In principle, this process can be controlled by changing the wavelength of the laser radiation. The accuracy of this procedure is determined by the spectral width of the laser pulse and could reach values of about 1 meV.

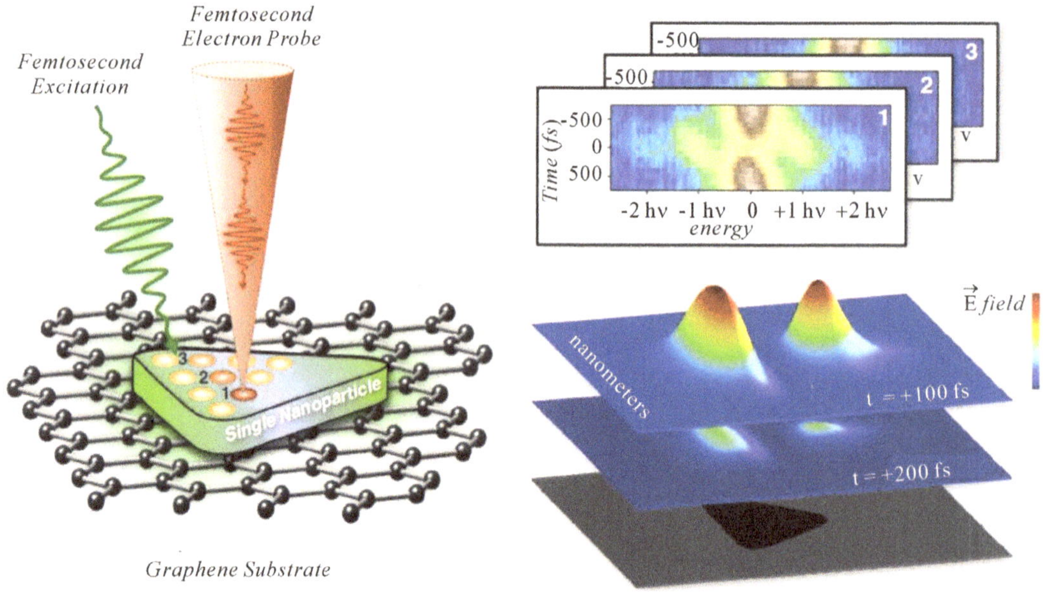

Figure 20. The process of ultrafast spectral imaging. A 10-nm electron beam was scanned over the silver particle that was previously excited by a femtosecond laser pulse. For each position of the probe, the change of the kinetic energy of the electrons (viz. electron spectrum) was measured as a function of the delay between the exciting, optical, and probing, electronic, pulses. The increase in the energy of the electrons is measured in units of the photon energy, being equal to hν = 2.4 eV [47].

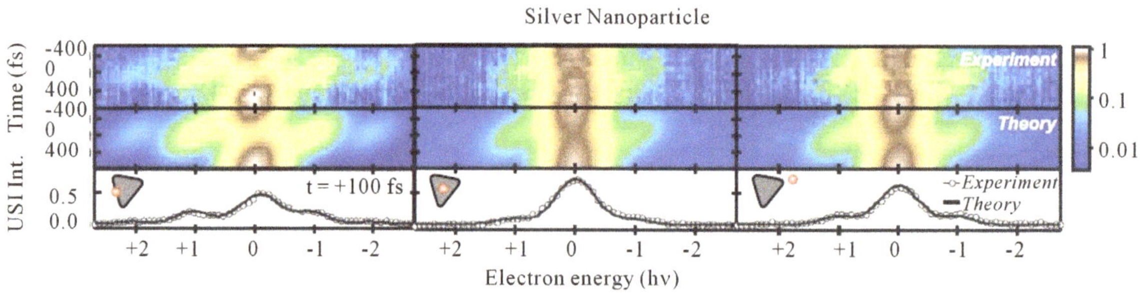

Figure 21. Spectral and temporal characteristics of silver nanoparticles. The electron spectrum, obtained at a delay of 100 fs after the femtosecond optical pulse, is shown below. It is schematically indicated by the position of the electron nanobeam with respect to the triangular silver particle, which is depicted in gray [47].

The results of the measurement of the plasmic texture, generated on the surface of the silver particle, is shown in **Figure 22**.

It is clearly seen in **Figure 22** that the field of plasma oscillations, excited by the laser radiation, is concentrated near the vertices (the cusps) of the triangular particle. There is a fairly good agreement with theory. Qualitatively similar results were obtained near the sharp edge of a copper surface irradiated by the laser [47].

13. Conclusions

Electron microscopy and diffraction with a high temporal resolution have opened the possibility of directly observing processes that occur in non-steady states of the studied substances. A temporal resolution on the order of 100 fs corresponds to the transition of a quantum system through an energy barrier of a potential surface, describing a chemical reaction in the processes of breaking and forming new bonds of the interacting agents. The advances thus opened the possibility of investigating the coherent nuclear dynamics of molecular systems and condensed matter [24,25,48-51].

In the past two decades it has been possible to observe the motions of nuclei in time intervals corresponding to the period of the oscillation of the nuclei. The observed coherent changes in the nuclear system at such time steps determine the fundamental transition from the standard kinetics to the dynamics of the phase trajectory of a single molecule, the molecular quantum state tomography [50-53].

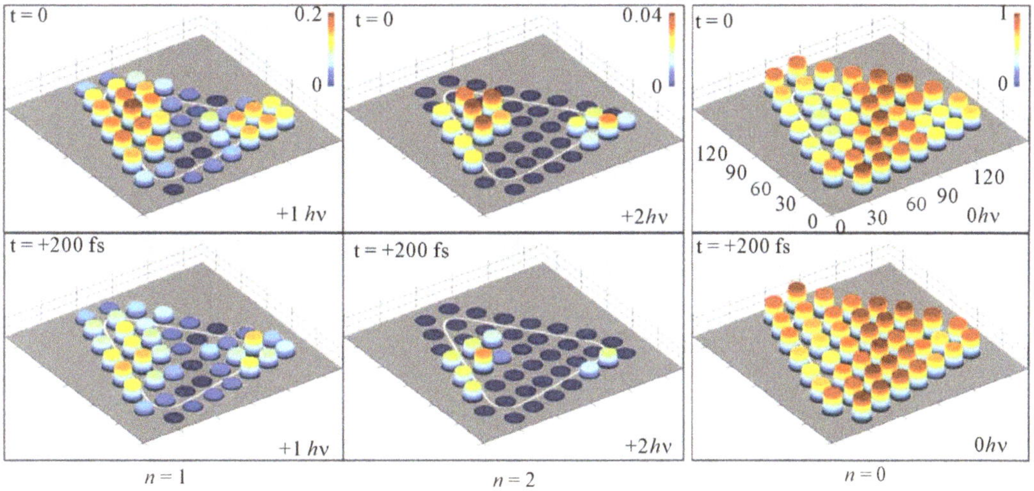

Figure 22. Plasma oscillations, optically excited in the silver nanostructure, were visualized by a 10 - nm electron beam. Here, the electron signal is proportional to the height of the corresponding cylinder. For each panel, the energy, acquired as a result of the interaction of the electrons with nanolocalized fields, is shown in the lower right corner [47].

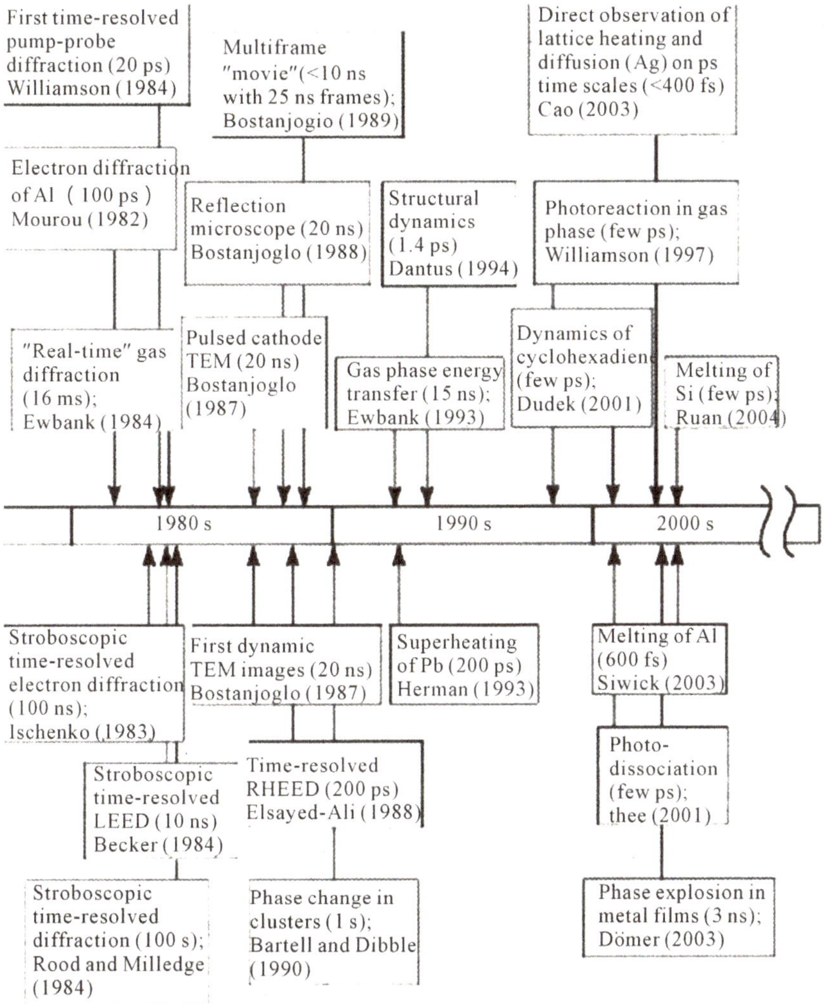

Figure 23. Chronology of the basic works (the first steps) in Dynamic Electron Microscopy, done in the first two decades (1982-2004) [21]. The parallel development of UED, UEM, UEC and DTEM until 2010 is described in [2]. The references [58-109] demonstrate the worldwide experimental activity in this field.

At the present time, the method of ultrafast diffraction continues its rapid development. The methods of ultrafast electron crystallography and electron microscopy with temporal resolution from micro- to subpicoseconds provide great opportunities to the study of the 4D structural dynamics (see the review articles [21,54] and the monograph [2]). Very recent advances in the formation of ultrashort electron pulses allow us to reach an attosecond temporal resolution and observe the coherent dynamics of the electrons in molecules [55-57].

The development of dynamic electron microscopy and the first steps devoted to the progress in the study of structural dynamics of ultrafast processes are illustrated in **Figure 23**. It is all but certain that exciting developments in this field will continue into the future.

14. Acknowledgements

Academician of the RAS, Prof. A. L. Buchachenko, Prof. E. A. Ryabov, Prof. V. N. Bagratashvili, Dr. A. E. Luk'yanov, A. A. Timofeev, Prof. V. R. Flid, and Dr. James Beckman (Therapon, Inc.) are acknowledged for the hepful discussions. This work was supported by grants: No. 2012-1.1-12-000-1009-055 (from the Ministry of Education and Science of the Russian Federation) and No. 10-02-92000-HHC_a, 11-02-00868-a, 12-02-00840-a (from the Russian Fund for Basic Research). PMW acknowledges support by the Division of Chemical Sciences, Geosciences, and Biosciences, the Office of Basic Energy Sciences, the U.S. Department of Energy, Grant No. DE-FG02-03ER15452.

REFERENCES

[1] A. A. Ischenko, V. V. Golubkov, V. P. Spiridonov, A. V. Zgurskii, A. S. Akhmanov, M. G. Vabishevich and V. N. Bagratashvili, "A Stroboscopical Gas-Electron Diffraction Method for the Investigation of Short-Lived Molecular Species," *Applied Physics B*, Vol. 32, No. 3, 1983, pp. 161-163.

[2] A. H. Zewail and J. M. Thomas, "4D Electron Microscopy. Imaging in Space and Time," Imperial College Press, London, 2010.

[3] B. K. Weinstein, "Atomic Resolution Electron Microscopy," *Soviet Physics Uspekhi*, Vol. 152, No. 3, 1987, pp. 75-122.

[4] J. Brandon and W. Kaplan, "Microstructure of Materials. Methods of Research and Monitoring," M. Technosphere, Bergenfield, 2004.

[5] A. Vlasov, K. A. Yelsukov and V. Panfilov, "Microscopy Techniques," Publishing House of Moscow State Technical University named after Bauman, Moscow, 2011.

[6] A. Vlasov, K. A. Yelsukov and I. A. Kosolapov, "Electron Microscopy," M. Bauman Bauman, Moscow, 2011.

[7] D. Shindo and T. Oikawa, "Analytical Transmission Electron Microscopy for Materials Science," M.: Mir, Moscow, 2006.

[8] Y. S. Umansky, J. N. Skakov, A. N. Ivanov and L. N. Rastorguev, "Crystallography, X-Ray and Electron Microscopy," Metallurgy, Moscow, 1982.

[9] R. Kheidenraikh, "Fundamentals of Transmission Electron Microscopy," Academic Press, Waltham, 1966.

[10] P. Hirsch, A. Howie, R. Nicholson, D. Pashley and M. Whelan, "Electron Microscopy of Thin Crystals," World, Moscow, 1968.

[11] M. V. Locquim and M. Langeron, "Handbook of Microscopy," Butterworths, London, 1983.

[12] J. Spence, "Experimental Electron Microscopy Razresheniya," M.: Science, Poole, 1986.

[13] I. Watt, "The Principles and Practice of Electron Microscopy," Cambridge University Press, Cambridge, 1985.

[14] G. Thomas and M. J. Gorindzh, "Transmission Electron Microscopy of Materials," Nauka, Moscow, 1983.

[15] E.A. Gudilin, "Electron Microscopy," 2008. http://www.nanometer.ru/2008/12/21/electron_microscopy_55002.html

[16] C. Amelinks, "Methods of Direct Observation of Dislocations," Wiley, Hoboken, 1968.

[17] M. Motosuke and S. Tetsuya, "Stroboscopic Scanning Electron Microscope," US Patent Number: 4538065, 1985.

[18] G. V. Spivak, V. V. Shakmanov, V. I. Petrov, A. E. Lukyanov and S. Yakunin, "On the Application of the Gates with the Deflection Plates in a Stroboscopic Electron Microscopy," *Mathematical USSR Physics Series*, Vol. 32, No. 7, 1968, pp. 1111-1114.

[19] A. E. Lukyanov, V. Galstyan and G. V. Spivak, "About Stroboscopic Scanning Electron Microscopy Four-Semiconductor Structures," *Technology and Electronics*, No. 11, 1970, pp. 2424-2427.

[20] M. L. Taheri, N. D. Browning and J. Lewellen, "Symposium on Ultrafast Electron Microscopy and Ultrafast Science, Special Section: Ultrafast Electron Microscopy," *Microscopy and Microanalysis*, Vol. 15, No. 4, 2009, pp. 271-271.

[21] W. E. King, G. H. Campbell, A. Frank, B. Reed, J. F. Schmerge, B. J. Siwick, B. C. Stuart and P. M. Weber, "Ultrafast Electron Microscopy in Materials Science, Biology, and Chemistry," *Journal of Applied Physics*, Vol. 97, No. 111101, 2005, pp. 1-27.

[22] V. A. Lobastov, R. Srinivasan and A. H. Zewail, "Four-Dimensional Ultrafast Electron Microscopy," *Proceedings of the National Academy of Sciences of the United States of America*, Vol. 102, No. 20, 2005, pp. 7069-7073.

[23] A. H. Zewail, "4D Ultrafast Electron Diffraction, Crystallography, and Microscopy," *Annual Review of Physical Chemistry*, Vol. 57, 2006, pp. 65-103.

[24] S. Williamson, G. Mourou and L. C. M. Li, "Time-Resolved Laser-Induced Phase Transformation in Aluminum," *Physical Review Letters*, Vol. 52, No. 26, 1984, pp. 2364-2367.

[25] S. A. Akhmanov, V. N. Bagratashvili, V. V. Golubkov, A. V. Zgurskaya, A. A. Ischenko, S. A. Krikunov, V. P. Spiridonov and A. G. Tunkin, "Generation in the EMR-100 Electron Diffraction Apparatus the Picosecond Pulses of Fast Electrons by Photoemission in the Laser Field," *Soviet Technical Physics Letters*, Vol. 11, No. 3, 1985, pp. 157-161.

[26] P. M. Weber, S. D. Carpenter and T. Lucza, "Reflectron Design for Femtosecond Electron Guns," *SPIE Proceedings*, Vol. 2521, No. 1, 1995, pp. 23-30.

[27] H. Domer and O. Bostanjoglo, "High-Speed Transmission Electron Microscope," *Review of Scientific Instruments*, Vol. 74, No. 10, 2003, pp. 4369-4372.

[28] O. Bostanjoglo, "High-Speed Electron Microscopy," *Advances in Imaging and Electron Physics*, Vol. 121, 2002, pp. 1-51.

[29] M. S. Grinolds, V. A. Lobastov, J. Weissenrieder and A. H. Zewail, "Four-Dimensional Ultrafast Electron Microscopy of Phase Transitions," *Proceedings of the National Academy of Sciences of the United States of America*, Vol. 103, No. 49, 2006, pp. 18427-18431.

[30] A. V. Gaponov and M. A. Miller, "Potential Wells for Charged Particles in a High-Frequency Electromagnetic Field," *Soviet Physics—JETP*, Vol. 7, No. 1, 1958, pp. 168-169.

[31] T. W. B. Kibble, "Refraction of Electron Beams by Intense Electromagnetic Waves," *Physical Review Letters*, Vol. 16, No. 23, 1966, pp. 1054-1056.

[32] C. T. Hebeisen, R. Ernstorfer, M. Harb, T. Dartigalongue, R. E. Jordan and R. J. D. Miller, "Femtosecond Electron Pulse Characterization Using Laser Ponderomotive Scattering," *Optics Letters*, Vol. 31, No. 23, 2006, pp. 3517-3520.

[33] C. T. Hebeisen, G. Sciaini, M. Harb, R. Ernstorfer, T. Dartigalongue, S. G. Kruglik and R. J. D. Miller, "Grating Enhanced Ponderomotive Scattering for Visualization and Full Characterization of Femtosecond Electron Pulses," *Optics Express*, Vol. 16, No. 5, 2008, pp. 3334-3341.

[34] I. R. Gekker, "Interaction of Strong Electromagnetic Fields with Plasmas," Clarendon Press, Oxford, 1982.

[35] M. V. Fedorov, "Atomic and Free Electrons in a Strong Light Field," World Scientific, London, 1997.

[36] P. H. Bucksbaum, M. Bashkansky and T. J. McIlrath, "Scattering of Electrons by Intense Coherent Light," *Physical Review Letters*, Vol. 58, No. 4, 1987, pp. 349-352.

[37] E. J. Muybridge, "Animal Locomotion, an Electrophotographic Investigation of Consecutive Phases of Animal Movement," J. B. Lippincott and Co., Philadelphia, 1887.

[38] G. H. Kassier, K. Haupt, N. Erasmus, E. G. Rohwer and H. Schwoerer, "Achromatic Reflectron Compressor Design for Bright Pulses in Femtosecond Electron Diffraction," *Journal of Applied Physics*, Vol. 105, No. 1, 2009, pp. 113111-1-113111-10.

[39] S. Tokita, M. Hashita, S. Inoue, T. Nishoji, K. Otani and S. Skabe, "Single-Shot Femtosecond Electron Diffraction with Laser-Accelerated Electrons: Experimental Demonstration of Electron Pulse Compression," *Physical Review Letters*, Vol. 105, No. 21, 2010, pp. 215004-215007.

[40] C. Guo, G. Rodriguez, A. Lobad and A. J. Taylor, "Structural Phase Transition of Aluminum Induced by Electronic Excitation," *Physical Review Letters*, Vol. 84, No. 19, 2000, pp. 4493-4496.

[41] B. J. Siwick, J. R. Dwyer, R. E. Jordan and R. J. D. Miller, "An Atomic-Level View of Melting Using Femtosecond Electron Diffraction," *Science*, Vol. 302, No. 5649, 2003, pp. 1382-1385.

[42] D. J. Flannigan, P. C. Samartzis, A. Yurtsever and A. H. Zewail, "Nanomechanical Motions of Cantilevers: Direct Imaging in Real Space and Time with 4D Electron Microscopy," *Nano Letters*, Vol. 9, No. 2, 2009, pp. 875-881.

[43] http://hqwall.net

[44] J. S. Baskin, H. S. Parkand A. H. Zewail, "Nanomusical Systems Visualized and Controlled in 4D Electron Microscopy," *Nano Letters*, Vol. 11, No. 5, 2011, pp. 2183-2191.

[45] J. Frank, "Electron Tomography: Methods for Three-Dimensional Visualization of Structures in the Cell," Springer, New York, 2010.

[46] O.-H. Kwon and A. H. Zewail, "4D Electron Tomography," *Science*, Vol. 328, No. 5986, 2010, pp. 1668-1673.

[47] A. Yurtsever, R. M. van der Veen and A. H. Zewail, "Subparticle Ultrafast Spectrum Imaging in 4D Electron Microscopy," *Science*, Vol. 335, No. 6064, 2012, pp. 59-64.

[48] A. A. Ischenko, V. P. Spiridonov, L. Schäfer and J. D. Ewbank, "The Stroboscopic Gas Electron Diffraction Method for Investigation of Time-Resolved Structural Kinetics in Photoexcitation Processes," *Journal of Molecular Structure*, Vol. 300, 1993, pp. 115-140.

[49] V. A. Lobastov, J. D. Ewbank, L. Schafer and A. A. Ischenko, "Instrumentation for Time-Resolved Electron Diffraction Spanning the Time Domain from Microseconds to Picoseconds," *Review of Scientific Instruments*, Vol. 69, No. 7, 1998, pp. 2633-2643.

[50] J. D. Ewbank, L. Schafer and A. A. Ischenko, "Structural and Vibrational Kinetics of Photoexcitation Processes Using Time-Resolved Electron Diffraction," *Journal of Molecular Structure*, Vol. 534, No. 1-3, 2000, pp. 1-49.

[51] A. A. Ischenko, V. N. Bagratashvili and A. S. Avilov, "Methods of Studying Coherent 4D Structural Dynamics of Free Molecules and Condensed Matter," *Crystallography*, Vol. 56, No. 5, 2011, pp. 805-828.

[52] A. A. Ischenko, "The Study of Coherent Dynamics of the Nuclei by Electron Diffraction with a Time Resolution. II.

Electron Scattering Coherently Excited Molecules," *Proceedings of the Universities: Chemistry and Chemical Technology*, Vol. 52, No. 5, 2009, pp. 62-67.

[53] A. A. Ischenko, "The Study of Coherent Dynamics of the Nuclei by Electron Diffraction with a Time Resolution. III. Molecular Quantum State Tomography," *Proceedings of the Universities: Chemistry and Chemical Technology*, Vol. 52, No. 8, 2009, pp. 58-63.

[54] J. R. Dwyer, C. T. Hebeisen, R. Ernstorfer, M. Harb, V. B. Deyirmenjian, R. E. Jordan and R. J. D. Miller, "Femtosecond Electron Diffraction: 'Making the Molecular Movie'," *Philosophical Transactions of the Royal Society A*, Vol. 364, No. 1840, 2006, pp. 741-778.

[55] M. Ben-Nun, J. Cao and K. Wilson, "Ultrafast X-Ray and Electron Diffraction: Theoretical Considerations," *Journal of Physical Chemistry A*, Vol. 101, No. 47, 1997, pp. 8743-8761.

[56] H.-C. Shao and A.F. Starace, "Detecting Electron Motion in Atoms and Molecules," *Physical Review Letters*, Vol. 105, No. 26, 2010, pp. 263201-263204.

[57] H.-C. Shao and A. F. Starace, "Ultrafast Electron Pulse (e,2e) Processes," *Bulletin of the American Physical Society 43rd Annual Meeting of the APS Division of Atomic, Molecular and Optical Physics*, Vol. 57, No. 5, 2012, Article ID: N3.00008.

[58] S. Williamson and G. Mourou, "Picosecond Electron Diffraction," *Applied Physics Letters*, Vol. 41, No. 1, 1982, pp. 44-45.

[59] V. V. Golubkov, A. V. Zgurskii, A. A. Ischenko, V. I. Petrov and V. P. Spiridonov, "Pulse-Resonance Method for Detecting a Signal in a Stroboscopic Electron Microscopy," Abstracts XII All-Union Conference on Electron Microscopy, Moscow, Science, 1982, p. 62.

[60] V. V. Golubkov, A. V. Zgurskii, A. A. Ischenko and V. I. Petrov, "New Methods of Signal Detection in Gas Electron Diffraction," *Proceedings of the USSR Academy of Sciences*, Vol. 47, No. 6, 1983, pp. 1115-1121.

[61] R. S. Becker, G. S. Higashi and J. A. Golovchenko, "Low-Energy Electron Diffraction during Pulsed Laser Annealing: A Time-Resolved Surface Structural Study," *Physical Review Letters*, Vol. 52, No. 4, 1984, pp. 307-310.

[62] A. P. Rood and J. Milledge, "Combined Flash-Photolysis and Gas-Phase Electron Diffraction Studies of Small Molecules," *Journal of the Chemical Society, Faraday Transactions*, Vol. 2, No. 80, 1984, pp. 1145-1151.

[63] J. D. Ewbank, L. Schafer, D. W. Paul, O. J. Benston and J. C. Lennox, "Real-Time Data Acquisition for Gas Electron Diffraction," *Review of Scientific Instruments*, Vol. 55, No. 10, 1984, pp. 1598-1603.

[64] A. A. Ischenko, V. N. Bagratashvili, V. V. Golubkov, V. P. Spiridonov, A. V. Zgurskii and A. S. Akhmanov, "Observation of Electron Diffraction from Free Radicals—Products IR Multiphoton Dissociation of CF_3I Molecules by Stroboscopic Gas Electron Diffraction," *Moscow University Chemistry Bulletin*, Vol. 26, No. 2, 1985, pp.140-143.

[65] O. Bostanjoglo, E. Endruschat, F. Heinricht, R. P. Tornow and W. Tornow, "Short-Interval Electron Microscopy and Pulsed Lasers," *European Journal of Cell Biology*, Vol. 44, No. 1, 1987, pp. 10-15.

[66] O. Bostanjoglo, R. P. Tornow and W. Tornow, "Nanosecond-Exposure Electron Microscopy of Laser-Induced Phase Transformations," *Ultramicroscopy*, Vol. 21, No. 4, 1987, pp. 367-372.

[67] O. Bostanjoglo and F. Heinricht, "A Laser Pulsed High Emission Thermal Electron-Gun," *Journal of Physics: Conference Series*, Vol. 93, No. 1, 1988, pp. 105-106.

[68] H. E. Elsayed-Ali and G. A. Mourou, "Picosecond Reflection High-Energy Electron Diffraction," *Applied Physics Letters*, Vol. 52, No. 2, 1988, pp. 103-104.

[69] O. Bostanjoglo and P. Thomsen-Schmidt, "Laser-Induced Multiple Phase Transitions in Ge and Te Films Traced by Time-Resolved TEM," *Applied Surface Science*, Vol. 43, No. 1-4, 1989, pp. 136-141.

[70] H. E. Elsayed-Ali and J. W. Herman, "Picosecond Time-Resolved Surface-Lattice Temperature Probe," *Applied Physics Letters*, Vol. 57, No. 15, 1990, pp. 1508-1510.

[71] L. S. Bartell and T. S. Dibble, "Observation of the Time Evolution of Phase Changes in Clusters," *Journal of American Chemical Society*, Vol. 112, No. 2, 1990, pp. 890-891.

[72] J. C. Williamson, M. Dantus, S. B. Kim and A. H. Zewail, "Ultrafast Diffraction and Molecular Structure," *Chemical Physics Letters*, Vol. 196, No. 6, 1992, pp. 529-534.

[73] J. D. Ewbank, J. Y. Luo, J. T. English, R. F. Liu, W. L. Faust and L. Schafer, "Time-Resolved Gas Electron Diffraction Study of the 193-nm Photolysis of 1,2-Dichloroethenes," *Journal of Physical Chemistry*, Vol. 97, No. 34, 1993, pp. 8745-8751.

[74] J. W. Herman, H. E. Elsayed-Ali and E. A. Murphy, "Time-Resolved Structural Study of Pb(100)," *Physical Review Letters*, Vol. 71, No. 3, 1993, pp. 400-403.

[75] M. Dantus, S. B. Kim, J. C. Williamson and A. H. Zewail, "Ultrafast Electron Diffraction. 5. Experimental Time Resolution and Applications," *Journal of Physical Chemistry*, Vol. 98, No. 11, 1994, pp. 2782-2796.

[76] J. C. Williamson, J. Cao, H. Ihee, H. Frey and A. H. Zewail, "Clocking Transient Chemical Changes by Ultrafast Electron Diffraction," *Nature*, Vol. 6, No. 386, 1997, pp. 159-162.

[77] A. A. Ischenko, L. Schafer and J. D. Ewbank, "Tomography of the Molecular Quantum State by Time-Resolved Electron Diffraction," *SPIE Proceedings*, Vol. 3516, 1998, pp. 580-595.

[78] R. C. Dudek and P. M. Weber, "Ultrafast Diffraction Imaging of the Electrocyclic Ringopening Reaction of 1,3-Cyclohexadiene," *Journal of Physical Chemistry A*, Vol. 105, No. 17, 2001, pp. 4167-4171.

[79] H. Ihee, V. A. Lobastov, U. M. Gomez, B. M. Goodson, R. Srinivasan and A. H. Zewail, "Direct Imaging of Transient Molecular Structures with Ultrafast Diffraction," *Science*, Vol. 291, No. 5503, 2001, pp. 458-462.

[80] B. J. Siwick, J. R. Dwyer, R. F. Jordan and R. J. D. Miller, "Ultrafast Electron Optics: Propagation Dynamics of Femtosecond Electron Packets," *Journal of Applied Physics*, Vol. 92, No. 3, 2002, pp. 1643-1648.

[81] H. Domer and O. Bostonjoglo, "High-Speed Transmission Electron Microscope," *Review of Scientific Instruments*, Vol. 74, No. 10, 2003, pp. 4369-4372.

[82] C.-Y. Ruan, F. Vigliotti, V. A. Lobastov, S. Chen and A. H. Zewail, "Ultrafast Electron Crystallography: Transient Structures of Molecules, Surfaces, and Phase Transitions," *Proceedings of the National Academy of Sciences of the United States of America*, Vol. 101, No. 5, 2004, pp. 1123-1128.

[83] C.-J. Ruan, D.-S.Yang and A. H. Zewail, "Structures and Dynamics of Self-Assembled Surface Monolayers Observed by Ultrafast Electron Crystallography," *Journal of the American Chemical Society*, Vol. 126, No. 40, 2004, pp. 12797-12799.

[84] W. E. King, G. H. Campbell, A. Frank, B. Reed, J. F. Schmerge, B. J. Siwick, B. C. Stuart and P. M. Weber, "Ultrafast Electron Microscopy in Materials Science, Biology, and Chemistry," *Journal of Applied Physics*, Vol. 97, No. 11, 2005, pp. 111101-111127.

[85] W. E. King, M. R. Armstrong and O. Bostonjoglo, "High-Speed Electron Microscopy," *Science of Microscopy*, Vol. 1, 2007, pp. 404-444.

[86] O.-H. Kwon, B. Barwick, H. S. Park, J. S. Baskin and A. H. Zewail, "Nanomechanical Motions of Cantilevers: Direct Imaging in Real Space and Time with 4D Electron Microscopy," *Nano Letters*, Vol. 8, No. 11, 2008, pp. 3557-3562.

[87] D. J. Flannigan, B. Barwick and A. H. Zewail, "Biological Imaging with 4D Ultrafast Electron Microscopy," *Proceedings of the National Academy of Sciences of the United States of America*, Vol. 107, No. 22, 2010, pp. 9933-9937.

[88] O.-H. Kwon, H. S. Park, J. P. Baskin and A. H. Zewail, "Nonchaotic Nonlinear Motion Visualized in Complex Nanostructures by Stereographic 4D Electron Microscopy," *Nano Letters*, Vol. 10, No. 8, 2010, pp. 3190-3198.

[89] V. Ortalan and A. H. Zewail, "4D Scanning Transmission Ultrafast Electron Microscopy: Single-Particle Imaging and Spectroscopy," *Journal of the American Chemical Society*, Vol. 133, No. 28, 2011, pp. 10732-10735.

[90] G. Sciaini and R. J. D. Miller, "Femtosecond Electron Diffraction: Heralding the Era of Atomically Resolved Dynamics," *Reports on Progress in Physics*, Vol. 74, No. 9, 2011, pp. 096101-096136.

[91] S. T. Park, D. J. Flannigan and A. H. Zewail, "Irreversible Chemical Reactions Visualized in Space and Time with 4D Electron Microscopy," *Journal of the American Chemical Society*, Vol. 133, No. 6, 2011, pp. 1730-1733.

[92] O.-H. Kwon, V. Ortalan and A. H. Zewail," Macromolecular Structural Dynamics Visualized by Pulsed Dose Control in 4D Electron Microscopy," *Proceedings of the National Academy of Sciences of the United States of America*, Vol. 108, No. 15, 2011, pp. 6026-6031.

[93] O. F. Mohammed, D.-S. Yang, S. K. Pal and A. H. Zewail," 4D Scanning Ultrafast Electron Microscopy: Visualization of Materials Surface Dynamics," *Journal of the American Chemical Society*, Vol. 133, No. 20, 2011, pp. 7708-7711.

[94] V. Ortalan and A. H. Zewail, "4D Scanning Transmission Ultrafast Electron Microscopy: Single-Particle Imaging and Spectroscopy," *Journal of the American Chemical Society*, Vol. 133, No. 28, 2011, pp. 10732-10735.

[95] M. M. Lin, D. Shorokhov and A. H. Zewail, "Structural Dynamics of Free Proteins in Diffraction," *Journal of the American Chemical Society*, Vol. 133, No. 42, 2011, pp. 17072-17086.

[96] S. Schäfer, W. Liang and A. H. Zewail, "Structural Dynamics of Nanoscale Gold by Ultrafast Electron Crystallography," *Chemical Physics Letters*, Vol. 515, No. 4-6, 2011, pp. 278-282.

[97] S. Schäfer, W. Liang and A. H. Zewail, "Structural Dynamics of Surfaces by Ultrafast Electron Crystallography: Experimantal and Multiple Scattering Theory," *Journal of Chemical Physics*, Vol. 135, No. 21, 2011, pp. 214201-214215.

[98] I.-R. Lee, A. Gahlmann and A. H. Zewail, "Structural Dynamics of Free Amino Acids in Diffraction," *Angewandte Chemie International Edition*, Vol. 51, No. 1, 2012, pp. 99-102.

[99] A. Yurtsever, J. S. Baskin and A. H. Zewail, "Entangled Nanoparticles: Discovery by Visualization in 4D Electron Microscopy," *Nano Letters*, Vol. 12, No. 9, 2012, pp. 5027-5032.

[100] S. T. Park and A. H. Zewail, "Relativistic Effects in Photon-Induced near-Field Electron Microscopy," *Journal of Physical Chemistry A*, Vol. 116, No. 46, 2012, pp. 11128-11133.

[101] A. Yurtsever, S. Schäfer and A. H. Zewail, "Ultrafast Kikuchi Diffraction: Nanoscale Stress-Atrain Dynamics of Wave-Guiding Structures," *Nano Letters*, Vol. 12, No. 7, 2012, pp. 3772-3777.

[102] A. Yurtsever and A. H. Zewail, "Direct Visualization of Near-Fields in Nanoplasmonics and Nanophotonics," *Nano Letters*, Vol. 12, No. 6, 2012, pp. 3334-3338.

[103] S. T. Park, D. J. Flannigan and A. H. Zewail, "4D Electron Microscopy Visualization of Anisotropic Atomic Motions in Carbon Nanotubes," *Journal of the American*

Chemical Society, Vol. 134, No. 22, 2012, pp. 9146-9149.

[104] S. T. Park, O.-H. Kwon and A. H. Zewail, "Chirped Imaging Pulses in Four-Dimensional Electron Microscopy: Femtosecond Pulsed Hole Burning," *New Journal of Physics*, Vol. 14, 2012, Article ID: 053046.

[105] S. T. Park and A. H. Zewail, "Enhancing Image Contrast and Slicing Electron Pulses in 4D Near-Field Electron Microscopy," *Chemical Physics Letters*, Vol. 521, 2012, pp. 1-6.

[106] D. J. Flannigan and A. H. Zewail, "4D Electron Microscopy: Principles and Applications," *Accounts of Chemical Research*, Vol. 45, No. 10, 2012, pp. 1828-1839.

[107] D.-S. Yang, O. F. Mohammed and A. H. Zewail, "Environmental Scanning Ultrafast Electron Microscopy: Structural Dynamics of Solvation at Interfaces," *Angewandte Chemie International Edition*, Vol. 52, No. 10, 2013, pp. 2897-2901.

[108] W. Liang, S. Schäfer and A. H. Zewail, "Ultrafast Electron Crystallography of Monolayer Adsorbates on Clean Surfaces: Structural Dynamics," *Chemical Physics Letters*, Vol. 542, 2012, pp. 1-7.

[109] W. Liang, S. Schäfer and A. H. Zewail, "Ultrafast Electron Crystallography of Heterogeneous Structures: Gold-Graphene Bilayer and Ligand-Encapsulated Nanogold on Graphene," *Chemical Physics Letters*, Vol. 542, 2012, pp. 8-12.

[110] J. B. Hastings, F. M. Rudakov, D. H. Dowell, J. F. Schmerge, J. Cardoza, J. M. Castro, S. M. Gierman, H. Loos and P. M. Weber, "Ultrafast Time-Resolved Electron Diffraction with Megavolt Electron Beams," *Applied Physics Letters*, Vol. 89, No. 18, 2006, pp. 184109-184111.

[111] F. M. Rudakov, J. B. Hastings, D. H. Dowell, J. F. Schmerge and P. M. Weber, "Megavolt Electron Beams for Ultrafast Time-Resolved Electron Diffraction," In: M. D. Furnish, M. Elert, T. P. Russel and C. T. White, Eds. *Shock Compression of Condensed Matter*—2005, American Institute of Physics, New York, 2006, pp. 1287-1292.

[112] J. D. Geiser and P. M. Weber, "High Repetition Rate Time-Resolved Gas Phase Electron Diffraction," *Proceedings of SPIE Conference on Time Resolved Electron and X-Ray Diffraction*, Vol. 2521, San Diego, July 1995, pp. 136-144.

Transmission of N-Atoms Produced by N_2 Flowing Microwave Afterglows through Hollow Tubes

Andre Ricard, Jean-Philippe Sarrette

Paul Sabatier University, Toulouse, France.

ABSTRACT

Transmission of N-atoms (T_N) through small diameters tubes (1.5 and 3 mm internal diameter (i.d) and 9, 50 and 80 cm length for silicone tubes, 1.5 mm i.d and 6.5 cm length for stainless steel tubes) has been measured in late N_2 and Ar-N_2 flowing afterglows of microwave plasmas in continuous and pulsed gas injection at a flow rate of 1 and 3 Standard liter by minute (Slm), a gas pressure from 2 to 4 Torr for N_2 and 20 Torr for Ar-1%N_2 and a plasma power from 150 to 300 Watt. From the experimental T_N values, it is deduced the γ-destruction probability inside the tube walls as being $\gamma = (1-1.6)\times 10^{-3}$ for the silicon tubes and $\gamma = (1.6-2)\times 10^{-2}$ for the stainless steel tubes.

Keywords: Spectroscopy; N-Atoms Transmission; Hollow Tubes; γ-Destruction Probability

1. Introduction

Production of N-atoms has been previously studied in flowing N_2 microwave post-discharges at medium gas pressures (1 - 10 Torr) [1]. Bacteria inactivation by N-atoms in such post-discharge reactors has been largely studied [1-3]. In these previous papers, it has been obtained a 6 log reduction of an initial bacteria population (*E. coli* 10^6 colony forming unit by milliliter, cfu/ml) deposited on planar surfaces with a 40 minutes exposure to a N_2 flowing afterglow at 350 K.

In the cited previous works, the N atoms are produced by microwave plasmas at power of 100 - 300 Watt. The plasma is located inside a quartz tube of internal diameter (i.d) 5 mm and extends along a few centimeters outside the cavity (surfatron) gap for a gas pressure between 1 and 30 Torr and a flow rate between 0.5 and 2 Slm.

In these conditions, the plasma gas temperature reached up to 1000 K near the surfatron gap [4], needing an air cooling of the quartz tube. After a few centimetres of post-discharge, it is found the room gas temperature.

After the plasma, early and late afterglows appear and the late afterglow extends in the post-discharge reactor where the sterilization of bacteria is studied.

It is presently reported the results obtained for the transmission of N-atoms produced in the N_2 late afterglow through small diameters hollow tubes.

These studies are of interest to appreciate the possibility of N-atoms sterilization inside hollow silicon and stainless steel tubes.

2. The Experimental Setup

The flowing microwave post-discharge reactor described in previous studies [1-3] is reproduced in **Figures 1(a)** and **(b)**. A two-stage oil rotary pump with a pumping speed of 48 sm^3/h (800 slm) is set up after a 5 litre post-discharge reactor. The N_2 flow rate is controlled from 1 to 3 slm. The pressure inside the reactor can be adjusted by means of a throttle valve above the pump (see **Figure 1**).

A 2.45 Ghz microwave plasma is produced by a surfatron cavity in a quartz tube of 5 mm i.d and 30 cm in length. The discharge tube is connected to a tube of 18 mm i.d and 30 cm in length setup before the 5 litres post-discharge reactor in Pyrex glass. The dia.18 mm tube is in quartz after the dia.5 mm discharge tube and in Pyrex before the 5 litre reactor. A Teflon junction connects the two quartz and Pyrex tubes.

At a pressure of 5 Torr and a flow rate of 1 slm, the flow times are calculated as being 2×10^{-3} sec and 10^{-2} sec in the tubes of 5 mm i.d and of 18 mm i.d., respectively.

The emission spectroscopy measurements are performed by moving an optical fibre along the tubes from the plasma gap and the external wall of the post-discharge reactor. The N atom density is obtained by NO titration [4] by injecting an Ar-2%NO gas mixture in the

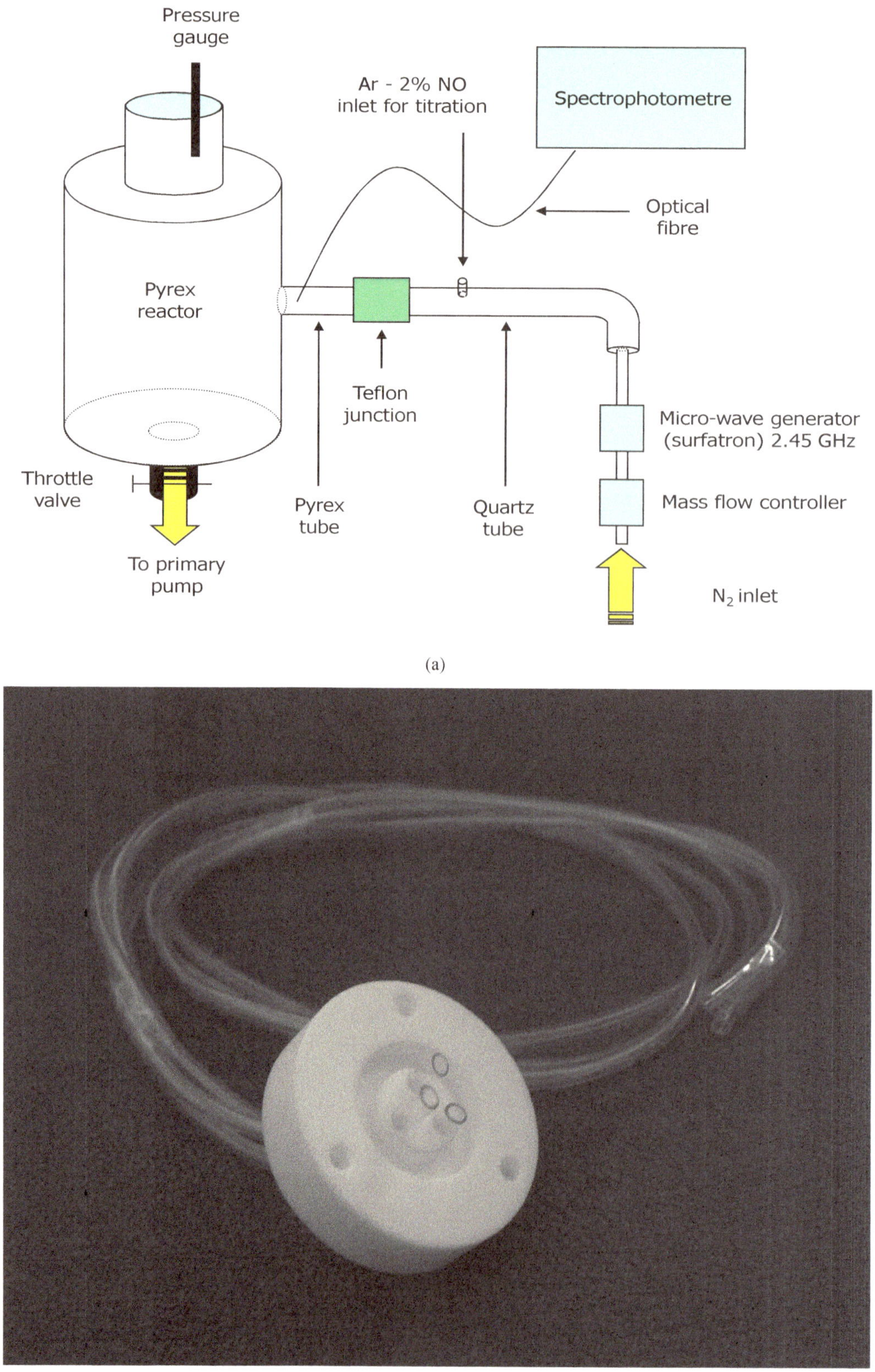

(b)

Figure 1. Post discharge set up with the emplacement of the Teflon junction connecting the quartz and the Pyrex parts of the 18 mm i.d. tubes (a). The picture shows details of the Teflon junction and the emplacement of the three silicone tubes used in the transmission experiments (b).

post-discharge tube (see **Figure 1(a)**).

The transmission of N-atoms through hollow tubes was studied by using the Teflon junction allowing to join the quartz and Pyrex parts of the 18 mm i.d. tube **Figure 1(b)**.

Teflon junctions were worked to receive between 3 to 9 tubes of dia. 1 - 3 mm across it.

The insertion of hollow tubes in the Teflon junction produces pressure variations between the plasma tube and the 5 litre reactor.

Such pressure variations were analysed from the afterglow emissions in the 18 mm i.d tube at 2 cm before the Teflon junction (named hereafter z_0) by keeping a given pressure in the 5 litre reactor.

The intensity I (580 nm) in z_0 versus the gas pressure and for several flow rates is reproduced in **Figure 2** without the silicon tubes.

With the silicon tubes, the I (580 nm) intensity in z_0 as reported in **Figure 2** allows to determine the upstream pressure for a given flow rate.

At a too high N_2 pressure ($p > 5$ Torr), it was not always possible to switch on the N_2 plasma at a power less than 300 Watt.

Also as described in ref.5 it has been found that the plasma can be switched on at higher pressure by pulsing the gas. For that purpose, a Parker E.M valve was set up between the flow controller and the microwave cavity (see **Figure 1**)

In such a way the plasma was switched on as the gas pressure decreased when the valve was closed and it was maintained at high pressure when the valve was opened (at a constant time of 2×10^{-2} sec).

An Ar-N_2 gas mixture was studied since the plasma is easier to sustain at high pressure [6].

3. Plasmas at High Gas Pressures

3.1. Plasmas in Pulsed N_2 Gas

The microwave cavity was first tuned with a continuous gas inlet of 1 slm and an injected power between 100 and 300 Watt. Then by pulsing the gas with a δt pulse varying from 0.1 to 2 sec and a Δt period of 1 and 2 sec, it was successively produced two plasmas at high pressure (2 - 3 Torr) and at low pressure (0.4 - 0.5 Torr) [5].

By checking the N_2^+/N_2 band intensity ratio near the surfatron gap, it was found that the microwave cavity tuning remained with the gas on, as for the continuous case with a negligible reflected power. At low pressure when the gas is off, the reflected power is important, varying from 30 to 90 Watt for an incident power from 100 to 150 Watt [5].

In these conditions, a strong late afterglow of N_2 1st pos. bands (with a specific emission at 580 nm, resulting of N+ N atoms recombination) was observed at high pressure and not at low pressure. As discussed in ref.4, the I(580 nm) afterglow intensity at high pressure (>2 Torr) is given by the following equation:

$$I(580\text{nm}) = k\cdot[\text{N}]^2, \quad (1)$$

where k is a constant.

By analysing the time varying gas pressure, afterglow intensity obtained with a photodiode (600 - 900 nm), it is

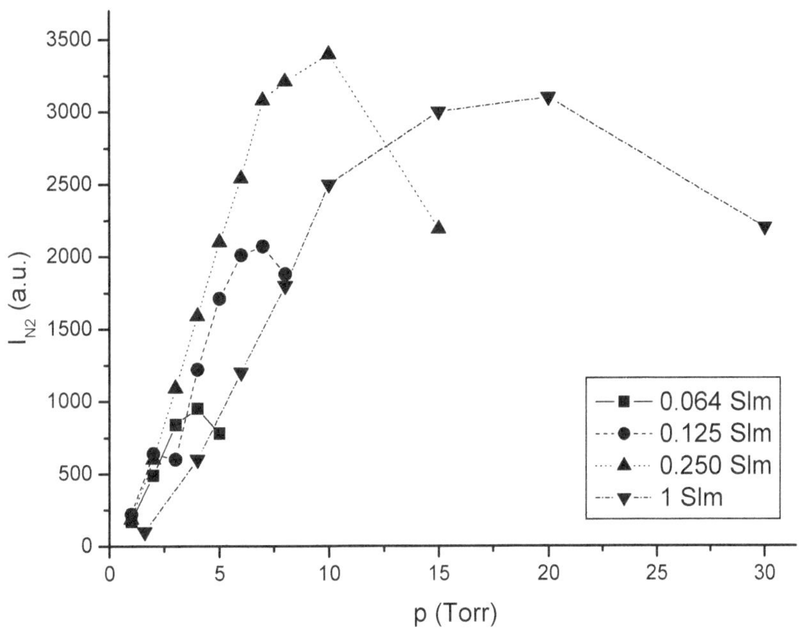

Figure 2. Variation of I (N_2, 580 nm) intensity in z_0 with N_2 gas pressure without the silicon tubes, for several flow rates at 100 Watt.

obtained the results as reproduced in **Figure 3**.

It was established that for long pulses (0.4 - 1 sec), the N-atom density increased up to the end of gas pulse as did the plasma length

A maximum of N-atom density is obtained for time period Δt, twice the pulse time δt. The maximum pressure variations are for long pulses. In the following experimental conditions: incident power 200 Watt, flow rate 1 slm, pump speed 800 slm, $\Delta t \sim 2\delta t = 2\,\text{sec}$, it is determined a mean N-atom density of 9×10^{14} cm^{-3} (10^{15} cm^{-3} in continuous flow) and a pressure variation $\Delta p = 2.3 - 0.5\,\text{Torr}$.

3.2. Plasmas in Ar-N_2 Gas Mixtures

Homogeneous N_2 microwave post-discharges can be obtained in the continuous mode at reduced gas pressures (2 - 10 Torr). When the pressure is increased, the plasma became unstable and a 300 W power is not sufficient to sustain the discharge. Adding Ar to N_2 allows to obtain a stable discharge from low pressure up to atmospheric gas pressure with N_2 percentage below 23% with a microwave power limited to 300 Watt [6].

In the Ar-N_2 gas mixtures, the plasma gas temperature was found to be lower than 1000 K. In pure N_2 plasma, the electron energy distribution function is depleted above 3 eV because of the low mean vibrational excitation of N_2 [7]. In contrast in Ar gas, the first excitation threshold being 11 eV, the extension of electron energy is higher. Consequently, by comparing to pure N_2 plasma the ionization in Ar-N_2 gas mixture is enhanced particularly at low N_2 percentage.

4. Transmission of N-Atoms in Hollow Tubes

4.1. Silicone Tubes

It was first experimented 9 tubes of silicones of int.dia. (i.d) 1.5 mm and out.dia. (o.d) 2 mm located inside the Teflon junction of **Figure 1** where 9 holes of dia. 2 mm were made inside a Teflon tap.

When the length of the tubes was 150 cm, it was necessary to pulse the gas with $\Delta t \sim 2\delta t = 2\,\text{sec}$ at 200 Watt to switch on the N_2 plasma (see 3.1). In continuous mode, it was not possible to produce the N_2 plasma at 300 Watt.

In these pulsed conditions the afterglow with mainly emission of N_2 (580 nm) extended up to 50 cm inside the 9 tubes.

The length of the tubes was then reduced to 50 cm and the afterglow was detected inside the 5 litre reactor.

In conditions of $\Delta t \sim 2\delta t = 2\,\text{sec}$, 200 Watt, 1 slm, the gas pressure was found varying from 0.8 to 2.0 Torr in the 5 litre reactor and the N-atom density was obtained from the I(580 nm) intensity (see Equation (1)) after cali-

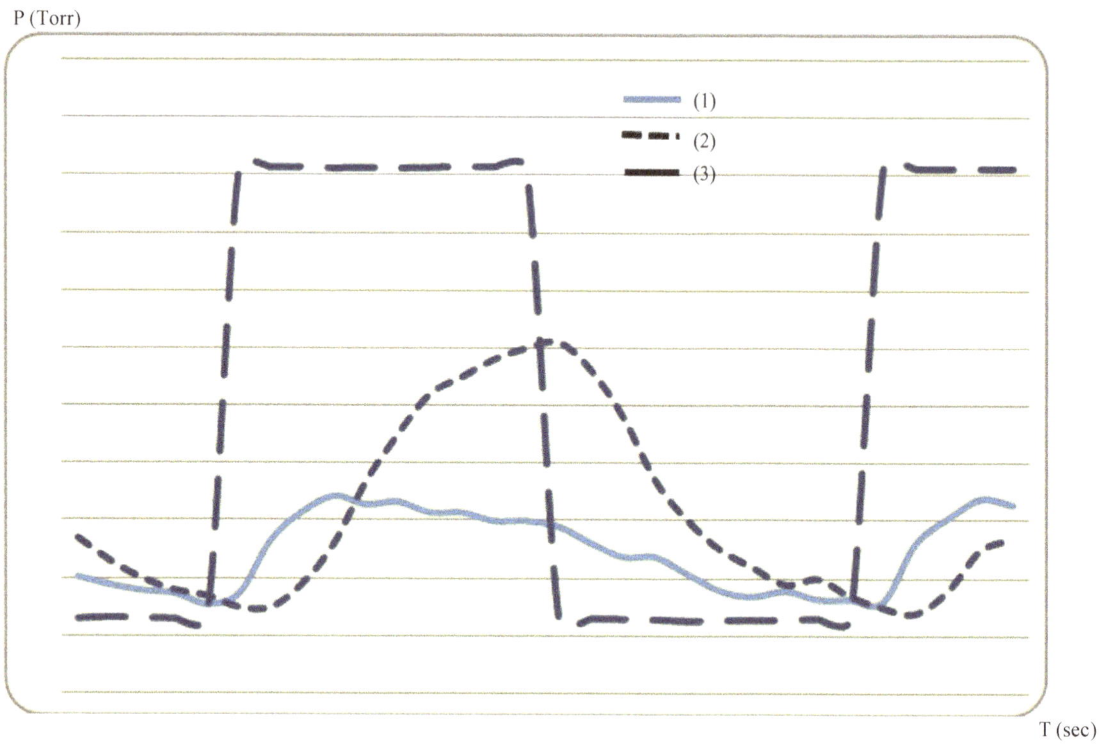

Figure 3. Variations with time of gas pressure (1) in Torr, of photodiode signal (2) and of electric field of the EM valve (3), obtained by the labview program. The photodiode and electric field signals are in arbitrary units. 150 Watt, 1 slm, Δt = 2 sec, δt = 1 sec .

bration by NO titration [1-4].

It was found $[N]=3.4\times10^{15}cm^{-3}$ at 1 cm before the Teflon junction. In the 5 litre reactor, the N-atom density was evaluated from the I (580 nm) intensity, corrected from the change of optical thickness between the Pyrex tube and reactor chamber. It was deduced $[N]=1.2\times10^{14}cm^{-3}$ after 50 cm of silicon tubes length. The N-atom transmission T_N was thus evaluated to 3.5%.

A second experiment was undertaken with a Teflon tap bored of 3 holes of dia. 4 mm connecting 3 silicon tubes of i.d. 3 mm and o.d. 4 mm, of length from 80 to 250 cm.

For these larger diameter tubes, the N_2 plasma was switched on in the continuous mode.

At a flow rate of 1 Slm, a pressure of 4 Torr in the 5 litre reactor, a power of 150 Watt and silicon tube lengths of 80 cm, it is determined a N-atom density of 2.7×10^{15} cm^{-3} before the Teflon tap and $T_N=8.5\%$ in the 5 litre reactor.

At a low flow rate of 125 Sccm, a power of 100 Watt and silicon tube lengths of 60 cm, it was obtained a slow increase of T_N from 3.9% at 1 Torr to 4.9% at 3 Torr.

To improve the plasma lighting at high flow rate thus at high gas pressure, it was experimented as discussed in 3 - 2, an Ar-1%N_2 gas mixture.

Then the plasma was obtained in continuous mode at 150W, 3slm, 20 Torr and length 80 cm of the silicon tubes. The N-atom density before the Teflon tap was found to be 8.2×10^{15} cm^{-3} and $T_N=8.5\%$ at 80 cm, that is an increase of 3 of N-atom density and the same T_N value in comparison to N_2 150 W, 1 slm.

4.2. Stainless Tubes

9 stainless tubes of i.d 1.5 mm and o.d 2 mm, of length 6.5 cm was located across the tap of the Teflon junction as for the silicon tubes in the 1st experiment in 4.1.

A continuous N_2 plasma was obtained at 300 W, 1 slm and 1.7 Torr in the 5 litre reactor. In these conditions, the N atom density was 8×10^{15} cm^{-3} at z_0, before the Teflon junction and decreased sharply to about 10^{14} cm^{-3} at the exit of the 9 stainless tubes of length 6.5 cm, that is an N-atom transmission of 1.3%.

In the pulse mode $\Delta t\sim2\delta t=2$ sec, 200 Watt, 1 slm, gas pressure varying from 0.9 to 2.3 Torr, the N-atom transmission was increased to 3.5% - 4%.

The T_N transmission through silicone tubes of int. dia. 1.5 and 3 mm and various lengths of 9, 10 and 80 cm in pulsed and continuous N_2 plasmas are reported in **Table 1(a)**. The result obtained with a continuous Ar-1%N_2 plasma is also reported. In **Table 1(b)**, it is reported the T_N values through stainless tubes of int.dia 1.5 mm and length 6.5 cm in pulsed and continuous N_2 plasmas.

4.3. γ-Probability of N-Atom Destruction on the Tube Walls

By estimating that the main loss of N-atoms is by colli-

Table 1. Transmission of N-atoms through: (a) Silicone tubes of int. dia. 1.5 and 3 mm and various lengths of 9, 10 and 80 cm.Pulsed and continuous N_2 plasmas. Continuous Ar-1%N_2 plasmas; (b) Stainless tubes of int.dia 1.5 mm and length 6.5 cm. Pulsed and continuous N_2 plasmas.

(a) Silicone tubes

Plasma gas	Number of tubes	Tube i.d. (mm)	Operating mode	Power (W)	Flow rate (slpm)	Pressure (Torr)	$[N]_{z0}$ (10^{15} cm^{-3})	Tube length (cm)	T_N (%)
N_2	9	1.5	Pulsed $\Delta t\approx2\delta t=2s$	200	1.0	0.9 - 2.3	3.5	10.0	30.0
N_2	9	1.5	Pulsed $\Delta t\approx2\delta t=2s$	200	1.0	0.9 - 2.3	3.5	50.0	3.0
N_2	3	3.0	Continuous	300	1.0	4.0	4.2	9.0	35.0
N_2	3	3.0	Continuous	300	1.0	4.0	4.2	80.0	8.5
Ar-1% N_2	3	3.0	Continuous	150	3.0	20.0	8.2	80.0	8.5

(b) Stainless steel tubes

Plasma gas	Number of tubes	Tube i.d. (mm)	Operating mode	Power (W)	Flow rate (slpm)	Pressure (Torr)	$[N]_{z0}$ (10^{15} cm^{-3})	Tube length (cm)	T_N (%)
N_2	9	1.5	Pulsed $\Delta t\approx2\delta t=2s$	200	1.0	0.9 - 2.3	8.3	6.5	3.5 - 4.0
N_2	9	1.5	Continuous	300	1.0	1.7	8.0	6.5	1.3

sions on tube wall, it is written:

$$[N]_z = [N]_0 \cdot \exp-\left[\gamma(c'/2R)(z/v)\right] \quad (2)$$

where z is the axial distance in the silicon tube, $[N]_0$ is the N-atom density at z_0, c' is the thermal gas velocity at 300K: $c' = 5\times10^4 \text{cms}^{-1}$, R is the tube radius and v the axial gas velocity.

The axial gas velocity is deduced from the flow rate Q by:

$$Q(p) = S\cdot v \quad (3)$$

where $Q(p)$ is the flow rate at pressure $p(\text{Torr}): Q(p) = Q_0\, 760/p$, Q_0 is the standard flow rate (760 Torr, 300K) and S is the tube internal surface.

To obtain a nearly constant gas pressure in the 3 silicon tubes, the flow rate was decreased to 0.125 Slm for a pressure of 1 - 3 Torr in the 5 litre reactor.

At $Q = 0.125\text{Slm}$, the I (580 nm) measured in z_0 with the silicon tubes at 2 Torr indicates a pressure before the Teflon junction of 2 - 3 Torr.

With Equation (3), it is calculated at 2 Torr an axial gas velocity of 3.7×10^3 cm·sec^{-1} corresponding to a flow time in the silicon tubes of 1.6×10^{-2} s at 60 cm.

Then, it is determined from Equations (2) and (3), a γ value varying from 1.2 to 0.8×10^{-3} as the pressure in the reactor increased from 1 to 3 Torr.

Such values are 3 times lower than published values for Nylon [8]: $\gamma = 3-2\times10^{-3}$ at 2 - 3 Torr.

The γ-value has been calculated with the Ar-1%N_2 gas mixture in conditions of 3 slm and 20 Torr in the 5 litre reactor. From the I(580 nm) intensity calibrated with the Ar-1%N_2 gas mixture as for N_2 in **Figure 2**, it is estimated a pressure of 45 Torr in front the Teflon tape. By taking the mean value of pressure p_m= 32 Torr in the tubes, it is calculated from Equation (3) an axial velocity of 5.6×10^3 cm·s^{-1}. At 80 cm, it is measured T_N = 8.5%. It is then calculated from Equation (1) that $\gamma = 10^{-3}$.

In N_2, 1 slm, 150 Watt, the pressure before the Teflon tape is found from **Figure 2** to be 10 Torr. For a pressure of 4 Torr in the 5 litre reactor, the mean pressure in the tubes is 7 Torr, the axial velocity is 8×10^3 cm·s$^{-1.}$ With T_N = 8.5% at 80 cm, it is deduced $\gamma = 1.6\times10^{-3}$ near the published value for Nylon at 4 Torr: $\gamma = 2\times10^{-3}$ [8].

It is concluded to about the same value of γ probability of N atoms in pure N_2 and in the Ar-1%N_2 gas mixture: $\gamma = (1-1.6)\times10^{-3}$. The argon gas has no influence on the N-atom destruction probability on the silicon tubes walls.

With such γ values, it can be calculated the N-atom density at the exit of the 3 silicon tubes of i.d. 3 mm and length 250 cm: 1.3×10^{12} cm^{-3} for N_2 at 1 Slm, 150 W and 4Torr in the 5 litre reactor and 5.7×10^{12} cm^{-3} for Ar-1% N_2 at 3 Slm, 150 W and 20 Torr. The increase of N-atoms density by 4.4 when N_2 is changed into the Ar-1%N_2 gas mixture comes from the increases by a factor 3 of N-atom density before the Teflon tape and by a factor 3 of flow rates in spite of increase of pressure in the tubes of length 250 cm from about 20 Torr in N_2 to 30 Torr in Ar-1%N_2.

Equations (2) and (3) are applied to the results of the stainless tubes in part 4.2.

The pressure before the tap of the 9 stainless tubes was determined from the I (580 nm) intensity as it has been experimented with the silicon tubes. It is estimated a pressure of 10 - 16 Torr for 4 Torr in the 5 litre reactor. The mean pressure in the tubes is taken between 7 - 10 Torr.

With a N-atom transmission of 1.3% in continuous mode after 6.5 cm of tube length, it is calculated $\gamma = (1.6-2)\times10^{-2}$.

This value is 3 - 4 times higher than the published values in pulsed RF discharge for stainless steel [9]: $\gamma = 5\times10^{-3}$ at 5 Torr and $\gamma = 6\times10^{-3}$ at 3 Torr.

5. Conclusions

Transmission of N-atoms through hollow tubes in the late flowing afterglow of a N_2 microwave discharge has been measured for silicon tubes of i.d. 3 mm and a length up to 250 cm and for stainless steel tubes of i.d 1.5 mm and 6.5 cm length. The hollow tubes were placed though holes in a Teflon tap across a Teflon junction of i.d.18 mm in the main N_2 flow before a 5 litre reactor. Depending on the gas flow, a gas pressure variation is produced before and after the Teflon junction. To switch on the plasma at high gas pressure, it was experimented pulsed N_2 gas modes and Ar-N_2 gas mixtures. The pulse gas mode was of interest to produce the plasma at lower gas pressure when the gas was off and to maintain it at high pressure when the gas was on.

In comparison to continuous discharges, the N-atom transmission across the 1.5 mm i.d stainless steel tubes increased with the pulsed gas mode.

An Ar-1%N_2 gas mixture has allowed to produce plasmas at high flow rates and gas pressures with higher N-atom density than in pure N_2.

The γ-destruction probability has been determined inside the tube walls as being $\gamma = (1-1.6)$ 10^{-3} for the silicon tubes and $\gamma = (1.6-2)\times10^{-2}$ for the stainless steel tubes. The γ-values are in good agreement with published values for the silicon tubes but not for the stainless steel tubes which appear to be 3 - 4 times higher.

With a value $\gamma = 10^{-3}$, the best result for N-atom injection in the silicon tubes of 3 mm i.d and length 250 cm is obtained with an Ar-1%N_2 gas mixture at 3 Slm, 150 W giving a measured density of 8×10^{15} cm^{-3} before and a calculated density of 6×10^{12} cm^{-3} after the tubes in the 5 litre reactor.

6. Acknowledgments

Thanks to the Acteon-Satelec company (Mérignac-France) to have supported the research on N_2 plasma—post-discharge sterilization.

REFERENCES

[1] S. Villeger, J. P. Sarrette and A. Ricard, "Synergy between N and O Atom Action and Substrate Surface Temperature in a Sterilization Process Using a Flowing N_2-O_2 Microwave Post Discharge," *Plasma Processes and Polymers*, Vol. 2, No. 9, 2005, pp. 709-714.

[2] S. Villeger, S. Cousty, A. Ricard and M. Sixou, "Sterilization of *E. coli* Bacterium in a Flowing N_2-O_2 Post-Discharge Reactor," *Journal of Physics D: Applied Physics*, Vol. 36, No. 13, 2003, pp. L60-L62.

[3] S. Villeger, J. P. Sarrette, B. Rouffet, S. Cousty and A. Ricard, "Treatment of Flat and Hollow Substrates by a Pure Nitrogen Flowing Post Discharge," *The European Physical Journal Applied Physics*, Vol. 42, No. 1, 2008, pp. 25-32.

[4] P. Merel, M. Tabbal, M. Chaker, M. Moisan and A. Ricard, "Influence of the Field Frequency on the Nitrogen Atom Yield in the Remote Plasma of an N_2 High Frequency Discharge," *Plasma Sources Science and Technology*, Vol. 7, No. 4, 1998, p. 550.

[5] A. Ricard, F. Moser, S. Cousty, S. Villeger and J. P. Sarrette, "Time Varying Afterglow Emission and Gas Pressure in a Pulsed N_2 Gas Microwave Flowing Discharge at Reduced Pressure," *The European Physical Journal Applied Physics*, Vol. 49, No. 1, 2010, pp. 13104-13107.

[6] A. Ricard, F. Gaboriau and C. Canal, "Optical Spectroscopy to Control a Plasma Reactor for Surface Treatments," *Surface and Coatings Technology*, Vol. 202, No. 22-23, 2008, pp. 5220-5224.

[7] M. Capitelli (Ed.), "Non-Equilibrium Vibrational Kinetics, Topics in Current Physics Volume 56," Springer, Berlin, 1986.

[8] J.-P. Sarrette, B. Rouffet and A. Ricard, "Determination of Nitrogen Atoms Loss Probabilities on Copper, Aluminium, Alumina, Brass and Nylon Surfaces," *Plasma Processes and Polymers*, Vol. 3, No. 2, 2006, pp. 120-126.

[9] S. F. Adams and T. A. Miller, "Surface and Volume Loss of Atomic Nitrogen in a Parallel Plate of Discharge Reactor," *Plasma Sources Science and Technology*, Vol. 9, No. 3, 2000, p. 248.

Direct Determination of Polychlorinated-Biphenyls in Automotive Shredder Residues by Gas Chromatography-Mass Spectrometry

María de los Desamparados Escribano Otero[1], Mohamed Nour Kayali Sayadi[1], Luis María Polo Díez[2]

[1]Mass Spectrometry Unit, Faculty of Chemistry, Complutense University of Madrid, Madrid, Spain; [2]Department of Analytical Chemistry, Faculty of Chemistry, Complutense University of Madrid, Madrid, Spain.

ABSTRACT

An easy and rapid method is proposed for the determination of PCBs in automotive shredder residues, using gas chromatography combined with low resolution mass spectrometry (GC-MS). It is based on direct n-hexane solid-liquid extraction, subtracting background of the lineal aliphatic hydrocarbon interferences and integration of chromatographic peaks containing selected ion PCBs masses (256, 292 and 326 m/z), which are common in all PCBs formulations. Recoveries were in the 80% - 120% range; PCBs were detected and quantified in shredder samples from an automotive shredder industry, thus indicating the validity of the method.

Keywords: PCBs; Solid-Liquid Extraction; Automotive Shredder Residues; Gas Chromatography-Mass Spectrometry (GC-MS)

1. Introduction

Recycling of materials from manufactured products has become of increasing interest due to environmental problems involved being a challenge in modern sustainable development. In particular, automotive industry in European Community Countries generates thousands of tons per year whose recycling continues to be an essential research subject [1,2].

The main basic steps of the automotive residue recycling process include crushing, flotation of plastics, rubber and glass, as well as aspiration of light particles; as a result of this process, two classic types of automotive residues have been differentiated; the metallic materials part has been largely recovered to be employed for different purposes; on the other hand, the so called "no metal or shredder organic residues part" has been less often recycled; it has sometimes been used to level land but more often it has been thrown on dumps [3]. However, it could be used as an energy source by combustion, for instance, in cement industries, which could increase recycling of these types of residues and moreover, it could replace fossil fuels [4]. For this purpose, it is necessary to know the potential presence of toxic precursor compounds in the shredder organic part, especially Polychlorinated Biphenyls (PCBs). These compounds have been used, as dielectric fluids in electrical capacitors, as stabilizers in engine oils and as plastic additives employed to manufacture a variety of automobile components [5]. They are very stable and resistant to chemicals and consequently, they are persistent in the environment, where their ubiquity is due to extensive use during the last century. These PCBs may give rise by combustion to even more toxic compounds such as benzofurans and dioxins, which are very unhealthy substances associated with serious illnesses [6]; these compounds have been detected in PCBs combustion at temperatures around 200˚C [7,8]. This fact makes it necessary for us to know the presence and concentration of such precursors in order to make an informed decision about the recycling of shredder organic residues by combustion.

Given the importance of PCBs, various matrixes have been analyzed to detect these compounds, including human food, accumulation in tissues, environmental pollution and oils [9-12]. To date, the analytical technique most often used to detect and quantify PCBs in environmental samples has been Gas Chromatography combined with high resolution Mass Spectrometry (GC-MS) in order to separate PCBs congeners. This technique obviously involves high instrumentation cost and technical skill [13]. In this research, a simple and rapid method based on direct solid-liquid extraction of PCBs from shredder organic residues and further total PCBs evaluation using low resolution GC-MS is proposed. It is based on background subtraction of the hydrocarbon noise over which PCBs chromatographic peaks appear and the use of selected characteristic PCBs ion masses (which are always present in chromatograms when shredder samples contains PCBs) along with integration of selected PCBs masses, without the need for a high resolution mass spectrometer. Method validity is established by recovery studies and shredder samples from a recycling automobile industry are analyzed.

2. Materials and Methods

2.1. Equipment

A Hewlett Packard gas chromatograph, model 5890 series 2 provided with a split/splitless injector and combined with a quadrupole mass spectrometer, model HP 5989A Engine as a detector; it is equipped with Hewlett-Packard Chemstation B.02.05 software which allows both correction of background over which chromatographic peaks appear and integration of selected ion masses. The column was a 5% dimethylphenylpolysiloxane, SLB-5MS, 30 m × 0.25 mm × 0.25 μm film thickness. Helium VIP, supplied by Carburos Metálicos S.A. (Madrid, Spain) was used as a carrier gas.

An Ultrasonic bath P-Selecta Precisterm (Barcelona, Spain), a rotary evaporator system Heildoph VV 2000 (Schwabach, Germany) and PTFE filters (13 mm, 0.5 μm) (Gloucestershire, UK) were used for sample preparation. Silica C18 cartridges (SUPELCO, Bellefonte, PA, USA) were used to optimize sample clean up.

2.2. Reagents

The PCBs solid-liquid extraction was undertaken using high purity n-Hexane ultra-trace analysis (99.6%) supplied by Scharlau (Barcelona, Spain). Other solvents used to optimize the extraction process, including methanol, ethanol and dichloromethane were supplied by Sigma Aldrich (St. Louis, USA).

2.3. Standards

Five representative Aroclors 1016, 1242, 1248, 1254 and 1260 solutions at a concentration level of 1000 μg/mL in isooctane were supplied by SUPELCO (Bellefonte, PA, USA); fresh working solutions were prepared by suitable dilution in n-hexane. The Calibration graph was obtained by the external standard method from solutions in the 17 - 500 ng/μL concentration range. Working solutions were stored at 4°C.

2.4. Samples

Automotive shredder samples were collected at an automotive residue industry; a scheme of the automotive residue recycling is showed in **Figure 1**. A representative sample of one kilogram was collected every two days for a period of 30 days to obtain about 12 kg. This raw sample was spread homogeneously on a clean surface (20 - 35 $g \cdot dm^{-2}$) and divided into quarters, discarding two opposite quarters, mixing the remaining ones and repeating the process several times, thus reducing the sample to less than one kilogram. This subsample was ground to obtain a maximum particle of about 4 mm as a stock sample from which the analytical sample was taken. The following five samples of differing size were analyzed as showed in **Table 1**.

2.5. Aroclors Solid-Liquid Extraction from the Organic Shredder Sample

About 0.5 g of the ground sample was weighed accurately and transferred into a 20 mL extraction screw tube; then, 10 mL of n-Hexane were added to the tube. Subsequently the top vial was covered with an aluminium foil and the tube was shaken during 40 minutes in the ultrasonic bath. Afterwards, the mixture was left to stand and the extract was transferred into an evaporating tube, suitable for the rotary evaporator system. The extracting process was repeated by adding another 10 mL of n-Hexane to the above residue and the tube was shaken during 10 minutes in the ultrasonic bath; the new extract was transferred into the same evaporation tube as before Combined extracts were evaporated in the rotary evaporator system up to 2 mL before being transferred to a conical evaporation tube where the volume was reduced to 0.3 mL by blowing an Argon flow. This final volume was filtered through a PTFE filter, transferred to a topaz crimp top vial, and stored at 4°C before analysis by GC/MS.

Table 1. Particle size samples analyzed.

Sample	Particle size
A	2 mm
B	1 mm
C	2 mm
D	1 mm
E	4 mm

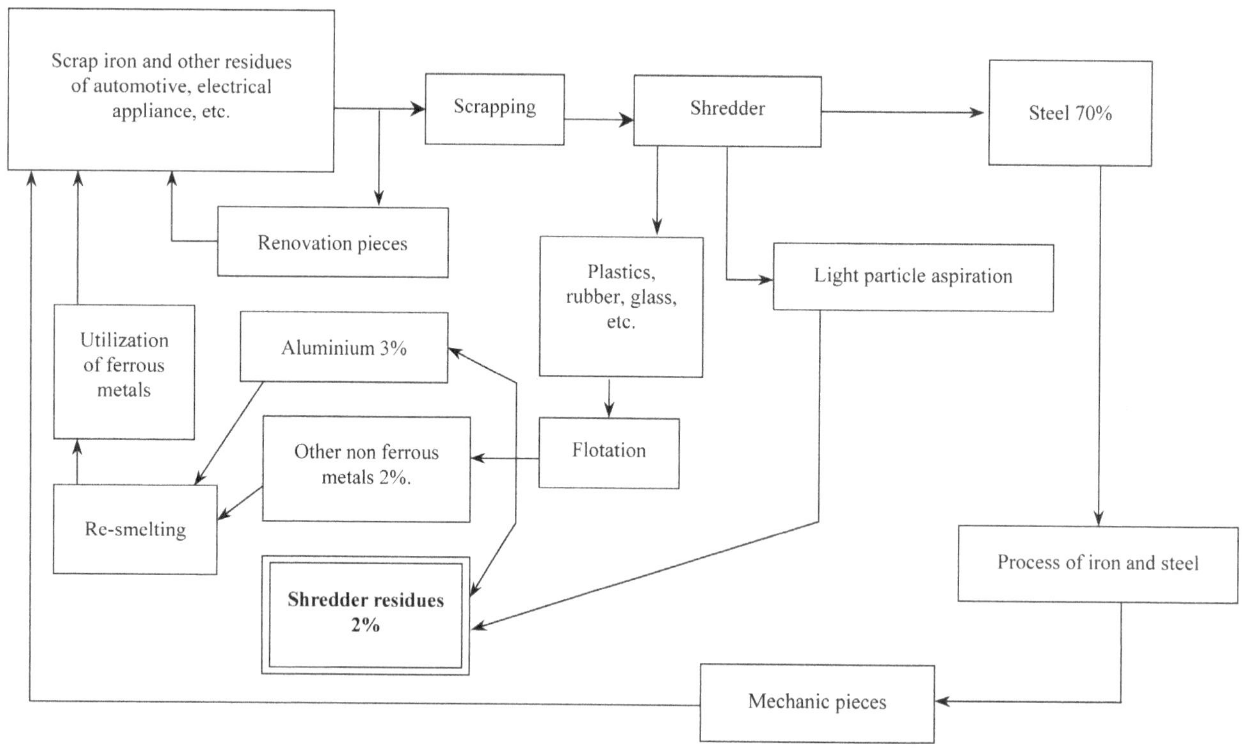

Figure 1. Automotive residue recycling process.

2.6. Chromatographic Method

One microliter of the above prepared analytical solution was injected into the gas chromatograph in splitless mode during 1 minute. The following temperature program was applied: injector temperature, 250°C; initial oven temperature, 80°C during 4 min, then a 5°C·min^{-1} ramp to 250°C was applied standing at this temperature for 5 minutes and then a 10°C·min^{-1} ramp to 280°C standing for 19 minutes. Interface temperature was 280°C; electronic impact ionization source temperature was 250°C and the potential between the filament and the anode was 70 eV; scan masses range was from 50 to 800 amu in total ion chromatogram mode. The background signal over which Aroclors chromatographic peaks appear was eliminated by subtracting lineal aliphatic hydrocarbons mass spectra background. Then, 256, 292 and 326 ion masses were selected to quantify PCBs and integrated chromatographic peak areas containing these masses were combined and used as a total area.

3. Results and Discussion

3.1. Selection of Common Ion Masses

In order to facilitate the detection of ion masses which are common to all PCBs, different chromatographic conditions were tested to achieve adequate chromatographic separation; for this purpose, the AROCLOR 1254 standard solution was analyzed in several chromatographic conditions; programmed temperature was used in all tests; the injector temperature was changed in the range 250°C - 280°C and different starting temperatures and ramps were tested; the best experimental conditions, which are specified in Section 2.6, were chosen and chromatograms of selected Aroclors were obtained. In all of them, 256, 292 and 326 m/z ion masses are present, corresponding to biphenyl-trichloro, biphenyl-tetrachloro and biphenyl-pentachloro, respectively (**Figure 2**); in order to quantify PCBs content as a whole, total integrated area of chromatographic peaks containing these ions was used.

The standard calibration was obtained using the Aroclor 1254 standard solution between 17 and 500 ng/µL; results and relevant data are showed in **Table 2**. Three standard solutions of 170 ng/µL concentration were analyzed to determine method precision and reproducibility; results are showed in **Table 2**.

3.2. Optimization of the PCBs Solid-Liquid Extraction Process

The B sample was enriched with 100 µL of Aroclor 1254 and variation of total PCBs area was taken for optimization purposes; extraction solvent, agitation time and other solid-liquid extraction conditions were studied.

Different extraction solvents including n-hexane, methanol, dichloromethane and n-hexane:ethanol (70:30; v:v)

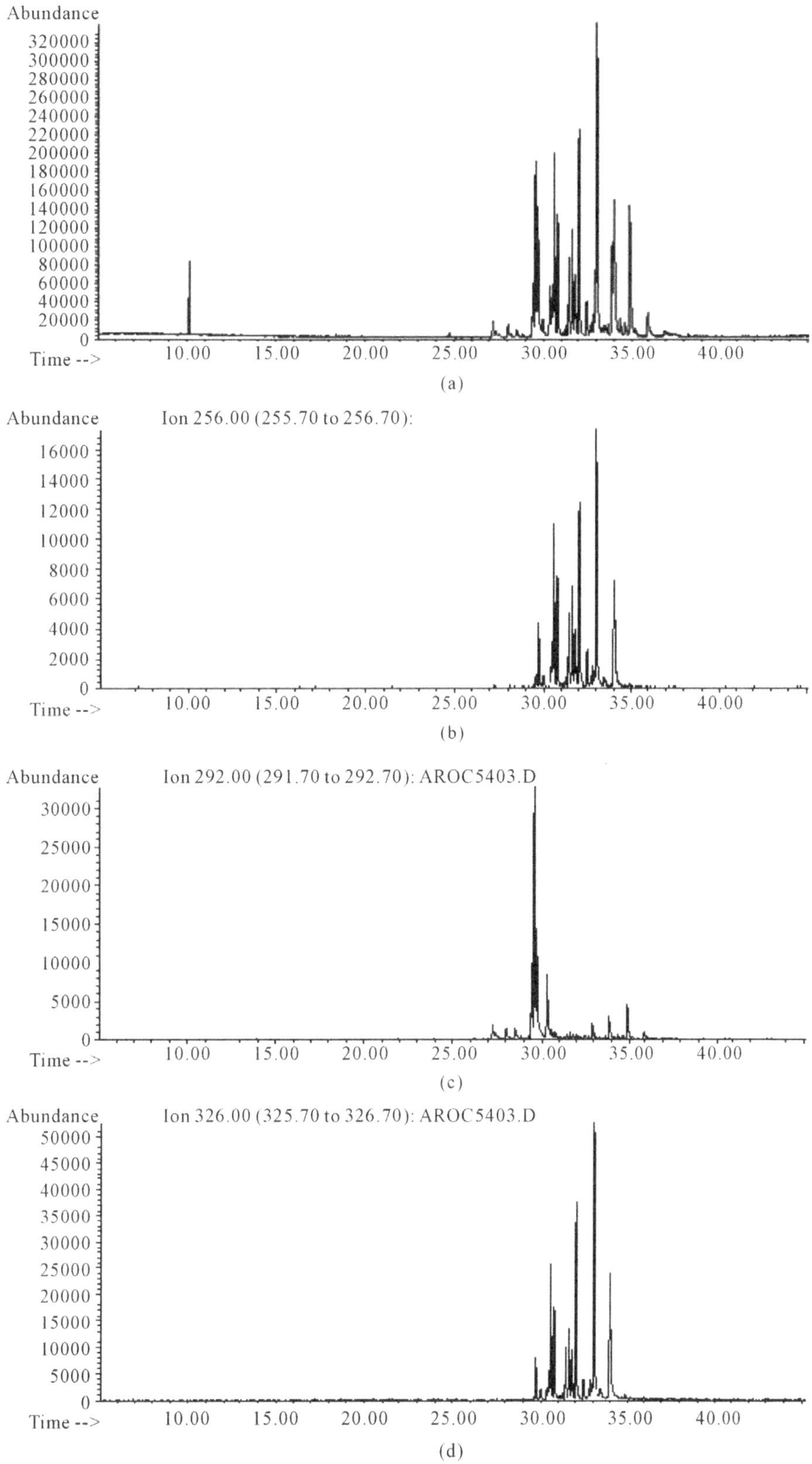

Figure 2. (a) Total ion chromatogram of AROCLOR 1254 standard solution; (b) Chromatogram obtained selecting 256 ion mass selected; (c) Chromatogram obtained selecting 292 ion mass selected; (d) Chromatogram obtained selecting 326 ion masses selected.

as well as different volumes and agitation times were tested; best results were found using n-hexane in two successive portions of 10 mL each, with agitation in an ultrasonic bath during 40 minutes for the first solvent portion and 10 minutes for the second one. A chromatogram obtained from this extract is showed in **Figure 3**; it can be seen that the typical unresolved lineal aliphatic hydrocarbons concave shape background appears, over which PCBs chromatographic peaks are detected. Consequently, a clean up by solid phase extraction using silica C18 cartridges was used, applying the protocol of M. M. Schantza *et al.* [14]; however, PCBs recoveries were quite low, indicating difficulties in separating PCBs from hydrocarbons efficiently. Therefore, a methodology was applied based on background subtraction of mass spectra over which selected PCBs ion masses appear. This background spectrum consisted of lineal aliphatic hydrocarbons (**Figure 4**) and it was common to the whole retention time range in which PCB peaks appear in the total ion chromatogram. When this hydrocarbon massspectra was subtracted, using the Hewlett-Packard Chemstation B.02.05 software with GC/MS Engine instrument, from the total ion chromatogram, the new chromatogram is showed in **Figure 5**; it can be seen that most of the initial background was eliminated. Over this clean chromatogram, common ion masses of PCBs (256, 292 and 326 m/z) were selected and chromatographic peaks were integrated on a chromatogram of these selected masses. This chromatogram is showed in **Figure 6**.

3.3. Analysis of Automotive Shredder Samples

The procedure specified in the experimental chapter was applied to samples from an automotive shredder industry detailed in **Table 1**. Results obtained are showed in **Table 3**; as can be seen, PCBs were detected in all samples; the highest concentration detected was in sample B. In any case these PCBs content should be taken into account in order to be used as a combustion fuel in cemetery industry.

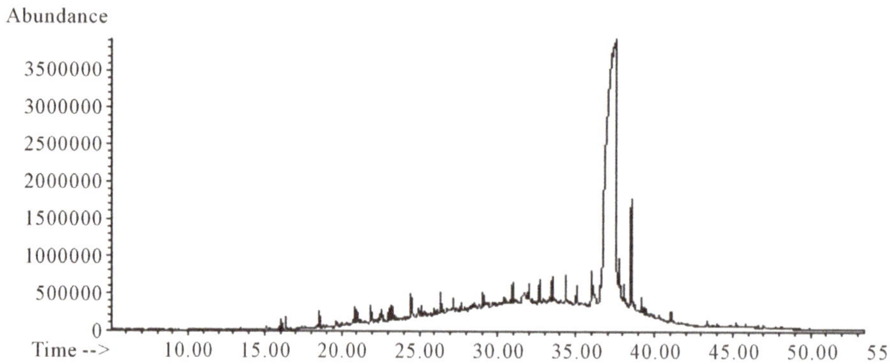

Figure 3. Total ion chromatogram of B sample enriched with 100 µL of AROCLOR 1254.

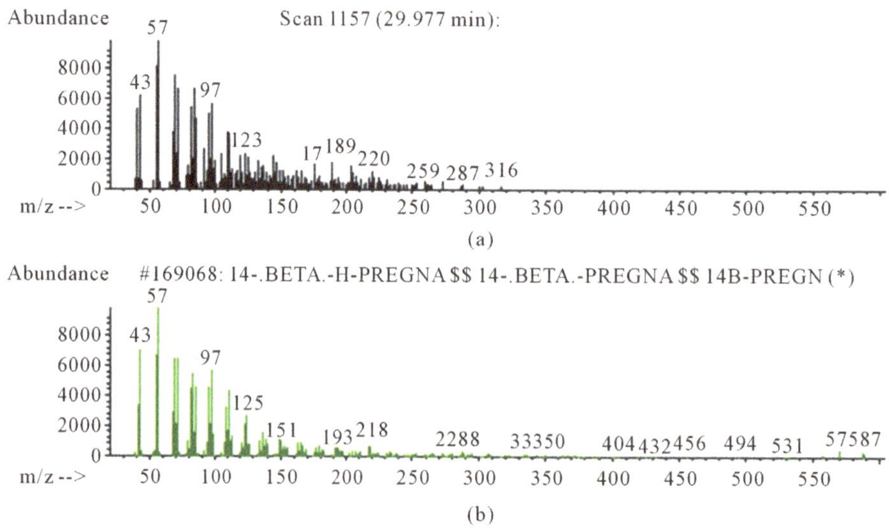

Figure 4. Mass spectra of B sample analysis: (a) Mass spectra of concave shape background obtained in B sample analysis

enriched with 100 μL of AROCLOR; (b) Identification by WILLEY library of background.

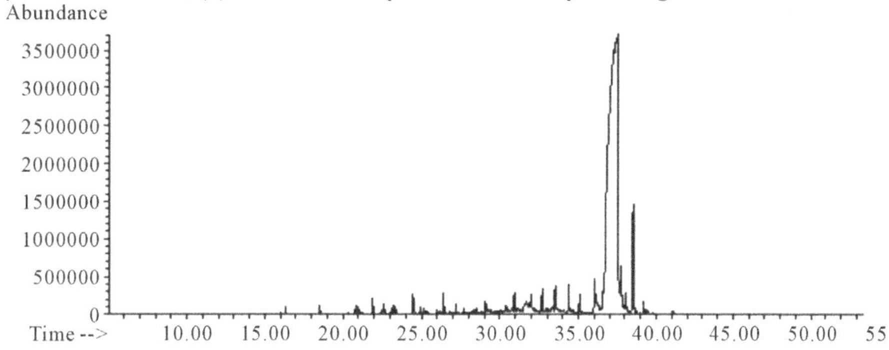

Figure 5. Total ion chromatogram of B sample subtracting background of the lineal aliphatic hydrocarbon interferences.

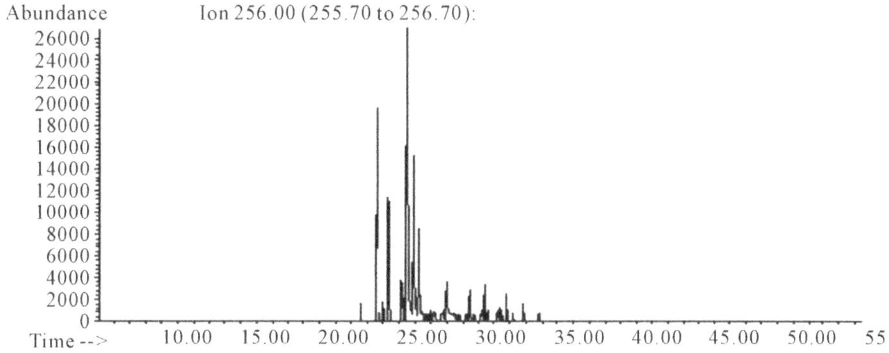

(a)

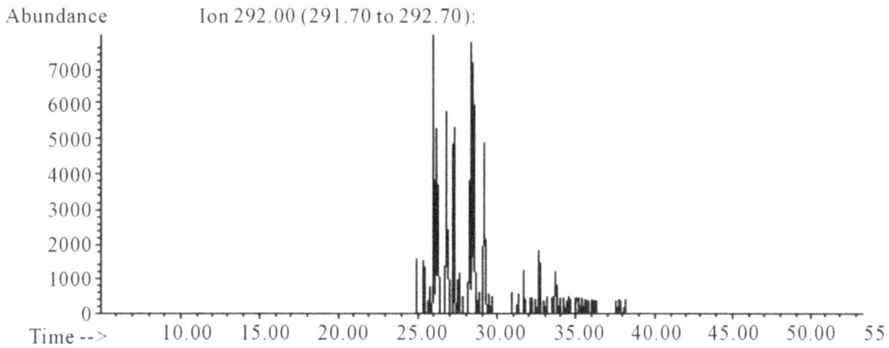

(b)

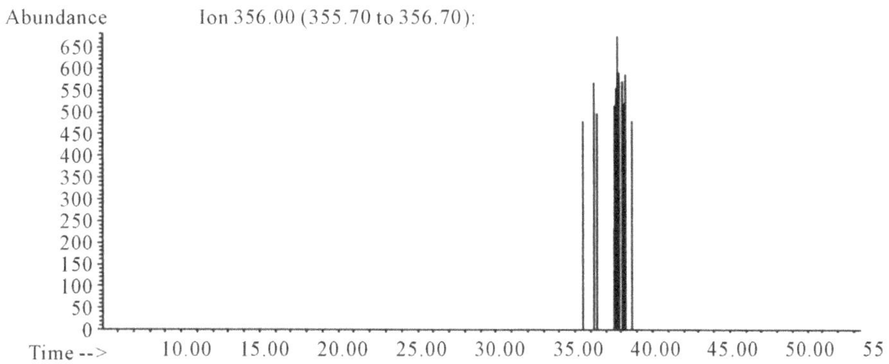

(c)

Figure 6. Total ion chromatogram obtained of sample B sample analysis: (a) Chromatogram obtained selecting 256 ion mass selected; (b) Chromatogram obtained selecting 292 ion mass selected; (c) Chromatogram obtained selecting 326 ion masses

selected.

3.4. Recovery Study

Because shredder reference samples were not available, recovery was determined to establish the validity of the proposed method. For this purpose, different volumes in the range of 25 - 100 µL of Aroclor 1254 were added to the sample and analyzed by following the same procedure as for samples. Results are showed in **Table 4**.

4. Conclusions

A direct method to identify and quantify PCBs in automotive shredder industry samples was proposed. The use of three selected ion masses (256, 292 and 326 m/z) which are present in all commercial Aroclors, making for an easy and fast method, was carried out in complex industry samples, without the need for high resolution Recovery results show the validity of the method. For this reason, a simple solid-liquid extraction process of PCBs and then their analysis by GC/MS is enough to conclude the presence or absence of these compounds in automotive shredder samples which could be used in recycled materials.

This method could be applied to other types of solid samples in order to conclude the presence or absence of PCBs.

Table 2. Standard calibration results.

Calibration graph equation	Total area = 8.68104C – 1127398.1
Regression coefficient	0.9991
*RSD	7.05%
Sensitivity	**L.D. = 3·[(S/N)/m] = 0.13 pg/µL **L.D. = (S/N)/m = 0.16 ng/µL

*RSD, Relative Standard Deviation; **L.D., Limit Detection.

Table 3. PCBs concentrations detected in automotive shredder samples.

*Samples	Caverage, µg/g	**RSD, %
A	116	21
B	191	24
C	118	22
D	94	12
E	102	23

*0.5 g weight samples; **RSD, Relative Standard Deviation.

Table 4. Results obtained in recovery study in B and C samples.

		PCBs, µg			
Sample Weight, g	Sample Content	Added	Found	*R%	**RSD
		25	80	119	14
B 0.5	51	50	108	116	11
		100	134	84	20
		25	78	108	12
C 0.5	47	50	101	104	11
		100	125	85	15

*R, Recovery; **RSD, Relative Standard Deviation for n = 3.

5. Acknowledgements

This article would have been impossible without the equipment and materials contributed by CAI-UEM, Research Support Centre-Mass Spectrometry Unit, University Complutense of Madrid

REFERENCES

[1] J. Péra, J. Ambroise, M. Chabannet, "Valorization of Automotive Shredder Residue in Building Materials," *Cement and Concrete Research*, Vol. 34, No. 4, 2004, pp. 557-562.

[2] K. J. Kima, D. Kima, S. H. Choia, K. Chunga, K. S. Shina, F. Barlatb, K. H. Oha and J. R. Youn, "Formability of AA5182/Polypropylene/AA5182 Sandwich Sheets," *Journal of Materials Processing Technology*, Vol. 139, No. 1-3, 2003, pp. 1-3.

[3] J. A. S. Williams, S. Wongweragiat, X. Qu, J. B. McGlinch, W. Bonawi-tan, J. K. Choi and J. Schiff, "An Automotive Bulk Recycling Planning Model," *Journal of Operational Research*, Vol. 177, No. 2, 2007, pp. 969-981.

[4] Börjeson, L. Börjeson, G. Löfvenius, M. Hjelt, S. Johansson and S. Marklund, "Characterization of Automotive Shredder Residues from Two Shredding Facilities with Different Refining Processes in Sweden," *Waste Management and Research*, Vol. 18, 2000, pp. 358-366.

[5] B. Heinzow, S. Mohr, G. Ostendorp, M. Kerst and W. Körner, "PCB and Dioxin-Like PCB in Indoor Air of Public Buildings Contaminated with Different PCB Sources—Deriving Toxicity Equivalent Concentrations from Standard PCB Congeners," *Chemosphere*, Vol. 67, No. 9, 2007, pp. 1746-1753.

[6] A. J. Pedersen and L. M. Ottosen, "Elemental Analysis of Ash Residue from Combustion of CCA Treated Wood Waste before and after Electrodialytic Extraction," *Chemosphere*, Vol. 65, No. 1, 2006, pp. 110-116.

[7] B. Colin, "Enviromental Chemistry," Freeman, pp. 252-272.

[8] H. T. Joung, Y. C. Seo, K. H. Kim and Y. C. Seo, "Effects of Oxygen, Catalyst and PVC on the Formation of PCDDs, PCDFs and Dioxin-Like PCBs in Pyrolysis Products of Automobile Residues," *Chemosphere*, Vol. 65, No. 9, 2006, pp. 1481-1489.

[9] S. Lorán, S. Bayarri, P. Conchello and A. Herrera, "Risk Assessment of PCDD/PCDFs and Indicator PCBs Contamination in Spanish Commercial Baby Food," *Food and Chemical Toxicology*, Vol. 48, No. 1, 2009, pp. 145-151.

[10] S. L. Wang, C. Y. Lin, Y. L. Guo, L. Y. Lin, W. L. Chou, L. W. Chang, "Infant Exposure to Polychlorinated Dibenzo-p-Dioxins, Dibenzofurans and Biphenyls (PCDD/Fs, PCBs)—Correlation between Prenatal and Postnatal Exposure," *Chemosphere*, Vol. 54, No. 10, 2004, pp. 1459-1473.

[11] S. Jursa, J. Chovancová, J. Petrík and J. Lokša, "Dioxin-Like and Non-Dioxin-Like PCBs in Human Serum of Slovak Population," *Chemosphere*, Vol. 64, No. 4, 2006, pp. 686-691.

[12] T. Takasuga, K. Senthilkumar, T. Matsumura, K. Shiozaki and S. Sakai, "Isotope Dilution Analysis of Polychlorinated Biphenyls (PCBs) in Transformer Oil and Global Commercial PCB Formulations by High Resolution Gas Chromatography-High Resolution Mass Spectrometry," *Chemosphere*, Vol. 62, No. 3, 2006, pp. 469-484.

[13] M. Ren, P. A. Peng, S. K. Zhang, Y. Y. Deng, B. X. Mai, G. Y. Sheng and J. M. Fu, "Determination of 2,3,7,8-Substitituted Polychlorinated Dibenzo-p-dioxins-dibenzofurans and Dioxin-Like Polychlorinated Biphenyls in Environmental Samples by Gas Chromatography/High Resolution Mass Spectrometry," *Chinese Journal of Analytical Chemistry*, Vol. 35, No. 2, 2007, pp. 176-180.

[14] M. M. Schantza, S. Bøwadt, B. A. Benner, S. A. Wisea and S. B. Hawthorneb, "Comparison of Supercritical Fluid Extraction and Soxhlet Extraction for the Determination of Polychlorinated Biphenyls in Environmental Matrix Standard Reference Materials," *Journal of Chromatography A*, Vol. 816, No. 2, 1998, pp. 213-220.

Determination of Inorganic Antimony Species Conversions during Its Speciation Analysis in Soil Using Isotope Dilution Techniques

Sameer Amereih[1], Thomas Meisel[2], Zaher Barghouthi[3], Wolfhard Wegscheider[2]

[1]Palestine Technical University-Kadoorie (PTUK), Tulkarm, Palestine; [2]General and Analytical Chemistry Department, University of Leoben, Leoben, Austria; [3]Institute of Chemistry, Karl-Franzens University, Graz, Austria.

ABSTRACT

A new method was developed in this work to account for inorganic Sb species interconversion during soil sample preparation and subsequent separation steps. The Sb(III) and Sb(V) concentrations at each investigated step in the analytical procedure were determined using species specific spikes (^{121}Sb(III) with 81.18% and ^{123}Sb(V) with 74.04% enrichment). The spiking of these enriched isotopes species solutions was done separately before soil sample extraction and before HPLC separation. Simply by subtracting the final concentration of each species done by on-line isotope dilution (ID) from its concentration at different stages of the analytical procedure done by species specific ID, the influence of each step on species transformation can be estimated. After optimization, the extraction procedure for inorganic Sb species 6% Sb(III) (1.3 RSD, n = 3) and 43.2% Sb(V) (2.9% RSD, n = 3) as percent of total Sb were detected in the examined soil sample using on-line ID. Using the above described methodology we found that there was no reduction of Sb(V) to Sb(III) during sample preparation or species separation. While about 9.3% of extractable Sb (4.6% of total) was converted from Sb(III) to Sb(V) during the extraction step, no conversion during HPLC separation step was observed. By compensating for Sb(III) transformation during the sample preparation step; the extractable Sb(III) and Sb(V) as percent of total Sb yielded 10.6% and 38.7%, respectively.

Keywords: Antimony Speciation Analysis; Species Interconversions; Isotope Dilution Techniques

1. Introduction

Antimony and its compounds are considered to be priority pollutants by Environmental Protection Agency of the United States (USEPA) and European Union (EU) [1] Sb toxicity depends on its oxidation state; Sb(III) compounds are ten times as toxic as Sb(V) compounds [2,3]. Therefore, accurate determination of inorganic antimony species distribution in environmental samples is of great importance.

Current measurement procedures like On-line coupling of liquid chromatography with ICP-MS can provide selective, sensitive and rapid measurements for inorganic metallic speciation analysis, but it doesn't necessarily guarantee accurate analytical results. Independent of the other effects, they are also presented in elemental speciation analysis, varying compositions in chromatographic peaks influence the ionization efficiency in the plasma and, therefore, hinder exact external calibration curve [4] Certified Reference Materials (CRMs) are a useful tool to test for accuracy. Unfortunately, at present they are scarce. Besides, a reliable extraction procedure for a certain material may not be so applied to another material because the adsorption/binding forces of the species of interest to the solid are strongly dependent on the matrix. An alternative approach to the use of CRMs for speciation could be resort to highly qualified primary methods such as Isotope Dilution-Inductively Coupled Plasma Mass Spectrometry (ID-ICP-MS) [5] which provides general accuracy and superior uncertainty over other calibration strategies, including external calibration and standard addition [6]. The high accuracy and precision provided by ID-ICP-MS in the species-specific spiking mode can used to correct for the majority of systematic errors occurred in speciation analysis (non-quantitative separation procedures, signal drifts, quantification measurements errors, etc.). Although isotope dilution ICP-MS has been widely employed for trace element analysis in a wide variety of sample matrixes, its application to species determinations has been limited by the non-availability of commercial species-specific enriched spikes. In species specific mode the original sample is spiked prior to separation with advantage that loss of substance has no effect on the analytical result [7].

However, the sample preparation is the most challengeable step of the analytical speciation procedures preventing that reliable enough speciation results can be achieved routinely. This statement refers specially to the speciation of elements in the solid samples where the major difficulty is not the quantification itself but the quantitative extraction of those compounds from the complex solid sample matrix without degradation or species transformation [8]. Rearrangement reactions between the species and degradation reactions cannot be controlled when more aggressive conditions are used in extraction in the search for quantitative recoveries. In such cases, the speciation information might be altered [9]. It is evident that the border line between quantitative extraction and alteration of the species is probably small and will depend on the matrix itself and on the labiality of the sought species [10].

Speciated isotope dilution mass spectrometry (SIDMS) takes a unique approach to speciated analysis that differs from traditional methods. Traditional speciation methods inherently measure the species after species conversion has occurred. SIDMS has been developed to address the correction for the species conversion. In SIDMS, each species is labeled with a different isotope-enriched spike in the corresponding species form. Thus, the interconversion that occur after spiking is traceable and can be corrected. While SIDMS maintains the advantages of IDMS, it is capable of correcting the degradation of the species or the interconversion between the species [11]. Evaluating and correcting for these species transformations in the sample preparation step could be attained if a spike solution containing inorganic antimony species isotopically labeled with different antimony isotopes. This can not be done in a similar way to previous studies which applied SIDMS successfully for the elements having more than two naturally isotopes like Cr [12], since Sb has only two isotopes. For this purpose a special approach is presented in this work. The Sb species distribution after the chromatographic separation was determined using on-line isotope dilution [13] after their extraction from a contaminated soil sample using 100 mmol·L^{-1} citric acid. The Sb(III) and Sb(V) concentrations at each investigated step in the analytical procedure were determined using species specific spikes [^{121}Sb(III) and ^{123}Sb(V)]. The influence of each step on species transformation can be estimated by subtracting the final concentration of each species done by on-line ID from its concentration done by species specific ID at the investigated step of analytical procedure.

2. Experimental

2.1. Sample

The soil sample examined in this work consisted of 20 subsamples and represented a road section between 50 - 100 m. The subsamples were taken from the soil surface (0 - 5 cm) near the edge of a highway (Vienna/A23) [14]. The collected soil sample was dried at ambient air temperature and then sieved to obtain the <2 mm soil fraction. The sieved sample was grinding using a mixer mill. The final fineness of samples particles after grinding was about 0.01 mm. The sample is an alkaline soil (pH of water extract (1:10) is 8.6) and the organic carbon was 1.71% m·m^{-1}. The soil sample contains 4.17 μg·g^{-1} as total Sb [13].

2.2. Instrumentation for Hyphenation

A standard Agilent Technologies HP4500 quadruple inductively coupled plasma mass spectrometry (QICP-MS) system was used with Babington for Sb speciation and a cooled quartz glass spray chamber was used. For the chromatographic module a quaternary HP 1050 series pump was used. Separation of Sb(III) and Sb(V) was carried out on a Hamilton PRP-X100 anion exchange column (150 mm × 4.6 mm i.d.) using 10 mmol·L^{-1} EDTA, 1 mmol·L^{-1} phthalic acid at pH 4.5 as a mobile phase of a flow rate of mL·min^{-1}. Samples and standards were injected using an autosampler (HP 1050 Series) with a 100 μL injection loop. The chromatographic system was interfaced with an ICP-MS instrument using 250 mm of PEEK tubing (0.54 mm i.d.) to connect the column to the nebulizer. On-line isotope dilution was preformed by a continuous addition (0.5 mL of min^{-1}) of the enriched spike was achieved by joining the spike solution line, pumped by a peristaltic pump equipped with a silicone tube (1.02 mm i.d.), with the HPLC effluent (1 mL·min^{-1}) via a three-way valve. The mixed solution passed directly to the nebulizer of the ICP-MS. The "time-resolved analysis" mode of the ICP-MS software was used for the data acquisition. The integration times of Sb isotopes were 0.9 s per mass.

2.3. Chemicals and Reagents

A Milli-Q (Millipore) purified water (18.2 MΩ·cm^{-1}) was used for the preparation of all solutions. Extraction solutions were prepared from citric acid monohydrate (99.5% - 100.5% mm^{-1}, Merck). The HPLC eluents were prepared with adequate concentrations by dissolving appropriate amounts phthalic acid (≥99.5% mm^{-1}, Fluka) and EDTA disodium salt (99.0% - 101.0% mm^{-1}, Baker). The pHs of these eluents were adjusted by adding ammonia solution (25%, Merck) or nitric acid (200 $mmol \cdot L^{-1}$). The resulting solutions were de-gassed before using. $^{nat}Sb(V)$ standard solution with natural isotopic abundance was prepared from antimony(V) oxide (99.995% mm^{-1}, Aldrich). Sb(V) stock solution (1 g ≈ 10 µg Sb) enriched with (94.2%) in ^{123}Sb was obtained as antimony oxide (Sb_2O_5) [lot number H3451] from Euriso-top France. Procedure describes antimony oxide as source material for $^{123}Sb(V)$ and $^{nat}Sb(V)$ illustrated elsewhere [13]. $^{nat}Sb(III)$ stock solution, 1 g = 0.5 mg Sb was prepared by dissolving 50 mg of Sb metal in 5 mL boiled concentric HCl. Then 2 mL of a reducing solution (40% KI + 5% ascorbic acid) were added to reduce all Sb to Sb(III). The solution was diluted to 100 g using highly purified water. The enriched isotope metal $^{121}Sb(III)$ (98.73% of 121) has been obtained from Euriso-top (France) [lot number H2841]. It was prepared in the same manner as the natural Sb(III). Reverse isotope dilution was used to find out the exact concentration of $^{121}Sb(III)$ by calibrating with the $^{nat}Sb(III)$ solution and independently, with the certified 10 $\mu g \cdot mL^{-1}$ ICP-MS standard solution after reduction of possible Sb(V) species using the reducing solution mentioned above. All standard and stock species solution were stored in dark polymeric containers in a refrigerator at a temperature 1°C - 4°C.

2.4. Analytical Procedures

2.4.1. Species Unspecific Isotope Dilution

100 µL of extraction solution was injected in the HPLC using the autosampler. From the measured intensities of the corresponding isotopes ^{121}Sb and ^{123}Sb the isotope ratio $R_{123/121(t)}$ for every measured time was calculated. The isotope ratio chromatograms were converted into mass flow chromatograms by calculating the absolute Sb amount for each $R_{123/121(t)}$ on the chromatogram by applying on-line isotope described on our previous work [13]. The background correction was done by subtracting the mean values of Sb mass before and after each peak from each point of the peaks in the mass chromatogram. The exact concentrations of Sb(III) and Sb(V) were determined by integration of the areas under the peaks after background subtraction. A special protocol for HPLC-ID-ICP-MS measurement (**Figure 1**) was followed in order to correct for mass bias which is the deviation of the measured ratio of a natural Sb solution from the agreed isotope ratio ($^{123/121}Sb$ 0.74785) [15] and to assume the blank contribution to the $R_{123/121(t)}$ of each measurement.

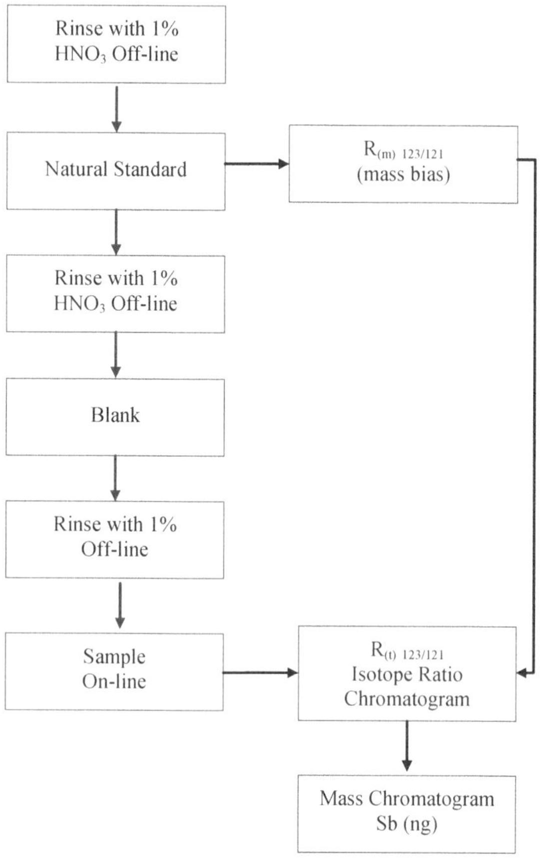

Figure 1. Protocol for HPLC-ID-ICP-MS measurements.

2.4.2. Species-Specific Isotope Dilution

Species-specific isotope dilution procedure which was used to determine the concentrations of inorganic Sb separately in two stages of the speciation analysis procedure is explained below:

1) Before extraction step

500 mg of a representative soil sample were weighed to the nearest 0.01 mg into a 10 mL screwed-cap polystyrene tube. This sample was spiked with 1.00 g of a calibrated isotope-enriched species $^{121}Sb(III)$ 0.93 $\mu g \cdot g^{-1}$ or $^{123}Sb(V)$ 2.91 $\mu g \cdot g^{-1}$ to quantify Sb(III) and Sb(V) at the spiking stage separately. Then soil and the spike solution were shaken manually and the suspension was left for 10 min in a refrigerator at less than 4°C. The spike species and the original species were extracted using 10 ml of 100 $mmol \cdot L^{-1}$ citric acid solution using an ultrasonic bath (power 120 W) for 45 min where equilibration can be achieved. The solution was then centrifuged at a speed of 4000 rpm for 5 min. and filtered using a 0.45 µm syringe filter. The filtrate was stored in polystyrene vials in a refrigerator at a temperature ≤ 4°C until the analysis was possible. Before analysis 3.0 g of spiked soil extract was diluted to 5.0 g using Milli-Q water. Then 100 µL of the diluted extraction solution was in-

jected in the HPLC using autosampler. From the measured intensities of the corresponding isotopes ^{121}Sb and ^{123}Sb the isotope ratios $R_{121/123}$ for Sb(III) and $R_{123/121}$ for Sb(V) were calculated from the measured of the peak area ratios after mass bias corrections. In order to correct for mass bias a soil extract obtained in the same way as the spiked soil portion was used before each measurement where the soil extract was assumed to have a natural Sb isotopic composition. The corrected isotope ratios $Sb(III)_{121/123}$ calculated from the chromatogram obtained for spiked soil sample with enriched ^{121}Sb(III) were used to calculate Sb(III) concentration at this step. While Sb(V) concentration was calculated using the corrected isotope ratio of $Sb(V)_{123/121}$ obtained from another portion of sample spiked with ^{123}Sb(V). Calculations for element species concentrations are illustrated elsewhere [16].

2) After extraction and before HPLC separation

Soil extraction was carried out in the same way as mentioned above for the spiked sample portions. 3.0 g of soil extract was spiked with 1.0 g of calibrated isotope-enriched species ^{121}Sb(III) 0.93 $\mu g \cdot g^{-1}$ or 1.0 g of ^{123}Sb(V) 2.91 $\mu g \cdot g^{-1}$ to quantify Sb(III) and Sb(V) and the spiked soil extract was diluted with highly purified water to 5.0 g. Then the determinations of species were done in a similar way to that used in case of the spiked soil portions (described above).

3. Results and Discussion

3.1. Interconversion of Inorganic Antimony Species during Its Speciation Analysis in Soil

3.1.1. Purity of ^{121}Sb(III) and ^{123}Sb(V) Spikes

Figure 2(a) represents a diluted solution (1:10) of ^{121}Sb(III) standard after the separation and determination using HPLC-ICP-MS. This solution was used for spiking of the soil portions and soil extracts. It is clear that only Sb(III) enriched with ^{121}Sb is found in this spike solution and no Sb(V). This is very important hence the presence of species specific spike is one of the most important steps in the methodology of species specific isotope dilution quantification. The source (metal) for this spike solution preparation has a certified enriched ratio of the isotope 121% of 98.73%. While this ratio was changed to 81.18% during the spike solution preparation and in the separation step where a lot of chemicals had been used: acids, reducing agents, mobile phase furthermore the complexing reagent. Thus it is important that the exact concentration and enriched isotope percent must be calibrated before spiking the sample. This was done using reverse isotope dilution. This diluted solution was spiked with a known amount of a ^{nat}Sb(III) species of known concentration. The original source which was used to prepare the enriched isotope species ^{123}Sb(V) was Sb_2O_5 where Sb is enriched with 123 isotope with a percent of 94.2%. The fraction of ^{123}Sb in Sb(V) species decreased after its preparation and separation to 74.05% **Figure 2(b)**. The exact concentration of the ^{123}Sb(V) spike was determined using reverse isotope dilution.

3.1.2. Calculating for Inorganic Antimony Species Interconversion during Analysis

Speciation analysis of Sb in the tested soil sample shows that the major difficulty is not the quantification of Sb species. The sample preparation is the most challenging step of the analytical speciation procedures preventing reliable speciation results. The final concentrations of the antimony species in the examined soil sample were determined using on-line isotope dilution after optimization of the extraction procedure for best separation of inorganic Sb species [17]. **Figure 3** represents the speciation of inorganic Sb species in the examined soil about 6% Sb(III) (1.3 RSD, n = 3) and 43.2% Sb(V) (2.9% RSD, n = 3) as percents of total Sb were determined. Interconversion between the species and degradation reactions

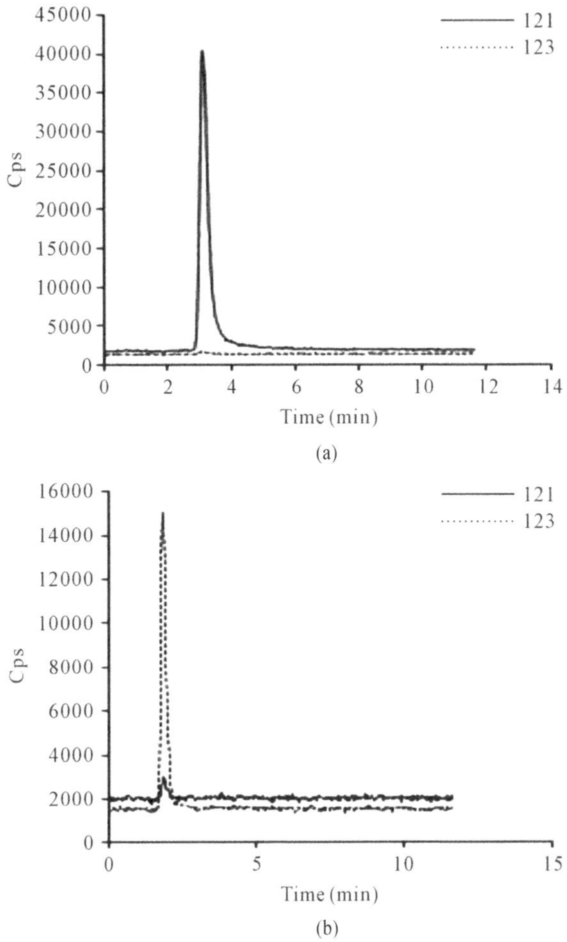

Figure 2. Chromatograms represent: (a) ^{121}Sb(III) and (b) ^{123}Sb(V) spike standards. HPLC flow rate was 1 $mL \cdot min^{-1}$.

cannot be discarded under the conditions which were used in extraction and further speciation analysis steps. In such cases, the speciation information might be altered and the pertinent question to be raised here: was there any species interconversion during the analysis and how much the data reflect the true speciation picture of Sb in the examined sample.

A novel method was developed in this work to calculate for inorganic Sb species interconversion during soil sample preparation and subsequent separation steps. The Sb species distribution after the chromatographic separation was determined using on-line isotope dilution after their extraction from the tested soil sample using the suitable extraction conditions for speciation analysis [13]. The separation of Sb(III) and Sb(V) was achieved using an anion exchange column (PRP-X100) and 10 $mmol \cdot L^{-1}$ EDTA, 1 $mmol \cdot L^{-1}$ phthalic acid at pH 4.5 as a mobile phase. The Sb(III) and Sb(V) concentrations at each investigated step in the analytical procedure were determined using species specific spikes (^{121}Sb(III) with 81.18% and ^{123}Sb(V) with 74.04% enrichment). The spiking of these enriched isotopes species solutions was done separately before soil sample extraction and before HPLC separation. The species interconversion during analysis can be calculated simply by subtracting the final concentration of each species done by on-line ID (speciation after species transformation) from its concentration at different stages of analytical procedure done by species specific ID (speciation before species transformation). The species transformations can be estimated as an absolute amount. This is illustrated in the following equations:

Sb(III) interconversion

$$= \left| C_x^{III}\left(\text{species-specific}\left[{}^{121}\text{Sb(III)}\right]\text{ID}\right) - C^{III}\left(\text{on-line ID}\right) \right|$$

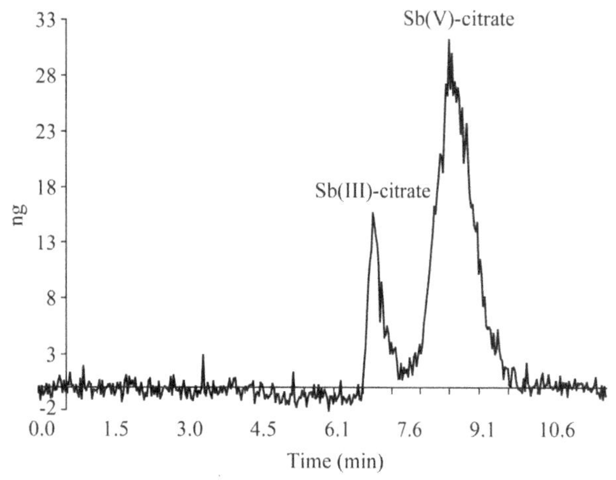

Figure 3. Mass chromatogram of Sb speciation in the examined soil sample done by on-line isotope dilution, HPLC flow rate was 1 $mL \cdot min^{-1}$.

Sb(V) interconversion

$$= \left| C_x^{V}\left(\text{species-specific}\left[{}^{123}\text{Sb(V)}\right]\text{ID}\right) - C^{V}\left(\text{on-line ID}\right) \right|$$

where C_x^{III} and C_x^{V} are the concentrations of Sb(III) and Sb(V) at the investigated step of analytical procedure done by species-specific isotope dilution. C^{III} and C^{V} are their concentrations done by on-line isotope dilution.

Two important steps of analytical procedure (extraction and separation) were examined for their influence on species interconversion. Three portions of the tested soil sample (unspiked, spiked with Sb(III) enriched with 121 and spiked with Sb(V) enriched with 123) were used to test for species transformation during extraction step. The first was an unspiked soil portion, which was used for three reasons:

- to determine the species concentration at the end of analytical procedure (after species transformation occurred) using on-line isotope dilution after species extraction following the procedure and the measurement protocol described elsewhere [17].
- The unspiked portion was also used before measurement of the spiked soil portions to correct the isotope ratios of the blends for the mass bias, since it has similar matrix liked spiked soil portions and it has the natural abundance of Sb isotopes.
- In addition it was used to control the influence of the enriched-isotopically spiked species on other species, since it represents the natural isotope ratio of unspiked species under the experimental conditions.

The second soil portion was spiked with Sb(III) enriched with 121 and also extracted using the same extraction procedure used for unspiked portion illustrated in details elsewhere [13]. Measuring the isotope ratio of Sb(III) and Sb(V) in this spiked portion will give information about Sb(III) species transformation to Sb(V). This can be determined if the natural isotope ratio (121/123) of Sb(V) was increased. Actually, there was a change in the natural isotope ratio of 121/123 of Sb(V) after spiking a soil portion with ^{121}Sb(III). The ratio of 121/123 of Sb(V) was increased (**Table 1**, part-A) compared with natural isotope ratio of Sb(V) in the unspiked soil portion. This means that part of Sb(III) was converted to Sb(V). On the other hand, no Sb(V) reduction could be detected, hence the isotope ratio 121/123 of Sb(III) did not decrease and it remained as the natural isotope ratio when another portion of soil was spiked with Sb(V) enriched with 123.

In the same manner, the second stage (after extraction and before HPLC separation) was examined for species interconversion. The difference here is that the soil extract (solution) was spiked instead of soil portions (solid) in first step. Results in **Table 1** part-B shows that there was no interconversion at HPLC separation step. Isotope ratio of 121/123 of Sb(V) did not change after spiking the soil extract with ^{121}Sb(III) and also the isotope ratio of 121/123 of Sb(III) did not change after spiking with ^{123}Sb(V).

The concentration of each Sb species was determined separately using species specific isotope dilutions equation [16] at the examined steps of analysis. This was done using the corrected isotopes ratios given in **Table 1**, which were obtained from previous spiking procedures. Then the percent of each species in the total extractable Sb was calculated and shown in **Table 2**. Comparing these results and results obtained for Sb species distribution at the final step of analysis done by on-line isotope dilution shows that the only significant difference was observed at the extraction step. The Sb(III) concentration before extraction is higher than Sb(III) at the final stage of analysis. This is in contrast, to Sb(V) concentration which is less than that found at the final stage of analytical procedure in the same percent as the increase of Sb(III) which is equal to 9.3% of extractable Sb (4.6% of total Sb). By compensating for Sb(III) transformation during sample preparation step; the extractable Sb(III) and Sb(V) as percents of total Sb yielded 10.6% and 38.7%, respectively.

Table 1. Mass bias corrected isotopes ratios of the inorganic Sb species after spiking with species specific at important stages of analytical procedure.

A—Mean of isotopes ratios results (n = 3) for spiking before extraction step		
Kind of sample	Rc121/123 Sb(III)	Rc121/123 Sb(V)
Spiked soil with ^{121}Sb(III)	1.798 ± 0.003	1.455 ± 0.010
Spiked soil with ^{123}Sb(V)	1.340 ± 0.001	1.078 ± 0.011
Unspiked soil	1.327 ± 0.006	1.341 ± 0.004
Natural (ICP-MS calibration solution)	1.339 ± 0.004	1.339 ± 0.004
B—Mean of isotopes ratios results (n = 3) after extraction and before HPLC separation		
Spiked soil extract with ^{121}Sb(III)	3.633 ± 0.101	1.354 ± 0.046
Spiked soil extract with ^{123}Sb(V)	1.348 ± 0.006	0.767 ± 0.021
Unspiked extract soil	1.334 ± 0.059	1.350 ± 0.046
IUPAC natural isotope ratio [18]	1.337 ± 0.001	1.337 ± 0.001

Table 2. Results of distribution of inorganic Sb species at important steps in analytical procedure using corrected isotope ratios.

Sb species	Before extraction step (1) %Sb[1], (RSD, n = 3) (SS)[2]	Before separation step (2) %, (RSD, n = 3) (SS)	Final step % Sb, (RSD, n = 3) (OL)[3]
Sb(III)	26.8 (0.1)	16.7 (4.1)	17.5 (3.44)
Sb(V)	73.2 (5.2)	83.3 (0.8)	82.5 (0.2)

[1]Percent of Sb species to total extractable Sb; [2]Species specific isotope dilution result; [3]Online isotope dilution result.

4. Conclusion

This is the first time that a combination of species unspecific and species specific isotope dilution results was used to calculate for species interconversion during the analytical procedure. The suggested methodology is applied for an element (Sb) having only two naturally isotope. Results show that only extraction step has an influence on Sb(III) transformations, even citric acid was used as an extractant. It was believed that citric acid can stabilize Sb(III) species, but it seems not in the presence of a soil matrix. There was no reduction of Sb(V) to Sb(III) during sample preparation or species separation and no conversion during HPLC separation step was observed.

REFERENCES

[1] M. Filella, N. Belzile and Y.-W. Chen, "Antimony in the Environment: A Review Focused on Natural Waters: I. Occurrence," *Earth-Science Reviews*, Vol. 57, No. 1-2, 2002, pp. 125-176.

[2] A. Guy, P. Jones and S. Hill, "Identification and Chromatographic Separation of Antimony Species with α-Hydroxy Acids," *Analyst*, Vol. 123, No. 7, 1998, pp. 1513-1518.

[3] M. Filella and N. Belzile, "Antimony in the Environment: A Review Focused on Natural Waters. III. Microbiota Relevant Interactions," *Earth Science Reviews*, Vol. 80, No. 3-4, 2007, pp. 195-217.

[4] K. Heumann, S. M. Gallus, G. Radlinger and J. Vogl, "Accurate Determination of Element Species by On-Line Coupling of Chromatographic Systems with ICP-MS Using Isotope Dilution Technique," *Spectrochimica Acta Part B: Atomic Spectroscopy*, Vol. 53, No. 2, 1998, pp. 273-287.

[5] A. Sanz-Medel, "Trace Element Analytical Speciation in Biological Systems: Importance, Challenges and Trends," *Spectrochimica Acta Part B: Atomic Spectroscopy*, Vol. 53, No. 2, 1998, pp. 197-211.

[6] G. Koellensperger, S. Hann, J. Nurmi, T. Prohaska and G. Stingeder, "Uncertainty of Species Unspecific Quantification Strategies in Hyphenated ICP-MS Analysis," *Journal of Analytical Atomic Spectrometry*, Vol. 18, No. 9, 2003, pp. 1047-1055.

[7] K. Heumann, L. Rottmann and L. Vogel, "Elemental Speciation with Liquid Chromatography-Inductively Coupled Plasma Isotope Dilution Mass Spectrometry,"

Journal of Analytical Atomic Spectrometry, Vol. 9, No. 12, 1994, pp. 1351-1355.

[8] P. Quevauviller, O. Donard, E. Maier and B. Griepink, "Improvements of Speciation Analyses in Environmental Matrices," *Mickrochimica Acta*, Vol. 109, No. 1-4, 1992, pp. 169-190.

[9] O. Donard, B. Lalere, F. Martin and R. Lobinski, "Microwave-Assisted Leaching of Organotin Compounds from Sediments for Speciation Analysis," *Analytical Chemistry*, Vol. 67, No. 23, 1995, pp. 4250-4254.

[10] F. Adams and S. Slaets, "Improving the Reliability of Speciation Analysis of Organometallic Compounds," *TrAC Trends in Analytical Chemistry*, Vol. 19, No. 2-3, 2000, pp. 80-85.

[11] H. M. Kingston, "Method of Speciated Isotope Dilution Mass Spectrometry," US Patent, 1995.

[12] H. M. Kingston and D. Hou, "Method 6800, Elemental and Speciated Isotope Dilution Mass Spectrometry," US Government Printing Office, Washington DC, 1998.

[13] S. Amereih, T. Meisel, E. Kahr and W. Wegscheider, "Speciation Analysis of Inorganic Antimony in Soil Using HPLC-ID-ICP-MS," *Analytical and Bioanalytical Chemistry*, Vol. 383, No. 7-8, 2005, pp. 1052-1059.

[14] J. Fritsche and T. Meisel, "Determination of Anthropogenic Input of Ru, Rh, Pd, Re, Os, Ir and Pt in Soils along Austrian Motorways by Isotope Dilution ICP-MS," *Science of the Total Environment*, Vol. 325, No. 1-3, 2004, pp. 145-154.

[15] S. Wehmeiser, R. Ellam and J. Feldmann, "Isotope Ratio Determination of Antimony from the Transient Signal of Trimethylstibine by GC-MC-ICP-MS and GC-ICP-TOF-MS," *Journal of Analytical Atomic Spectrometry*, Vol. 18, No. 9, 2003, pp. 1001-1007.

[16] H. M. Kingston, D. W. Huo, Y. S. Lu and S. Chalk, "Accuracy in Species Analysis: Speciated Isotope Dilution Mass Spectrometry (SIDMS) Exemplified by the Evaluation of Chromium Species," *Spectrochimica Acta Part B: Atomic Spectroscopy*, Vol. 53, No. 2, 1998, pp. 299-309.

[17] S. Amereih, T. Meisel, R. Scholger and W. Wegscheider, "Antimony Speciation in Soil Samples along Two Austrian Motorways by HPLC-ID-ICP-MS," *Journal of Environmental Monitoring*, Vol. 7, No. 12, 2005, pp. 1200-1206.

[18] T.-L. Chang, Q.-Y. Qian, M.-T. Zhao and J. Wang, "The Isotopic Abundance of Antimony," *International Journal of Mass Spectrometry and Ion Processes*, Vol. 123, No. 1, 1993, pp. 77-82.

A New SPME Thermal Desorption Interface for HPLC

John Jairo Ríos Acevedo[1], Mohamed Nour Kayali Sayadi[2], Luis María Polo Díez[1]

[1]Department of Analytical Chemistry, Faculty of Chemistry, Complutense University of Madrid, Madrid, Spain; [2]Mass Spectrometry Unit, Faculty of Chemistry, Complutense University of Madrid, Madrid, Spain.

ABSTRACT

A new SPME/HPLC interface is developed. It is based on thermal desorption from the SPME fiber and organic solvent collection of the desorbed analytes by sweeping them with an argon flow into a small organic solvent volume which is further injected into the HPLC chromatograph. Extraction and desorption parameters were investigated using five PAHs of different volatilities (naphthalene, acenaphthalene, fluorene, phenanthrene and anthracene) as test compounds and fluorimetric detection. Regression coefficients closed to 0.99 with RDS < 14% (n = 3) and detection limits in the range 0.07 - 0.99 $\mu g \cdot L^{-1}$ were found. A method was applied to determine the above PAHs in water samples. The results were compared with those supplied by the 550.1 EPA method showing the agreement of both methods at the 0.05 significance level.

Keywords: SPME-HPLC-FLD; SPME-HPLC Interface; Polynuclear Aromatic Hydrocarbons

1. Introduction

Development of rapid and efficient sample preparation methods for chromatographic analysis continues to be a challenge today. Ideally, sample preparation involves preconcentration and clean up with minimum steps to prevent contamination, to decrease costs, analysis time and lost of the analytes [1]. In this context, solid phase extraction (SPE) is now a general and well established technique introducing clear advantages over old liquid-liquid extraction [2-5]. However, this technique is generally time-consuming and requires large amounts of organic solvents [6,7].

More recently, Pawliszyn introduced solid phase microextraction (SPME) which is a quite elegant technique; it essentially integrates preconcentration and clean up in a single step [8-10]. Two basic types of sampling extraction modes into the fiber can be performed using SPME; direct extraction and headspace extraction (HS), which is also named HS-SPME [11-14]. In direct sampling extraction, the fiber is directly immersed into the liquid or gaseous sample while in HS-SPME, the fiber is suspended in the head space above the sample [14].

Initially, analytes desorption from the fiber was made thermally in the gas chromatograph (GC) injection port and then GC was continued in the usual way [1,9,15-18]. Consequently, most of the SPME methods involve GC and GC combined with Mass Spectrometry (MS). Desorption could also be made by solvent extraction which gave rise to an off-line interface with high performance liquid chromatograph (HPLC) to analyze compounds of different volatility [11,14,15,19]; or a heated liquid flow through interface has also been used [20]. These interfaces consisted of a chamber installed in place of the HPLC six-port valve injection loop; the analytes retained in the fiber were desorbed in static and/or dynamic mode. In the static mode an organic solvent was added into the desorption chamber settling the solvent in contact with the fiber during a suitable time; the analytes extracted into the solvent were further injected into the HPLC analytical column. Alternatively, the dynamic desorption involves passing the mobile phase through the desorption chamber at the working flow rate [21] and further injection into HPLC, an exhaustive review about this subject has been published [22]; some applications of this interface has been described [5,23]. The fiber should be carefully handled because of its fragility, and it can be easily broken, mainly during insertion into the desorption chamber [11]; we have also detected problems due to the introduction of air bubbles when the fiber was placed inside the interface; moreover, the organic solvents, or the mobile phase, damage the fiber coating decreasing its life time [14,24]. To date, no interface device design has proven to be optimal [25].

In this paper, an off-line SPME/HPLC interface is developed using old model of SPME fiber cleaner Nurka 390 [26]; analytes desorption from the fiber is made thermally preventing the contact of the fiber with the organic solvent. Desorbed analytes were flowed into a micro tube containing a small organic solvent volume which was then injected into the HPLC chromatograph. Several polynuclear aromatic hydrocarbons (PAHs) of different volatilities (considered as semivolatile compounds) were used as test analytes; the reason for this test compounds selection was that their sensibility by HPLC with fluorimetric detection can compete with GC techniques.

2. Experimental

2.1. Instrumental

PAHs analysis was carried out on a HPLC system which included an Agilent 1100 series HPLC-pump, a column oven with a temperature control module that allowed thermostatation at 35°C (± 0.5°C); a six-port Rheodyne 7125 manual injector with a 50 μl loop for partial filling injection and an Agilent 1100 series programmable fluorescence detector were also used. All the separations were carried out on a bonded phase specific for PAHs, Phenomenex Envirosep PP column (125 × 4.6 mm). The gradient started with 55:45 water:acetronitrile (v:v) during 2 minutes; next the acetonitrile content was increased to 90% in 25 minutes and returned to the initial composition after 15 minutes; the mobile phase flow rate was 1.0 ml·min^{-1}.

The column temperature was maintained at 35°C. The following excitation (λ_{ex}) and emission (λ_{em}) wavelength program was used for detection: naphthalene (λ_{ex} 274 nm, λ_{em} 335 nm), acenaphthene, fluorene (λ_{ex} 266 nm, λ_{em} 315 nm), phenanthrene (λ_{ex} 249 nm, λ_{em} 365 nm) and anthracene (λ_{ex} 251 nm, λ_{em} 420 nm).

Manual SPME commercially available 65 μm film thickness polydimethylsiloxane/divinylbencene (PDMS/DVB) StableFlex fibers, from Supelco, were used; other fibers were also tested (PDMS-100 μm, PDMS-7 μm, PDMS/DVB-65 μm). An agitation platform (PSelecta Agimatic 243) and stirrer bars were used to stir the water sample. An ultrasonic bath (PSelecta 300512 Barcelona, Spain) was used to prepare standard solutions. An ice bath was employed to help during the analytes solvent collection step. The fiber holder was provided of rotation with a rotor device (Nurka 390, Madrid, Spain).

2.2. Thermal Desorption/Organic Solvent Collection Interface

A scheme of this interface (Nurka 390, Madrid, Spain) to desorb the analytes from the fiber and to dissolve them further into a small amount of an organic solvent is showed in **Figure 1**. Essentially, it consisted of a thermal desorption chamber, which is a hole in a hotter block heated electrically; temperature inside this hole can be controlled in the range 30°C - 320°C by means of a thermostat; a GC glass liner (NK 390 0171) was placed inside this hole in which the fiber is adjusted. An argon flow is passed through this glass liner which flow rate is controlled by a manometer. This glass liner is connected to a solvent collection cone bottom shaped micro glass tube (50 × 3 mm) using a 1/16" stainless steel tube ending close to its bottom; this micro tube was placed inside an ice bath.

2.3. Chemicals and Reagents

Methanol and acetonitrile were of HPLC purity obtained from Scharlau, Barcelona, Spain. Water was purified with a Milli-Q system (Millipore; Milford, MA). The 5 PAHs standards were from Sigma Aldrich (naphthalene, bp 218°C; acenaphthene, bp 278°C; fluorene, bp 295°C; phenanthrene, bp 338°C; anthracene, bp 340°C) their purity being higher than 99.9%. Stock standard solutions were prepared from each PAH in methanol, at the following concentrations: naphthalene 676.8 mg·L^{-1}, acenaphthene 807.3 mg·L^{-1}, fluorene 498.9 mg·L^{-1}, phenanthrene 437.1 mg·L^{-1} and anthracene 160.2 mg·L^{-1}; this standard solutions were stored at 4°C in 100 mL amber flasks. Fresh standard solutions containing the 5 PAHs mixture were prepared by diluting 250.0 μl naphthalene, 180.0 μl acenaphthene, 160.0 μl fluorene, 60.0 μl phenanthrene and 30.0 μl anthracene of the above PAH standards solutions into a 2.0 ml flask with methanol; these diluted standard solutions were homogenized in the ultrasonic bath for five minutes.

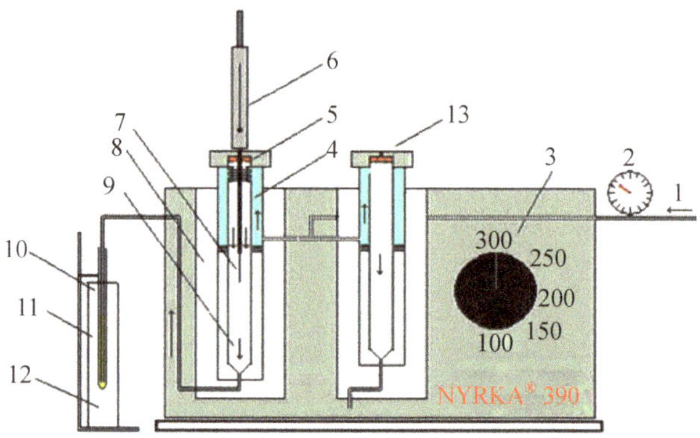

Figure 1. Scheme of the thermal desorption solvent collection interface; 1. Argon flow, 2. Manometer, 3. Temperature control, 4. Glass liner, 5. Septum, 6. SPME holder, 7. fiber, 8. Hotter block, 9. Argon flow direction, 10. Cone bottom shaped microtube, 11. Collecting solvent and 12. Ice baht.

2.4. Samples

A Milli-Q and a tap water sample from the Complutense University of Madrid (Faculty of Chemistry, Spain) as well as a drinking mineral water sample (available at the supermarket) were analyzed. Two water samples were taken from the Valdemayor reservoir and from Guadalix de la Sierra River in Madrid, respectively, in Pyrex borosilicate amber glass bottles and stored at 4˚C, protected from light, immediately after sampling.

2.5. SPME Procedure

2.5.1. Conditioning of the Fiber

Before being used, the 65 µm PDMS/DVB StableFlex fiber was conditioned for 1 h in the thermal desorption chamber under an argon stream at 280˚C and a flow rate of 30 $mL \cdot min^{-1}$. A fiber blank was registered daily.

2.5.2. Extraction Process

PAHs were extracted by direct immersion of the fiber into 100.0 ml of the water sample contained a beaker. The sample was stirred magnetically with a glass-coated stirrer at 3000 rpm during 90 min; the fiber was also rotated in opposite direction to the magnetic stirrer at 600 rpm using a rotor device. All extractions were carried out at room temperature (24˚C - 26˚C). After extraction the fiber was ready for thermal desorption.

2.5.3. Thermal Desorption/Analytes Collection

The fiber was placed inside the thermal desorption glass liner passing through it an argon flow of 7 ml/min at a temperature of 280˚C for 30 min. Afterwards, the fiber was removed and four aliquots of 10 µl methanol each 5 min were injected into this glass liner at the same temperature to sweep any PAHs residue condensed inside the connection tube. The analytes were collected in 50 µl of a 50:50 methanol:acetonitrile mixture placed inside the collection micro tube which was immersed in the ice bath. The volume of the analytical PAHs solution was decreased to 20 µl by using the same argon flow and the resulting total volume was then injected slowly into the HPLC injection valve.

2.5.4. Calibration Graphs

The external standard calibration method was applied as follow: 1000 ml of Milli-Q water were spiked with 20, 50, 100, 150 and 200 µl of the fresh PAHs standard solution mixture at concentration levels in the range 0.48 - 169.2 $\mu g \cdot L^{-1}$. The solid-phase microextraction and HPLC procedures were applied to analyze these standard samples. The results were adjusted by linear regression using the integrated peak areas. PAHs concentrations of PAHs in the water samples were determined by interpolation in the respective PAH calibration graphs.

3. Results and Discussion

3.1. Extraction Step

Extraction variables were studied by dipping the fiber in a milli-Q water sample spiked with a fresh standard solution PAHs mixture. A summary of the studied variables, changing systematically each one while keeping constant the rest of them, is shown in **Table 1**. The procedure specified in experimental was always applied. As **Figure 2** shows, among the tested fibers, the one yielding the highest signal for this studied PAHs was based on PDMS/DVB StableFlex 65 µm, which suggested that hydrophobic and π - π interactions must be predominant; this one was chosen for further studies.

A sample volume of 100 mL was enough to give the highest PAHs signals; **Figure 3(a)** shows a typical saturation graph above this sample volume. Regarding the time necessary to get the fiber-sample equilibrium, times about 90 min were necessary; above these, maximum saturation signals seem to be obtained, **Figure 3(b)**. Mass transfer into the fiber was favored by stirring the PAHs sample and rotting the fiber simultaneously in appositive way, as **Figure 3(c)** shows.

Table 1. Optimization of SPME extraction conditions.

Studied variables	Range	Selected
Fiber film thickness & phase	CAR/PDMS 85-µm, PDMS/DVB StableFlex 65-µm, PDMS/DVB 65-µm, PDMS 100-µm	PDMS/DVB StableFlex 65-µm
Sample volume, mL	10, 50, 100, 250, 500	100
Extraction time, min	30, 50, 60, 70, 80, 90, 120	90
Extraction temperature, ˚C	4, 26	26
Solution stirring/fiber rotting	One way Opposite way	Opposite way

CAR: Carboxen; CW: Carbowax; PDMS: Polydimethylsiloxane; DVB: Divinylbenzene.

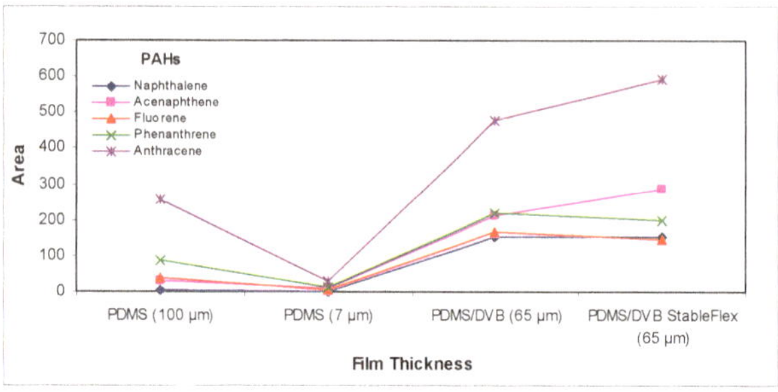

Figure 2. SPME fiber Selection.

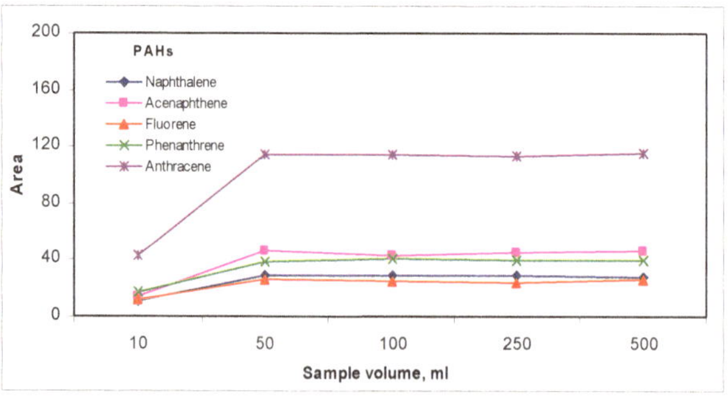

(a)

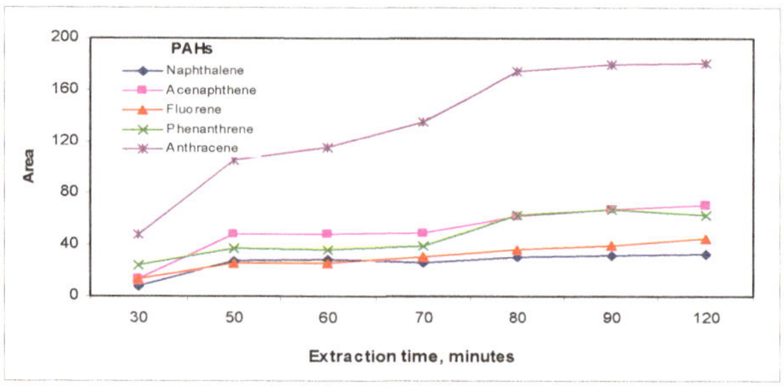

(b)

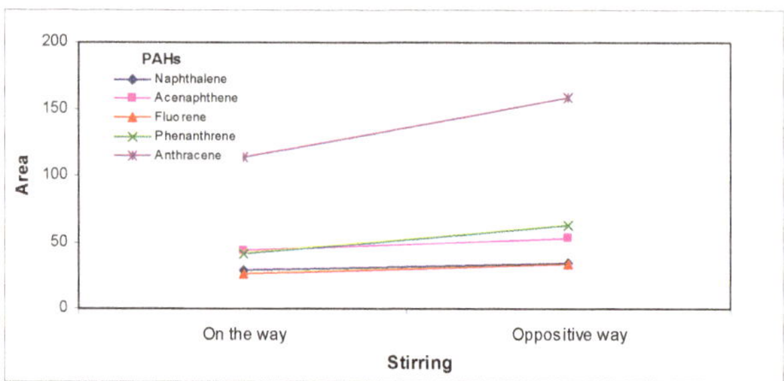

(c)

Figure 3. Extraction parameters effect: (a) Sample volume, (b) Time and (c) Sample solution stirring/fiber rotting.

3.2. Analytes Thermal Desorption and Organic Solvent Collection

After the extraction step, analytes extracted into the fiber were desorbed thermically using the new device and collected in a small organic solvent volume following the procedure specified in the experimental section. The effects of the different variables involved were studied; the range in which these variables were changed is showed in **Table 2**.

As **Figure 4(a)** shows signal increased slightly in the temperature range studied; so, a temperature of 280°C was chosen to get the highest signal near to complete desorption.

Regarding desorption time, the shape of the profiles showed in **Figure 4(b)** seems to indicate that after reaching a steady state, small losses of analytes appear for high desorption times, which is can be due to analyte losses during the whole process.

The effect of argon flow rate in **Figure 4(c)** shows that an optimum value for flow rate about 7 mL/min appeared, indicating analytes losses for higher argon flow rates, which is quite obvious because high flow rates may both sweep the analytes and carry them away from the micro tube.

Regarding the shape of the collection tube, small and long tubes yielded the highest signals because they allowed better contact between the argon gas which

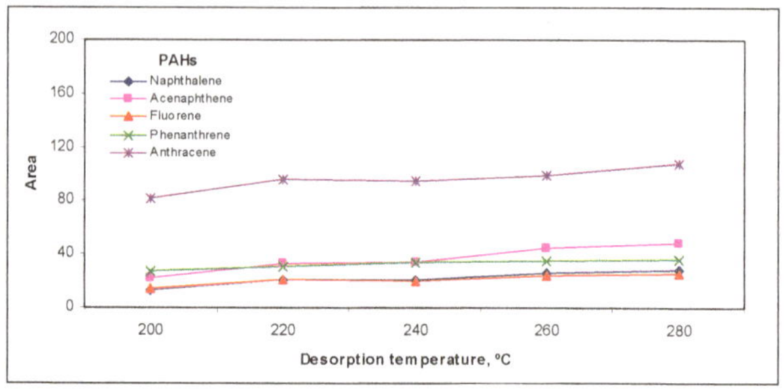

(a)

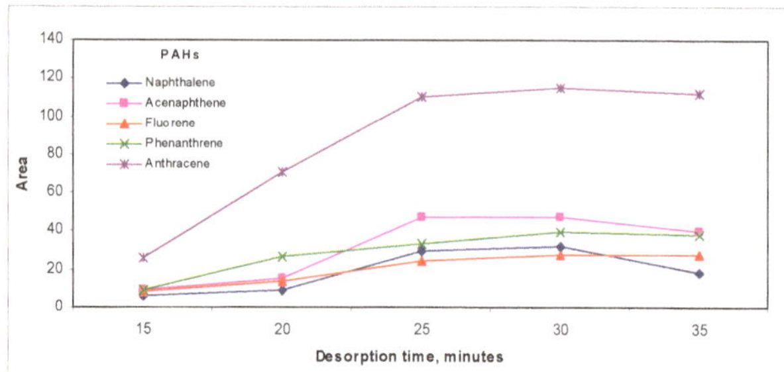

(b)

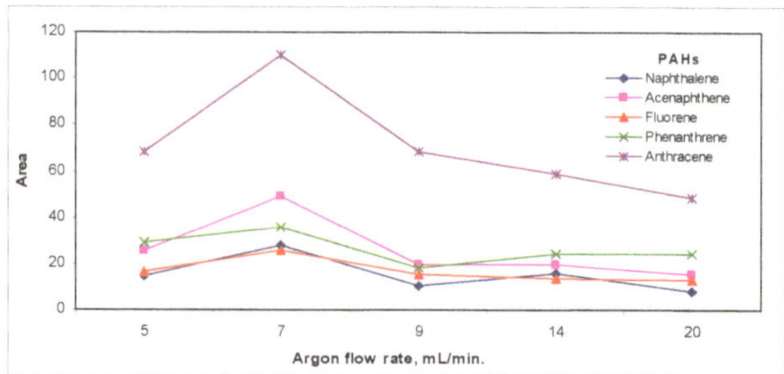

(c)

Figure 4. Desorption parameters effect: (a) Temperature, (b) Time and (c) Argon flow rate.

Table 2. Optimization of SPME thermal desorption/solvent collection interface.

Studied variables	Range	Selected
Desorption temperature, °C	200, 220, 240, 260, 280	280
Desorption time, min	15, 20, 25, 30, 35	30
Argon Flow, mL·min^{-1}	5, 7, 9, 14, 20	7
Collection vial volume, µL	400, 500, 1000	500
Nature of collection solvent, %	MeOH, ACN, 50:50 MeOH:ACN, 50:50 MeOH:Water, 50:50 ACN:Water	50:50 MeOH:ACN
Collection solvent volume, µL	50, 75, 100, 150, 200	50
Collection solvent temperature, °C	4, 24, 40	4
Sweeping solvent (MeOH), × 10 µL	2, 4, 5, 6, 7	4

sweeps the analytes and the collection solvent, **Figure 5(a)**.

The bottom of this tube was cone shaped to allow maximum transference of the collected PAHs solution using an HPLC syringe. When the volume of this tube increased, signals decreased for volumes above 500 µL.

The nature of the collection organic solvent in **Figure 5(b)** shows that the optimum mixture was based on MeOH:ACN; 1:1; v:v; the elution strength of this mixture is very high for injection in HPLC, although in practice, this effect was small in the resulting HPLC peaks.

The volume of the collecting organic solvent was optimized; **Figure 5(c)** shows that in 50 µL of the collecting solvent the retention of the analytes was the highest. Decreasing the temperature of the solvent also contributes to retain the analytes by increasing the solvent/gas distribution constant, **Figure 5(d)**; so the collection tube was placed inside an ice bath.

The volume of methanol used for sweeping was optimized to insure that all the analytes desorbed reached the collection solvent; the results are shown in the **Figure 5(e)**.

3.3. Evaluation SPME-HPLC Off-Line Method Performance

3.3.1. Analytical Characteristics for Standards

Linearity ranges, regression coefficients, precision (RSDs) and detection limits (LOD) for standards (3 S/N) are shown in **Table 3**. Precision below 14% seems to be acceptable taking into account the complexity of the whole procedure. Detection limits at the low µg·L^{-1} level were found. Regression coefficients showed an acceptable linearity taking into account the complexity of the procedure.

3.3.2. Analysis of Water Samples

PAHs were not detected in the Milli-Q, the tap water and the drinking mineral water samples analyzed. Four PAHs were detected in the "Valdemayor" reservoir samples (Acenaphtene, fluorene, phenanthrene and anthracene); in the "Guadalix de la Sierra River" samples, three PAHs (Naphthalene, phenanthrene and anthracene) were also detected (**Figure 6**); results obtained in the quantitationare given in **Table 4**. In order to validate the proposed SPME method, its accuracy was evaluated by comparing the results with those found by the 550.1 EPA method [28], based essentially in SPE and HPLC with fluorimetric detection. The results obtained from the SPME method seemed to be slightly lower than those found from the EPA method; however, differences were not significant between both methods using the student's t test at the significance level of 0.05 for all of the PAHs detected.

Table 3. Analytical characteristics of the method for standards.

PAH	Linear range µg·L^{-1}	r^2	LOD µg·L^{-1}	RSD % n = 3
Naphthalene	16.9 - 169.2	0.9752	0.99	9.9
Acenaphthene	14.5 - 145.4	0.9928	0.53	11.3
Fluorene	8.0 - 80.0	0.9827	0.33	10.8
Phenanthrene	2.6 - 26.0	0.9577	0.12	11.5
Anthracene	0.4 - 4.8	0.9945	0.07	13.1

4. Conclusions

The new SPME/HPLC thermal desorption/organic solvent collection interface showed to be an alternative to the existing one from Supelco. Some limitations of Supelco interface was overcame; in particular, the contact of the fiber with the organic solvent was avoided.

Validation of the new interface for some PAHs test compounds by comparing their results with those found from the 550.1 EPA method showed agreement at a 95% probability level; these results indicated the potentiality

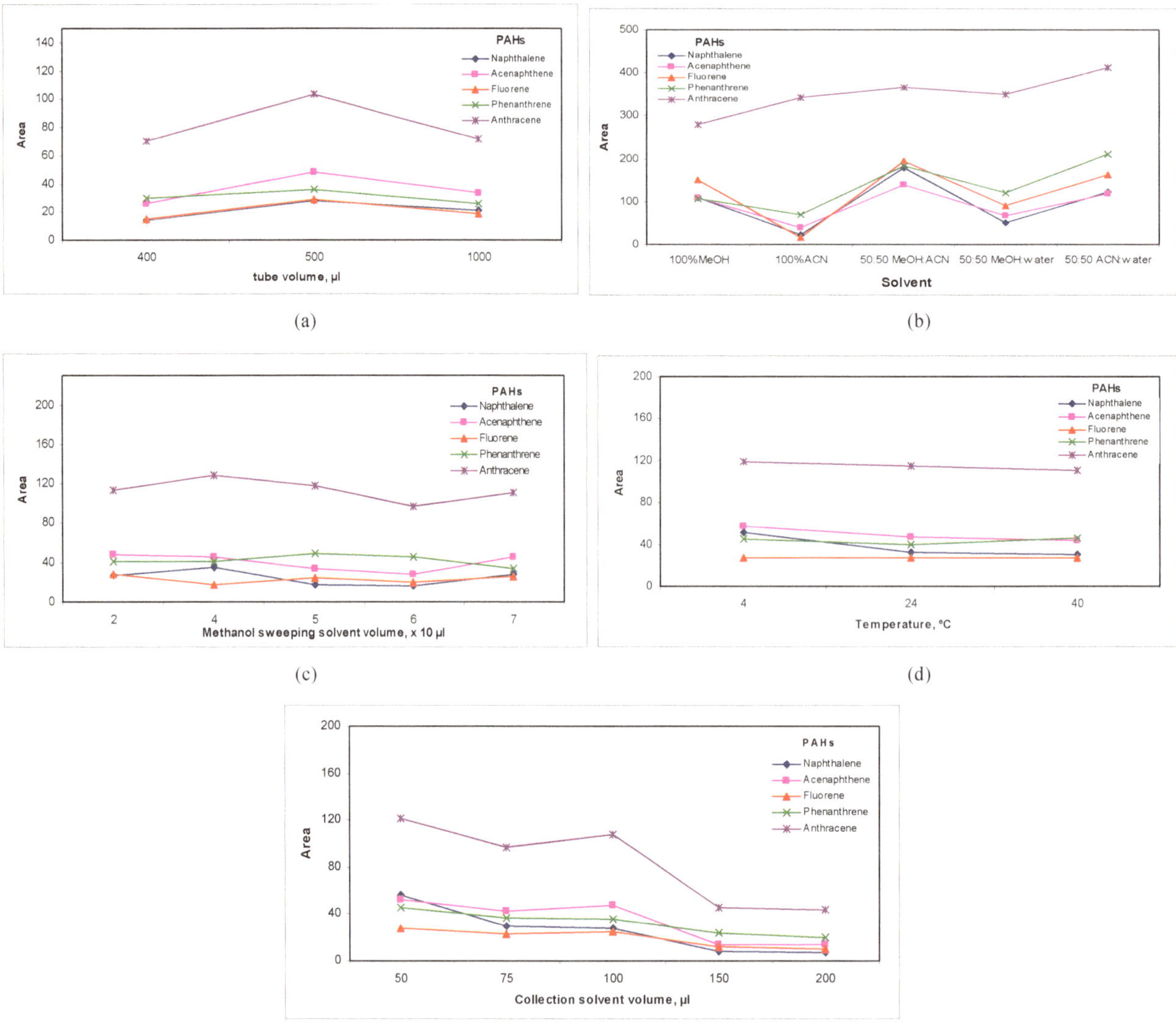

Figure 5. Collection parameters: (a) Tube volume, (b) Solvent, (c) Sweeping solvent volume, (d) Temperature and (e) Collection solvent volume.

Table 4. Determination of PAHs by SPME thermal desorption/solvent collection interface and SPE (EPA Method 550.1).

	Water sample													
	Valdemayor Reservoir							Guadalix de la Sierra River						
	SPME interface method			EPA 550.1 method			t_{exp}	SPME interface method			EPA 550.1 method			t_{exp}
	Mean	s**	RDS%	Mean	s	RDS%		Mean	s	RDS%	Mean	s	RDS%	
Naphthalene	n.d.	-	-	n.d.	-	-	-	89.9	5.3	5.8	93.4	8.8	7.4	**0.6**
Acenaphthene	38.3	4.8	9.5	43.3	1.9	4.3	**2.3**	n.d.	-	-	n.d.	-	-	-
Fluorene	50.6	3.5	6.9	57.5	3.1	5.3	**2.6**	n.d.	-	-	n.d.	-	-	-
Phenanthrene	25.6	2.4	13.3	27.9	1.3	6.9	**2.1**	34.9	2.5	10.0	37.6	1.7	6.4	**1.5**
Anthracene	19.1	0.4	9.9	19.8	0.5	7.9	**2.4**	10.8	0.9	10.4	11.4	0.8	7.5	**2.2**

n.d., no detected; **n = 3; Tabulated t value = 2.78 [27].

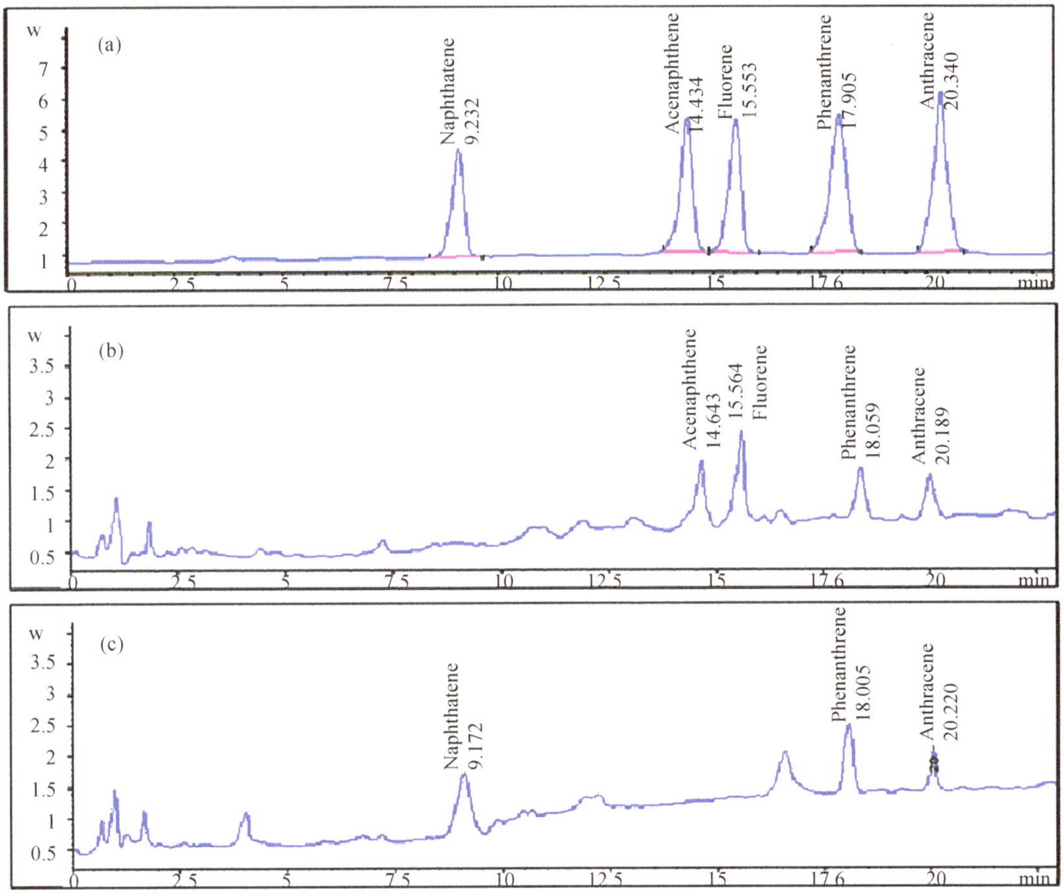

Figure 6. HPLC Chromatograms obtained from (a) A standard mixture of five PAHs, (b) Valdemayor reservoir sample and (c) Guadalix de la Sierra River sample.

of this interface to determine semivolatile compounds and others of different volatility.

REFERENCES

[1] Z. Mester, R. Sturgeon and J. Pawliszyn, "Solid Phase Microextraction as a Tool for Trace Element Speciation," *Spectrochimica Acta Part B: Atomic Spectroscopy*, Vol. 56, No. 3, 2001, pp. 233-260.

[2] Y. Fan, Y. Feng, T. Zhang, S. Da and M. Zhang, "Poly (Methacrylic Acid-Ethylene Glycol Dimethacrylate) Monolith In-Tube Solid Phase Microextraction Coupled to High Performance Liquid Chromatography and Analysis of Amphetamines in Urine Samples," *Journal of Chromatography A*, Vol. 1074, No. 1, 2005, pp. 9-16

[3] R. P. Belardi and J. Pawliszyn, "The application of chemically modified fused silica fibers in the extraction of organics from water matrix samples and their rapid transfer to capillary columns," *Water Pollution Research Journal of Canada*, Vol. 24, 1989, pp. 179-191.

[4] M. N. Kayali, S. Rubio, M. P. Cuesta and L. M. Polo, "Rapid Determination of Polycyclic Aromatic Hydrocarbons in Tea Infusion Samples by High-Performance Liquid Chromatography and Fluorimetric Detection Based on Solid-Phase Extraction," *Analyst*, Vol. 123, No. 10, 1998, pp. 2145-2148.

[5] A. Peñalver, E. Pocurull, F. Borrull and R. M. Marcé, "Method Based on Solid-Phase Microextraction—High-Performance Liquid Chromatography with UV and Electrochemical Detection to Determine Estrogenic Compounds in Water Samples," *Journal of Chromatography A*, Vol. 964, No. 1-2, 2002, pp. 153-160.

[6] Y. Wang, J. Zhang, *et al*., "Quantitative Determination of 16 Polycyclic Aromatic Hydrocarbons in Soil Samples Using Solid-Phase Microextraction," *Journal of Separation Science*, Vol. 32, No. 22, 2009, pp. 3951-3957.

[7] Gaurav, A. K. Malik and P. K. Rai, "Development of a New SPME-HPLC-UV Method for the Analysis of Nitro Explosives on Reverse Phase Amide Column and Application to Analysis of Aqueous Samples," *Journal of Hazardous Materials*, Vol. 172, No. 2-3, 2009, pp. 1652-1658.

[8] R. J. Bartelt, "Calibration of a Commercial Solid-Phase Microextraction Device for Measuring Headspace Concentrations of Organic Volatiles," *Analytical Chemistry*, Vol. 69, No. 3, 1997, pp. 364-372.

[9] C. Arthur and J. Pawliszyn, "Solid Phase Microextraction with Thermal Desorption Using Fused Silica Optical Fibers," *Analytical Chemistry*, Vol. 62, No. 19, 1990, pp. 2145-2148.

[10] M. R. Negrão and M. F. Alpendurada, "Solvent-Free Method for the Determination of Polynuclear Aromatic Hydrocarbons in Waste Water by Solid-Phase Microextraction-High-Performance Liquid Chromatography with Photodiode-Array Detection," *Journal of Chromatography A*, Vol. 823, No. 1-2, 1998, pp. 211-218.

[11] K. Hiroyuki, "Recent Advances in Solid-Phase Microextraction and Related Techniques for Pharmaceutical and Biomedical Analysis," *Current Pharmaceutical Analysis*, Vol. 1, No. 1, 2005, pp. 65-84.

[12] M. E. Miller and J. D. Stuart, "Comparison of Gas-Sampled and SPME-Sampled Static Headspace for the Determination of Volatile Flavor Components," *Analytical Chemistry*, Vol. 71, No. 1, 1999, pp. 23-27.

[13] O. Ezquerro and M. Tena, "Determination of Odour-Causing Volatile Organic Compounds in Cork Stoppers by Multiple Headspace Solid-Phase Microextraction," *Journal of Chromatography A*, Vol. 1068, No. 2, 2005, pp. 201-208.

[14] W. Zhang and C. Zlin, "Mussel Inspired Polydopamine Functionalized Poly (Ether Ether Ketone) Tube for Online Solid-Phase Microextraction—High Performance Liquid Chromatography and Its Application in Analysis of Protoberberine Alkaloids in Rat Plasma," *Journal of Chromatography A*, Vol. 1278, 2013, pp. 29-36.

[15] A. Bordagaray, R. García and E. Millán, "Development and Application of a Screening Method for Triazole Fungicide Determination in Liquid and Fruit Samples Using Solid-Phase Microextraction and HPLC-DAD," *Analytical Methods*, Vol. 5, No. 10, 2013, pp. 2565-2571.

[16] S. Panseri, S. Soncin, L. M. Chiesa and P. A. Biondi, "A Headspace Solid-Phase Microextraction Gas-Chromatographic Mass-Spectrometric Method (HS-SPME-GC/MS) to Quantify Hexanal in Butter during Storage as Marker of Lipid Oxidation," *Food Chemistry*, Vol. 127, No. 2, 2011, pp. 886-889.

[17] M. Tankiewicz, C. Morrison and M. Biziuk, "Application and Optimization of Headspace Solid-Phase Microextraction (HS-SPME) Coupled with Gas Chromatography-Flame-Ionization Detector (GC-FID) to Determine Products of the Petroleum Industry in Aqueous Samples," *Microchemical Journal*, Vol. 108, 2013, pp. 117-123.

[18] Q. L. Ma, N. Hamid, A. E. D. Bekhit, J. Robertson and T. F. Law, "Optimization of Headspace Solid Phase Microextraction (HS-SPME) for Gas Chromatography Mass Spectrometry (GC-MS) Analysis of Aroma Compounds in Cooked Beef Using Response Surface Methodology," *Microchemical Journal*, Vol. 111, 2013, pp. 16-24.

[19] Gaurav, V. Kaur, A. Kumar, A. Kumar and P. K. Rai, "SPME-HPLC: A New Approach to the Analysis of Explosives," *Journal of Hazardous Materials*, Vol. 147, No. 3, 2007, pp. 691-697.

[20] C. Fernandes, A. J. dos Santos Neto, J. C. Rodrigues, C. Alves and F. Mauro Lanças, "Solid-Phase Microextraction-Liquid Chromatography (SPME-LC) Determination of Fluoxetine and Norfluoxetine in Plasma Using a Heated Liquid Flow through Interface," *Journal of Chromatography B*, Vol. 847, No. 2, 2007, pp. 217-223.

[21] L. Cárdenes, J. Ayala, M. A. Afonso and V. González, "Solid-Phase Microextraction Coupled with High-Performance Liquid Chromatography for the Analysis of Heterocyclic Aromatic Amines," *Journal of Chromatography A*, Vol. 1030, No. 1-2, 2004, pp. 87-93.

[22] H. Kataoka and K. Saito, "Recent Advances in SPME Techniques in Biomedical Analysis," *Journal of Pharmaceutical and Biomedical Analysis*, Vol. 54, 2011, pp. 926-950.

[23] J. S. Aulakh, A. K. Malik, V. Kaur and P. M. Kopplin, "A Review on Solid Phase Micro Extraction—High Performance Liquid Chromatography (SPME-HPLC) Analysis of Pesticides," *Critical Reviews in Analytical Chemistry*, Vol. 35, No. 1, 2005, pp. 71-85.

[24] N. Kayali, F. G. Tamayo and L. M. Polo, "Determination of Diethylhexyl Phtalate in Water by Solid Phase Microextraction Coupled to High Performance Liquid Chromatography," *Talanta*, Vol. 69, No. 5, 2006, pp. 1095-1099.

[25] H. L. Lord, "Strategies for Interfacing Solid-Phase Microextraction with Liquid Chromatography," *Journal of Chromatography A*, Vol. 1152, No. 1, 2007, pp. 2-13.

[26] M. N. Kayali-Sayadi, J. M. Bautista, L. M. Polo-Díez and I. Salazar, "Identification of Pheromones in Mouse Urine by Head-Space Solid Phase Microextraction Followed by Gas Chromatography-Mass Spectrometry," *Journal of Chromatography B*, Vol. 796, No. 1, 2003, pp.55-62.

[27] J. N. Miller and J. C. Miller, "Statistics and Chemometrics for Analytical Chemistry," 4th Edition, Prentice-Hall, Harlow, 2000.

[28] L. H. Keith, "Compilation of EPA's Sampling and Analysis Methods," 2nd Edition, CRC Press, Boca Raton, 1996.

Silicotungstic Acid Modified Bentonite: An Efficient Catalyst for Synthesis of Acetal Derivatives of Aldehydes and Ketones

Reshu Chaudhary, Monika Datta*

Department of Chemistry, University of Delhi, New Delhi, India.

ABSTRACT

Acetals and ketals are among the important materials of organic synthesis and as protecting agent of carbonyl functionality. A milder, efficient and green synthesis of acetals and ketals has been developed using Silicotungstic acid modified Bentonite (STA-Ben) as a catalyst. STA-Ben has been synthesized and characterized by various analytical techniques. It has been found to be an efficient and reusable catalyst for the synthesis of acetyl derivatives in excellent yields. In order to elucidate the efficiency of the STA-Ben as catalyst, reaction has also been performed using various catalysts. The reaction conditions (time and amount of catalyst) have been optimized using various catalysts. The products of the various reactions have been characterized by FT-IR, NMR.

Keywords: Silicotungstic Acid; Bentonite; Aldehydes; Ketones

1. Introduction

Acetals and ketals are one of the most important perfume materials and industrial materials of organic synthesis. The protection of carbonyl functionality as an acetal or thioacetal derivative is a common practice in organic synthesis owing to their stability under mild acidic and basic conditions. They are generally obtained by the condensation of carbonyl compound with acetic anhydride [1], diols, dithiols and orthoformates [2-5], using Lewis acids as catalysts. A variety of catalysts are employed for this purpose like $CoCl_2$ [6], $NiCl_2$ [2], HCl [7], $HClO_4$ [8], sulfated zirconia [9] etc. But the most recent methods employ MMT-K10 [10], ionic liquids [11], metal cation exchanged MMT [12], natural kaolinite clay [3], silica sulfuric acid [13], $TaCl_5$-silicagel [14] as catalysts. However, many of these methods have some drawbacks such as low yields of the products, long reaction times, harsh reaction conditions, difficulties in work-up. Moreover, the main disadvantage of almost all existing methods is that the catalysts are destroyed in the work-up procedure and cannot be recovered or reused. Therefore, there is further scope to explore mild and efficient methods for acetalization of carbonyl compounds.

Solid acids offer many advantages by their nature, over soluble counterparts such as aluminium chloride and hydrogen fluoride. The substitution of liquid acids by solids as catalysts for organic synthesis offers a potential for superior effectiveness and environmental integrity. Although they differ in structure from liquid acids, solid acid catalysts work by the same principle. Among the solid acids, heteropoly acids (HPAs) have been extensively studied as acid catalysts for many reactions and found industrial applications in several processes [15]. The Keg- gin-type HPAs typically represented by the formula $H_{8-x}[XM_{12}O_{40}]$, where X is the heteroatom (e.g., P^{5+} or Si^{4+}), x is its oxidation state and M is the addenda atom (usually Mo^{6+} or W^{6+}), are the most important catalysts, especially $H_3PW_{12}O_{40}$ (PW), $H_3PMo_{12}O_{40}$ (PMo) and $H_4SiW_{12}O_{40}$ (SiW) [16]. They are more active than the conventional catalysts, such as mineral acids, ion-exchange resins, zeolites, SiO_2/Al_2O_3 and H_3PO_4/SiO_2 [17-19].

HPAs have several advantages over liquid acid catalysts, including being non-corrosive and environmen-

*Corresponding author.

tally benign. Thus, they present fewer disposal problems and more economically and environmentally attractive. Heteropoly acids (HPAs) due to their strong acidity have attracted much interest as the catalysts mostly in homogeneous systems [20,22]. It's not easy to separate HPA from reaction mixture and reuse it unless it is supported on some catalyst. Clays have also been proposed as suitable acid catalysts [23,24]. Clays are furthermore used as adsorbents, decoloration agents, ion exchangers, and catalysts [25]. The use of clays, as heterogeneous catalysts, offers many advantages over homogeneous acid catalysts such as ease of separation, mild reaction conditions, better selectivity of the desired product, and elimination of waste disposal problems. In a large number of organic reactions clays have been used as catalysts on laboratory/industrial scales. The properties of clay can further be improved by making pillared clays. Pillared clays are clays with high permanent porosity obtained by separating the clay sheets by a molecular prop or pillaring agent. These pillaring agents can be organic, organometallic, or inorganic complexes, preferably of a high positive charge. Pillared clay (PILC) possesses several interesting properties, such as large surface area, high pore volume and tunable pore size (from micropore to mesopore), high thermal stability, strong surface acidity and catalytic active substrates/metal oxide pillars. These unique characteristics make PILC an attractive material in catalytic reactions. It can be made either as catalyst support or directly used as catalyst [26-28].

In the present work silicotungstic acid (a heteropoly acid) modified bentonite have been synthesized to explore the properties of both clay and silicotungstic acid (a heteropoly acid). This makes it possible to carry out the catalytic process at a lower catalyst concentration and/or at a lower temperature. Further, catalysis of heteropoly acid lacks side reactions.

2. Experimental

2.1. Catalyst Preparation

Na-Bentonite

1.0 g of Bentonite clay was added into a 250 ml conical flask containing 50 ml of 1.0 M NaCl this clay suspension was stirred for 16 hours. The residue after centrifugation was washed several times with double distilled water till complete removal of chloride ions. The residue thus obtained after above procedure was dried in an oven at 100°C to generate the Na form of the Bentonite (Na-Ben).

Al-Pillared Bentonite (Al-Ben)

In a 250 ml conical flask containing 1.0 g of Na-Ben in 100 ml of double distilled water, 50 ml of the pillaring solution (keggin ion) was gradually added with vigorous stirring for 16 hours. The residue was washed several times with double distilled water till the complete removal of chloride ions was confirmed by $AgNO_3$ test. The residue, thus obtained was dried in an oven at 100°C is referred to as Al-Pillared Bentonite (Al-Ben) [29,30].

Acid Activated Bentonite (H^+-Ben)

5.0 g of Na-Ben was added into a 100 ml beaker containing 50 ml of 3N H_2SO_4, this mixture was exposed to microwave radiation for 30 minutes. The residue was washed several times with double distilled water till the complete removal of SO_4^{2-} ions was confirmed by $BaCl_2$ test. The residue thus obtained was dried in an oven at 100°C to generate the Acid activated Bentonite (H^+-Ben) [28].

Pillared Acid Activated Bentonite (PAA-Ben)

1.0 g of H^+ Ben, was added into a 250 ml conical flask containing 100 ml of double distilled water. 50 ml of pillaring solution was gradually added with vigorous stirring for 16 hours. The residue was washed several times with double distilled water till the complete removal of chloride ions was confirmed by $AgNO_3$ test. The residue thus obtained was dried in an oven at 100°C and is referred to as Pillared Acid Activated Bentonite (PAA-Ben).

Silicotungstic Acid Modified Bentonite (STA-Ben)

In a 250 ml conical flask, 1.0 g Na-Ben was suspended in 50 ml double distilled water. To this, aqueous solution of silicotungstic acid (100 mg) was added drop wise and then stirred for 16 hours. After this, the mixture was filtered and washed with double distilled water to remove the excess of silicotungstic acid. The product was dried at 100°C.

2.2. General Procedure for the Synthesis of Acetal Derivative of Aldehydes and Ketones

In order to elucidate the role of the STA-Ben as catalyst, a controlled reaction was carried out using STA, bentonite, Al-Ben, H^+-Ben, PAA-Ben and STA-Ben as catalyst with benzaldehyde as reactant. The best results were obtained with STA-Ben in microwave.

A number of aromatic/aliphatic aldehydes or ketones (10 mmol) and ethylene glycol (10 mmol) were mixed with the STA-Ben (50 mg). The mixture was exposed to microwave raditions. After irradiation of the mixture for a specified period, the contents were gradually cooled to room temperature. The completion of the reaction was checked by TLC. After confirming the completion of the reaction, the catalyst was recovered by filtration. The catalyst thus separated by filtration was washed 2 - 3 times with absolute ethanol to remove organic matter. The residue obtained after evaporation of the solvent from the filtrate was purified using TLC/column chromatography with $CHCl_3$:MeOH with increasing polarity.

Optimization of Reaction Conditions

For the optimization of reaction conditions 10 mmol of ethylene glycol and 10 mmol of Benzaldehyde have been selected as reactants.

Optimization of the Catalysts Amount

Reactions have been performed using varying amount of different catalysts under uniform conditions (**Table 1**).

Optimization of Time

The reaction time has been optimized by performing reactions at regular intervals of time. After a certain time period there is not much increase in the yield. That time has been selected as optimum time (**Table 2**).

After optimization of time and amount of catalyst a number of acetal derivatives have been synthesized **Table 3**.

3. Results and Discussion

3.1. Catalyst Characterization

X-Ray Diffraction

The XRD pattern of Ben [33], Na-Ben and H^+-Ben show a sharp peak at 2θ = 5.93, 6.83 and 9.67 respectively corresponding to a d-value of 14.9 Å, 12.9 Å and 9.1 Å. A decrease in the d-value of 2 Å and 5.8 Å is observed when interlayer cations (Ca^{2+}) are replaced by the smaller ions, Na^+ and H^+ respectively. PAA-Ben, shows a further shift in the peak position, 2θ = 5.30 and a d-value of 16.6 Å corresponding to an increase in the interlayer region by 7.5 Å w.r.t H^+-Ben confirms the intercalation of Al Keggin ion into the interlayer region of H^+-Ben. The Al-Ben shows a peak at 2θ = 5.20 and a d-value of 16.9 Å corresponding to an increase in the d-value of 4 Å w.r.t Na-Ben thus confirming the intercalation of Al Keggin ion in the interlayer region of Na-Ben. STA-Ben, show a peak at 2θ = 5.47 corresponding to a d-value of

Table 1. Optimization of catalyst amount.

S. No.	catalyst (mg)	Time, Sec.	% Yield of Benzaldehyde acetal with various catalysts					
			Ben	Al-Ben	H^+-Ben	PAA-Ben	STA	**STA-Ben**
1	20	180	30	40	45	49	47	62
2	30	180	36	47	52	57	55	72
3	40	180	45	54	60	65	65	80
4	**50**	180	50	60	65	70	68	**88**
5	60	180	50	60	65	70	68	88

Table 2. Optimization of time.

S.No.	Time, Sec.	Yield, %
01	90	66
02	105	67
03	120	70
04	135	75
05	150	80
06	165	85
07	180	88
08	200	88

Table 3. Products synthesized.

S. No.	Reactant	Product	Time, Sec	Yield, %
01	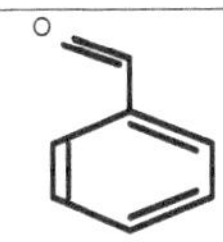	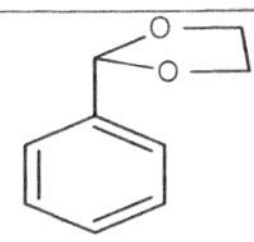	180	82
02	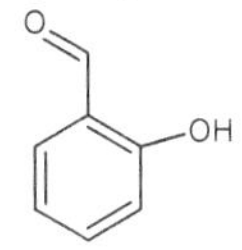	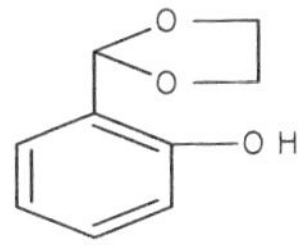	150	86
03	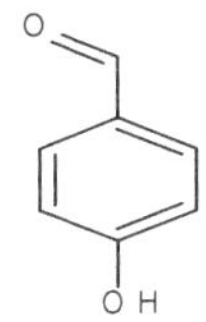	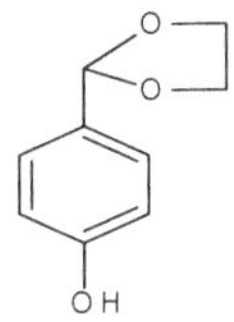	150	88
04	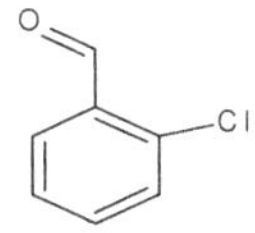	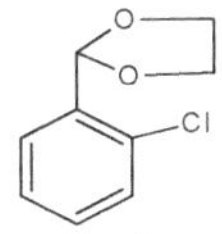	110	78
05	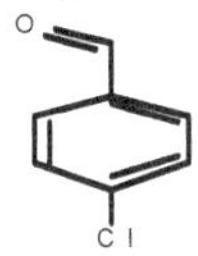	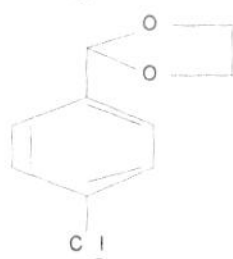	110	79
06	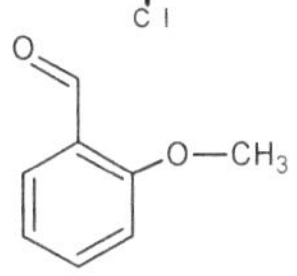	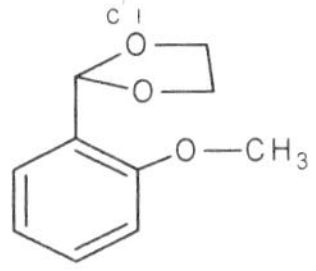	150	85
07	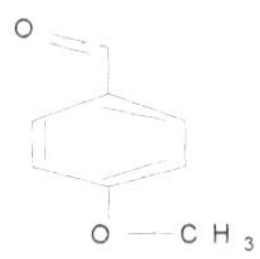	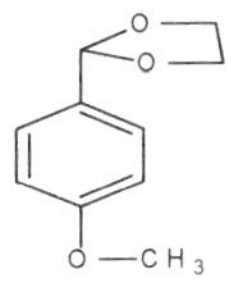	150	88
08	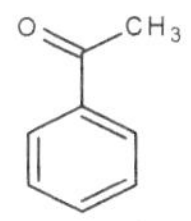	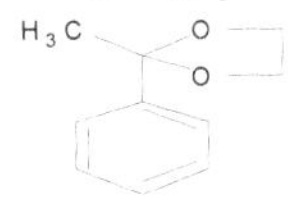	270	77
09	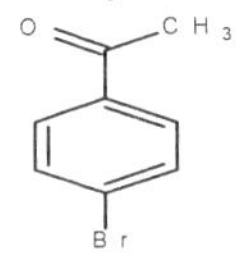	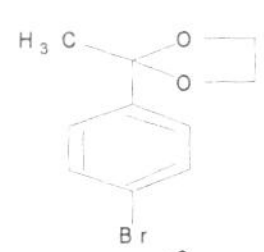	300	72
10#	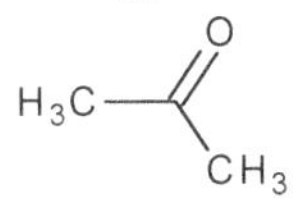	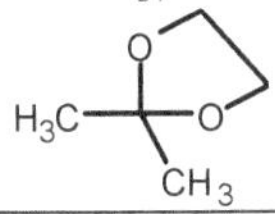	90	50

#Reactant is highly volatile. Therefore, the yield is very low.

16.1 Å. This increase in the d-value 3.2 Å w.r.t Na-Ben shows that STA is in the interlayer region of Na-Ben.

Total Surface Area

Total surface area studies have been performed by EGME method. Surface area values of Ben, Na-Ben, H^+-Ben, Al-Ben, PAA-Ben, STA-Ben are represented in the **Table 4**. Total surface area of Ben has been found to be 215 m^2/g. After conversion to Na-Ben total surface area has been found to be 121 m^2/g which may be due to the decrease in d-spacing as indicated by XRD data. Total surface area of H^+-Ben has been found to be 327 m^2/g. Although Na ions are replaced by smaller H^+ ions, but acid activation also causes the formation of small pores, consequently increasing the surface area [34,35]. Total surface area of Al-Ben has been found to be 295 m^2/g. Due to the intercalation of Al-Keggin ion the inner surface also becomes assessable. PAA-Ben has been found to have a total surface area of 350 m^2/g which is again due to intercalation of Al-Keggin ion in H^+-Ben. Total surface area of STA-Ben has been found to be 298 m^2/g which again shows intercalation of STA.

FT-IR

Vibrational spectra of Ben, Na-Ben, H^+-Ben, Al-Ben and PAA-Ben (**Figures 1** and **2**) indicate two strong absorption bands ~3626 cm^{-1} and ~3436 cm^{-1} corresponding to the stretching vibrations of the O-H group originating from the surface adsorbed and interlayer water. The ~1642 cm^{-1} band in these samples has been assigned to the H-O-H bending vibrations of water. The ~1043 cm^{-1} and ~796 cm^{-1} bands are attributed to the stretching vibration of the Si-O bond. The ~522 cm^{-1} and ~466 cm^{-1} bands have been assigned to the Si-O-Al and Si-O-Si deformation vibrations respectively [31,32]. The STA-Ben (**Figure 3(b)**) also shows the band in the similar region. The FT-IR spectra of STA (**Figure 3(a)**) shows an absorption band in the region of the ~3141 cm^{-1} and have been assigned to the stretching vibrations originating from the O-H groups present in the Keggin structure of STA. The absorption band at ~1711 cm^{-1} has been assigned to the bending vibrations of the O-H group. The fingerprint bands of the STA Keggin anion at 1081, 989, 890 and 793 cm^{-1}, which could be assigned to the typical antisymmetrical stretching vibrations of W = O, Si-O, W-O_c-W (Stretching Vibration involving corner sharing Oxygen atoms) and W-O_e-W (Stretch Vib. involving edge sharing Oxygen atoms) respectively.

Table 4. Surface area of catalysts.

Catalysts	Total Surface Area, (m^2/g)
Ben	215.0
Na-Ben	121.3
Al-Ben	295.2
H^+-Ben	327.2
PAA-Ben	350.1
STA-Ben	298.2

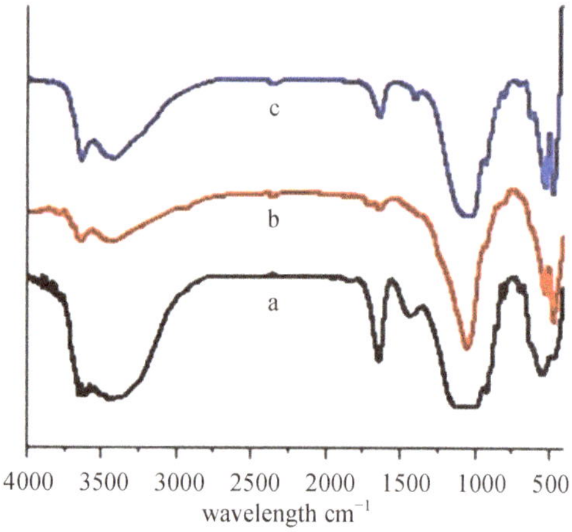

Figure 1. FTIR spectra of (a) Ben, (b) Na-Ben, (c) H^+-Ben.

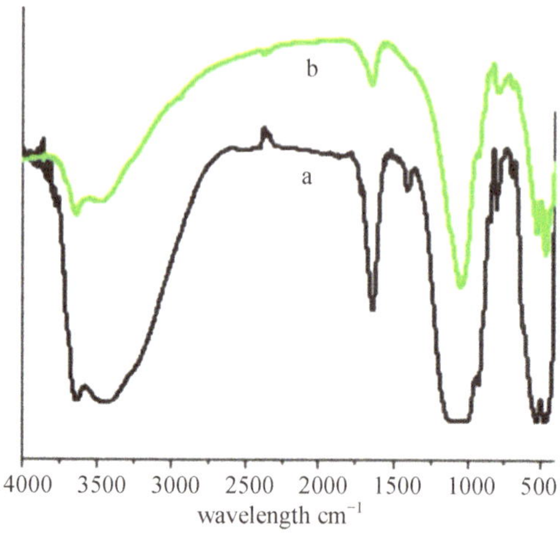

Figure 2. FTIR spectra of (a) Al-Ben, (b) PAA-Ben.

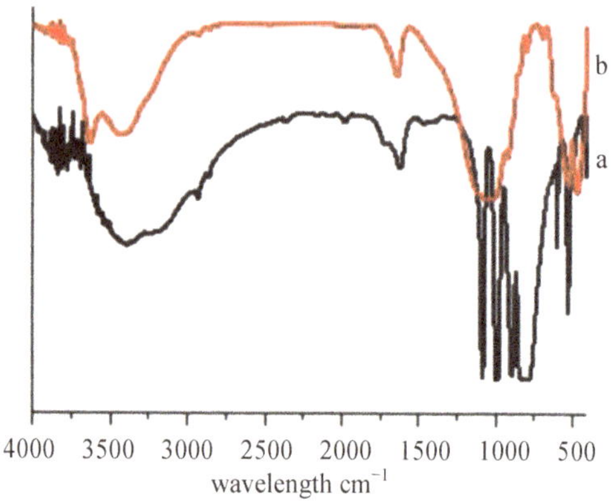

Figure 3. FTIR spectra of (a) STA, (b) STA-Ben.

Thermal Method of Analysis

The thermogram (TGA) of Ben, Na-Ben, H^+-Ben (**Figure 4**) shows four step weight loss pattern. The first step, up to 180˚C, corresponds to dehydration of surface adsorbed water. The second step, 180˚C to 550˚C, is attributed to dehydration of interlayer water. The third step 550˚C to 675˚C, is attributed to the gradual dehydroxylation of clay layers. Beyond 675˚C the clay loses its structure and practically shows no weight loss [31].

The thermogram (TGA) of PAA-Ben, Al-Ben (**Figure 5**) shows a similar four step weight loss pattern. The first step, up to 200˚C, corresponds to dehydration of surface adsorbed water. The second step, 200˚C to 350˚C, is attributed to dehydration of interlayer water. The third step, 350˚C to 600˚C, is attributed to the gradual dehydroxylation of clay layers. Beyond 600˚C the clay loses its structure and practically shows no weight loss [31].

The TGA of pure STA (**Figure 6(a)**) shows two step weight losses. The first step, up to 300˚C, corresponds to the loss of water of crystallization. The second step, 300˚C to 500˚C, shows weight loss corresponding to the decomposition of keggin unit of STA [33].

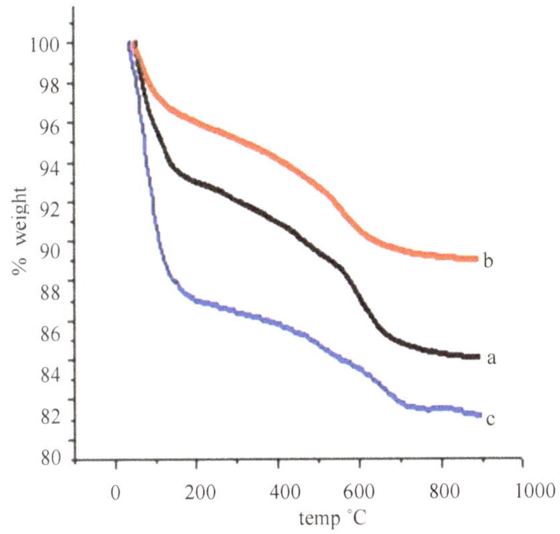

Figure 4. TGA studies of (a) Ben, (b) Na-Ben, (c) H^+-Ben.

The TGA of STA-Ben (**Figure 6(b)**) shows four step weight loss patterns, the first step, up to 150˚C, corresponds to dehydration of surface water. The second step, 150˚C to 350˚C, is attributed to dehydration of interlayer water. The third step, 350˚C to 750˚C, is attributed to the dehydroxylation of clay layers and the loss of keggin ion structure The heteropoly acids are stable up to 500˚C but after intercalation the catalyst becomes stable up to 750˚C.

SEM with EDX Analysis

The SEM-EDX pictures (magnification of 832X) of Ben and STA-Ben are represented in (**Figures 7(a)** and **(b)**). The SEM pictures of Ben and STA-Ben show no distinct change in the surface morphology and appear to have the layered structure. Therefore, it appears that the STA is not adsorbed on the surface of the clay but is in the layers of the clay. This fact is also supported by the XRD and IR studies. The SEM results are also supported by the EDX analysis. In both the cases surface composition consist of Al, Si, O, Mg, Fe, Ca & K. No signal corresponding to W was supported by the EDX data which further shows that STA may be in the interlayers region.

3.2. Characterization of Acetal Derivatives

The acetal derivatives synthesized during the reaction has been characterized using FT-IR and NMR spectroscopy techniques.

1) Benzaldehyde acetal IR (υ in cm^{-1}) 3141 (aromatic C-H), 1632, 1511, 1456 (aromatic ring), 1050 & 1026 (C-O), 720 and 763 (mono. Subs.); ^{1}H NMR δ 7.9 (d, 2H, $H_{2,6\ Aromatic}$), 7.6 (t, 1H, $H_{4\ Aromatic}$), 7.5 - 7.4 (t, 2H, $H_{3,5\ Aromatic}$), 4.2 (t, 2H, CH_2), 3.6 (t, 2H, CH_2), 2.4 (s, 1H,

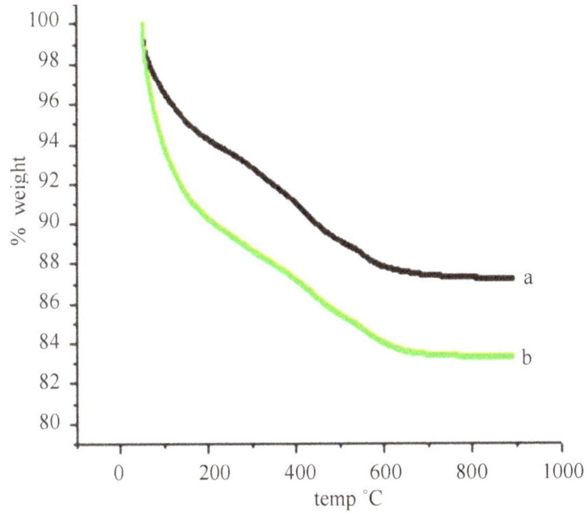

Figure 5. TGA studies of (a) Al-Ben, (b) PAA-Ben.

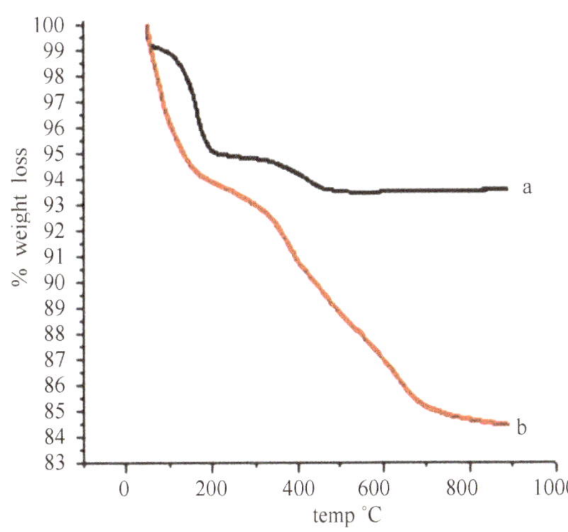

Figure 6. FTIR spectra of (a) STA, (b) STA-Ben.

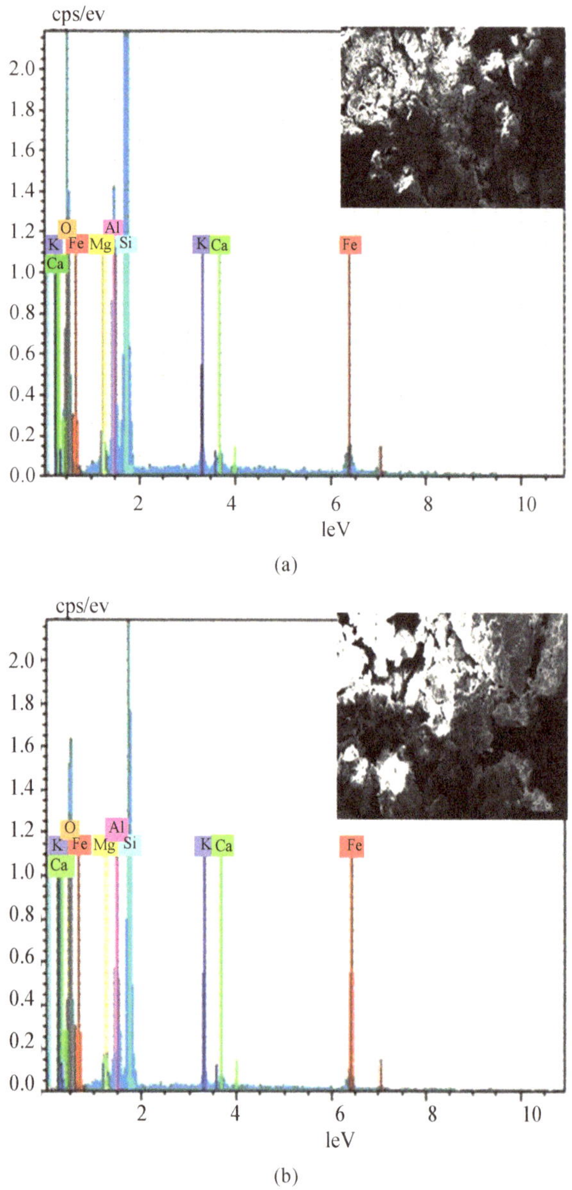

Figure 7. SEM-EDX of (a) Ben (b) STA-Ben.

CH).

2) 2-hydroxy Benzaldehyde acetal IR (υ in cm^{-1}) 3411 (O-H strech), 1551, 1456 (aromatic ring), 1003 & 1026 (C-O), 814 (p Subs.); ^{1}H NMR δ 10.1 (s, 1H, OH), 7.6 (d, 1H, $H_{6\ Aromatic}$), 7.1 (d, 1H, $H_{3\ Aromatic}$), 6.6 (m, 2H, $H_{4,5\ Aromatic}$), 3.6 (t, 2H, CH_2), 3.5 (t, 2H, CH_2), 3.4 (s, 1H, CH).

3) 4-hydroxy Benzaldehyde acetal IR (υ in cm^{-1}) 3400 (O-H strech), 1593,1464,1439 (aromatic ring), 1177, 1051 (C-O), 814 (p Subs.); ^{1}H NMR δ 10.3 (s, 1H, OH), 7.4 (d, 1H, $H_{2,6\ Aromatic}$), 6.8 - 6.9 (d, 2H, $H_{3,5\ Aromatic}$), 3.8 (t, 2H, CH_2), 3.2 (t, 2H, CH_2), 2.5 (s, 1H, CH).

4) 2-chloro Benzaldehyde acetal IR (υ in cm^{-1}) 3179 (aromatic C-H), 1589, 1491, 1463, 1400 (aromatic ring), 1273, 1093 (C-O), 750 (o-Subs.); ^{1}H NMR δ 7.8 (d, 1H, $H_{6\ Aromatic}$), 7.4 (d, 1H, $H_{3\ Aromatic}$), 6.8 - 6.9 (m, 2H, $H_{4,5\ Aromatic}$), 3.3 (m, 4H, CH_2), 2.6 (s, 1H, CH).

5) 4-chloro Benzaldehyde acetal IR (υ in cm^{-1}) 3157 (aromatic C-H), 1595, 1486, 1401 (aromatic ring), 1272, 1175 (C-O), 849 (p Subs.); ^{1}H NMR δ 7.7 (d, 2H, $H_{2,6\ Aromatic}$), 7.4 (d, 2H, $H_{3,5\ Aromatic}$), 4.42 (m, 4H, CH_2), 4.44 (s, 1H, CH).

6) 2-methoxy Benzaldehyde acetal IR (υ in cm^{-1}) 3100 (aromatic C-H), 1606, 1511, 1459 (aromatic ring), 1258, 1169 (C-O), 770 (o-Subs.); ^{1}H NMR δ 7.8 (d, 1H, $H_{6\ Aromatic}$), 7.4 (d, 1H, $H_{3\ Aromatic}$), 6.8 - 6.9 (m, 2H, $H_{4,5\ aromatic}$) 3.7 (m, 4H, CH_2), 3.6 (s, 3H, OCH_3), 2.6 (s, 1H, CH).

7) 4-methoxy Benzaldehyde acetal IR (υ in cm^{-1}) 3041 (aromatic C-H), 1511, 1457 (aromatic ring), 1276, 1259 (C-O), 848 (P-Subs.); ^{1}H NMR δ 7.9 (d, 2H, $H_{2,6\ Aromatic}$), 6.9 (d, 2H, $H_{3,5\ Aromatic}$), 3.8 (m, 4H, CH_2), 3.8 (s, 3H, OCH_3), 2.4 (s, 1H, CH).

8) Acetophenone acetal IR (υ in cm^{-1}) 3127 (aromatic C-H), 1568, 1464, 1473 (aromatic ring), 1026, 1006 (C-O), 692, 763 (monoSubs.); ^{1}H NMR δ 7.6 (m, 3H, $H_{3,4,5\ Aromatic}$), 7.1 (d, 2H, $H_{2,6\ Aromatic}$), 3.5 (m, 4H, CH_2), 2.5 (s, 3H, CH_3).

9) 4-bromo acetophenone acetal IR (υ in cm^{-1}) 3107 (aromatic C-H), 1598, 1498 (aromatic ring), 1027 (C-O), 824 (p-Subs.), 628 (C-Br); ^{1}H NMR δ 7.5 (d, 2H, $H_{2,6\ Aromatic}$), 7.1 (d, 2H, $H_{3,5\ Aromatic}$), 4.3 (t, 2H, CH_2), 4.2 (t, 2H, CH_2), 2.3 (s, 3H, CH_3).

10) Acetone acetal IR (υ in cm^{-1}) 2923, 2853 (C-H strech), 1256, 1204 (C-O); ^{1}H NMR δ 3.6 (m, 4H, CH_2), 2.3 (s, 6H, CH_3).

4. Reusability of STA-Ben

The reusability of STA-Ben has been investigated up to six repeated cycles using synthesis of Acetal derivative of Benzaldehyde. The catalyst, STA-Ben, was washed with MeOH after every cycle and was characterized using FT-IR, TGA, DSC, XRD, SEM and EDX etc. techniques. No noticeable changes were observed even after six cycles, which not only indicates the stability of the catalyst, but also confirms that the catalyst can be reused. The product/s, after separation and isolation, was characterized and the yield in each case was calculated (**Figure 8**). The variation in the yield was found to be in the range of 82% to 62%. The catalyst was characterized after the 6th cycle of reaction and following observations were made.

The SEM picture (**Figure 9**) does not show any change in the surface morphology, the layered structure is maintained. The EDX data (**Figure 9**) indicates the same elemental composition as earlier, *i.e.* Al, Si, O, Mg, Fe, Ca, K and Na.

The thermal stability of the catalyst was found to be unaffected after reuse. (**Figure 10**). XRD and FTIR data (**Figure 11**) of STA-Ben has been found to have no significant change.

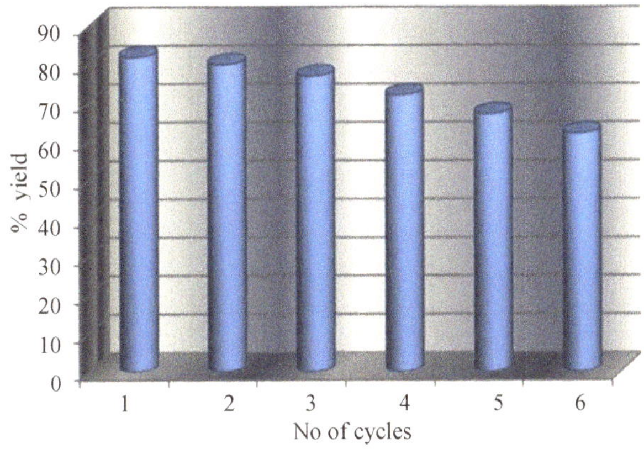

Figure 8. Reusability of STA-Ben up to 6 repeated cycles.

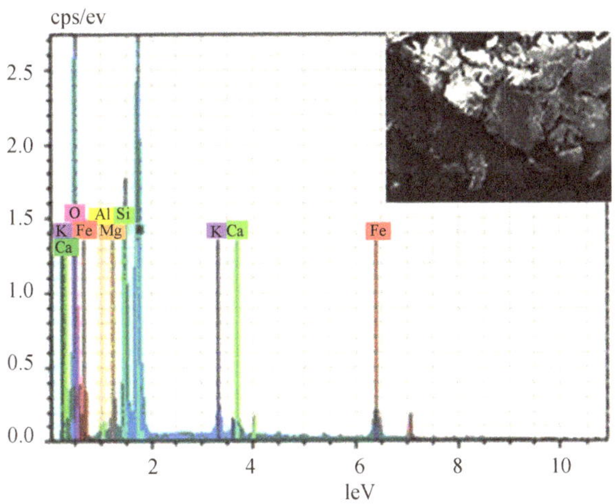

Figure 9. SEM-EDX of STA-Ben after 6th cycle.

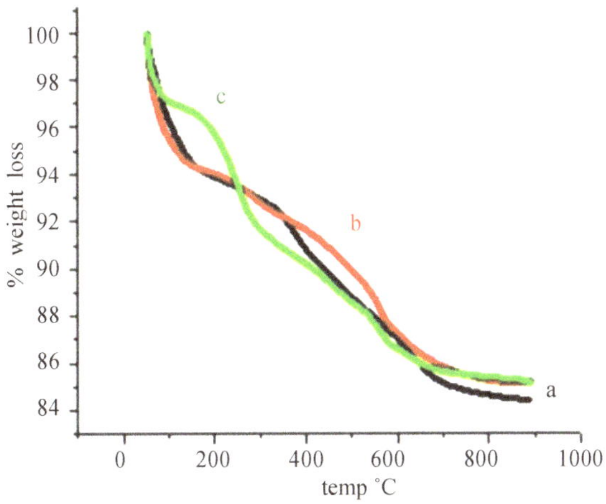

Figure 10. TGA studies of (a) STA-Ben, (b) STA-Ben after 1st cycle, (c) STA-Ben after 6th cycle.

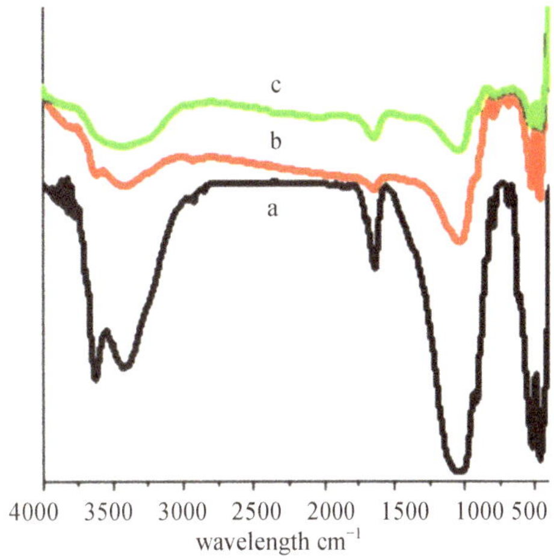

Figure 11. FTIR spectra of (a) STA-Ben, (b) STA-Ben after 1st cycle, (c) STA-Ben after 6th cycle.

5. Conclusion

We have synthesized STA intercalated Bentonite which proved to be a highly efficient, reusable and environmental benign catalyst for synthesis of acetal derivatives of aldehydes and ketones.

REFERENCES

[1] J. Cossy, H. Rakotoarisoa, P. Kahnand and J. R. Desmurs, "A One-Pot Synthesis Of Functionalized 2,2-Disubstituted 2H-1-Benzopyrans," *Tetrahedron Letters*, Vol. 39, No. 52, 1998, pp. 9671-9674.

[2] A. T. Khan, E. Mondal, P. R. Sahu and S. Islam, "Nickel(II) Chloride as an Efficient and Useful Catalyst for Chemoselective Thioacetalization of Aldehydes," *Tetrahedron Letters*, Vol. 44, No. 5, 2003, pp. 919-922.

[3] P. Datta, H. B. Borate, A. Sudalai, T. Ravlndranathan and V. H. Deshpande, "Natural Kaolinitic Clay: A Remarkable Reusable Solid Catalyst for the Selective Functional Protection of Aldehydes and Ketones," *Tetrahedron Letters*, Vol. 37, No. 26, 1996, pp. 4605-4608.

[4] B. K. Banik, M. Chapa, J. Marquez and Magda Cardona, "A Remarkable Iodine-Catalyzed Protection of Carbonyl Compounds," *Tetrahedron Letters*, Vol. 46, No. 13, 2005, pp. 2341-2343.

[5] B. Perio, M. J. Dozias, P. Jacquault and J. Hamelin, "Solvent Free Protection of Carbonyl Group under Microwave Irradiation," *Tetrahedron Letters*, Vol. 38, No. 45, 1997, pp. 7867-7870.

[6] S. K. De, "Cobalt(II)Chloride Catalyzed Chemoselective Thioacetalization of Aldehydes," *Tetrahedron Letters*, Vol. 45, No. 5, 2004, pp. 1035-1036.

[7] J. W. Ralls, R. M. Dodson and B. Riegel, "Addition of Mercaptans to Unsaturated Steroid Ketones," *Journal of the American Chemical Society*, Vol. 71, No. 10, 1949, pp. 3320-3325.

[8] E. Mondal, P. R. Sahu and A. T. Khan, "A Useful and Catalytic Method for Protection of Carbonyl Compounds into the Corresponding 1,3-Oxathiolanes and Deprotection to the Parent Carbonyl Compounds," *Synlett*, Vol. 3, No. 3, 2002, pp. 463-467.

[9] P. K. Mandal and S. C. Roy," Ceric Ammonium Nitrate as a Convenient Catalyst for Chemoselective Thioacetalisation," Tetrahedron, Vol. 51, No. 28, 1995, pp. 7823-7828.

[10] S. Gogoi, J. C. Borah and N. C. Barua, "Montmorillonite K-10 Clay as an Efficient Solid Catalyst for Chemoselective Protection of Carbonyl Compounds as Oxathiolanes with 2-Mercaptoethanol," *Synlett*, Vol. 9, No. 9, 2004, pp. 1592-1594.

[11] J. S. Yadav, B. V. S. Reddy and G. Kondaji, "Eco-Friendly and Highly Chemoselective 1,3-Oxathio-and 1,3-Dithioacetalization of Aldehydes Using Ionic Liquids," *Chemistry Letters*, Vol. 32, No. 8, 2003, pp. 672-673.

[12] L. Jankovic and P. Komadel , "Metal Cation-Exchanged Montmorillonite Catalyzed Protection of Aromatic Aldehydes with Ac2O," *Journal of Catalysis*, Vol. 218, No. 1, 2003, pp. 227-233.

[13] U. V. Desai, T. S. Thopate, D. M. Pore and P.P. Wadgaonkar, "An Efficient, Solvent-Free Method for the Chemoselective Synthesis of Acylals from Aldehydes and Their Deprotection Catalyzed by Silica Sulfuric Acid as a Reusable Solid Acid Catalyst," *Catalysis Communications*, Vol. 7, No. 7, 2006, pp. 508-511.

[14] S. Chandrasekhar, M. Takhi, Y. R. Reddy, S. Mohapatra, C. R. Rao and K. V. Reddy, "Tacl5-Silicagel and Tacl5 as New Lewis Acid Systems for Selective Tetrahydropyranylation of Alcohols and Thioacetalisation, Trimerisation and Aldolisation of Aldehydes," *Tetrahedron*, Vol. 53, No. 44, 1997, pp. 14997-15004.

[15] M. khenkin, L. Weiner and R. Neumann, "Selective Ortho Hydroxylation Of Nitrobenzene with Molecular Oxygen Catalyzed by the H5pv2mo10o40 Polyoxometalate," *Journal of the American Chemical Society*, Vol. 127, No. 128, 2005, pp. 9988-9989.

[16] A. M. Khenkin and R. Neumann, "Oxygen Transfer from Sulfoxides: Selective Oxidation of Alcohols Catalyzed by Polyoxomolybdates," *The Journal of Organic Chemistry*, Vol. 67, No. 20, 2002, pp. 7075-7079.

[17] T. Okuhara, N. Mizuno and M. Misono, "Catalytic Chemistry of Heteropoly Compounds," *Advances in Catalysis*, Vol. 41, 1996, pp. 113-252.

[18] I. V. Kozhevnikov, "Catalysis by Heteropoly Acids and Multicomponent Polyoxometalates in Liquid-Phase Reactions," *Chemical Reviews*, Vol. 98, No. 1, 1998, pp. 171--198.

[19] I. V. Kozhevnikov, "Fine Organic Synthesis with the Aid of Heteropolycompounds," *Russian Chemical Reviews*, Vol. 62, No. 5, 1993, p. 473.

[20] I. V. Kozhevnikov, "Catalysts for Fine Chemicals, Catalysis by Polyoxometalates," Wiley, Chichester, 2002.

[21] M. N. Timofeeva, "Acid Catalysis by Heteropoly Acids," *Applied Catalysis A*: *General*, Vol. 256, No. 1-2, 2003, pp. 19-35.

[22] J. M. Adams, "Synthetic Organic Chemistry Using Pillared, Cation-Exchanged and Acid-Treated Montmorillonite Catalysts—A Review," *Applied Clay Science*, Vol. 2, No. 4, 1987, pp. 309-342.

[23] G. Brain, A. Loupy and D. Villemin, "Microwave Activation of Reactions on Inorganic Solid Supports," In: K. Smith, Ed., *Solids Supports and Catalysis in Organic Synthesis*, Ellis Horwood, Chichester, 1992, pp. 100-122.

[24] L. Fowden, R. M. Barrer and P. B. Tinker, "Clay Minerals: Their Structure, Behaviour and Use," The Royal Society, London, 1984.

[25] E. González and A. Moronta, "The Dehydrogenation of Ethylbenzene to Styrene Catalyzed by a Natural and an Al-Pillared Clays Impregnated with Cobalt Compounds: A Comparative Study," *Applied Catalysis A*: *General*, Vol. 258, No. 1, 2004, pp. 99-105.

[26] M. Kurian and S. Sugunan, "Characterisation of the Acid--Base Properties of Pillared Montmorillonites," *Microporous and Mesoporous Materials*, Vol. 83, No. 1-3, 2005, pp. 25-34.

[27] H. J. Chae, I.-S. Nam, S. W. Ham and S. B. Hong, "Physicochemical Characteristics of Pillared Interlayered Clays," *Catalysis Today*, Vol. 68, No. 1-3, 2001, pp. 31-40.

[28] B. Tyagi, C. D. Chudasama and R. V. Jasra, "Characterization of Surface Acidity of an Acid Montmorillonite Activated with Hydrothermal, Ultrasonic and Microwave Techniques," *Applied Clay Science*, Vol. 31, No. 1-2, 2006, pp. 16-28.

[29] P. Salerno, M. B. Asenjo and S. Mendioroz, "Influence of Preparation Method on Thermal Stability and Acidity of Al–PILCs," *Thermochimica Acta*, Vol. 379, No. 1-2, 2001, pp. 101-109.

[30] W. O. Parker and I. Kiricsi Jr., "Aluminum Complexes in Partially Hydrolyzed Aqueous $AlCl_3$ Solutions Used to Prepare Pillared Clay Catalysts," *Applied Catalysis A*, Vol. 121, No. 1, 1995, pp. L7-L11.

[31] E. M. Araujo, T. J. A. Melo, L. N. L. Santana, G. A. Ne-

ves, H. C. Ferreira, H. L. Lira, L. H. Carvalho, M. M. Avila Jr., M. K. G. Ponte and I. S. Araujo, "The Influence of Organo-Bentonite Clay on The Processing and Mechanical Properties of Nylon 6 and Polystyrene Composites," *Materials Science and Engineering: B*, Vol. 112, No. 2-3, 2004, pp. 175-178.

[32] J. L. Valverde, P. Canizares, M. R. Sun Kou and C. B. Molina, "Enhanced Thermal Stability of Al-Pillared Smectites Modified with Ce and La," *Clays and Clay Minerals*, Vol. 48, No. 4, 2000, pp. 424-432.

[33] M. M. M. Abd El-Wahab and A. A. Said, "Phosphomolybdic Acid Supported on Silica Gel and Promoted with Alkali Metal Ions as Catalysts for the Esterification of Acetic Acid By Ethanol," *Journal of Molecular Catalysis A: Chemical*, Vol. 240, No. 1-2, 2005, pp. 109-118.

[34] N. Yildiz, Z. Aktas and A. Calimli, "Sulphuric Acid Activation of a Calcium Bentonite," *Particulate Science and Technology*, Vol. 22, No. 1, 2004, pp. 21-33.

[35] J. Temuujin, Ts. Jadambaa, G. Burmaa, Sh. Erdenechimeg and J. Amarsanaa, "Characterisation of Acid Activated Montmorillonite Clay from Tuulant (Mongolia)," *Ceramics International*, Vol. 30, No. 2, 2004, pp. 251-255.

10

A Rapid Method for Determination of Some Phenolic Acids in Brazilian Tropical Fruits of Mangaba (*Hancornia speciosa* Gomes) and Umbu (*Spondias tuberosa* Arruda Camara) by UPLC

Edelvio de Barros Gomes[1,2], Suyare Araújo Ramalho[1], Nayjara Carvalho Gualberto[1], Rita de Cássia Mendonça de Miranda[1], Nisha Nigam[3*], Narendra Narain[1]

[1]Laboratory of Flavor and Chromatographic Analysis, Federal University of Sergipe, São Cristóvão, Brazil; [2]Instituto Federal de Educação, Ciência e Tecnologia da Bahia, Canela, Brazil; [3]Department of Chemistry, R. D. National College, Bandra, Mumbai, India.

ABSTRACT

A rapid chromatographic method for the determination of six phenolic acids (chlorogenic, ferulic, gallic, *p*-coumaric, protocatechuic and vanillic acids) by ultra performance liquid chromatography (UPLC), was developed and applied for Brazilian tropical fruits mangaba (*Hancornia speciosa* Gomes) and umbu (*Spondias tuberosa* Arruda Camara). A multivariate statistical experimental design was employed to optimize analytical conditions (solvent A, solvent B concentrations and flow rates). Samples were cleaned-up by solid-phase extraction (SPE) with different solvents (methanol and acetone) employing SPE cartridges (amine and octadecyl-silane). The method using dihydrogen potassium phosphate 5 mmoL as solvent A and 8% acetonitrile as solvent B presented limits of detection varying from 14 to 94 $ng \cdot mL^{-1}$, limits of quantification from 39 to 277 $ng \cdot mL^{-1}$ with 2 μL of injection volume while total run time for all six compounds was only 9.6 minutes. Higher recovery was obtained by extraction with methanol-acetone of 69.51% to 72.59% for protocatechuic acid and 69.58% to 126.31% for the chlorogenic acid. The concentrations of chlorogenic, *p*-coumaric and ferulic acids in mangaba extracts were 113.4, 32.1 and 1.5 $\mu g \cdot g^{-1}$, respectively while concentrations of chlorogenic, protocatechuic, gallic, vanillic, ferulic and *p*-coumaric acids present in umbu fruit were 118.9, 141.3, 3.5, 2.5, 2.2 and 1.8 $\mu g \cdot g^{-1}$, respectively.

Keywords: Antioxidants; Bioactive Compounds; Tropical Fruits; SPE; UPLC-DAD

1. Introduction

Evidences suggest that a diet rich in fruits and vegetables may decrease the risk of chronic diseases, due to the presence of high concentrations of antioxidant compounds, such as phenolic acids [1]. These compounds are widely spread throughout the plant kingdom, occurring in the forms of esters, ethers or free form, and are well-known for their bioactive properties such as antimutagenic, anticarcinogenic, anti-inflammatory and antimicrobial. In particular, some typical low molecular weight phenolic acids such as gallic, chlorogenic, protocatechuic (3,4 dihidroxy benzoic), *p*-coumaric, vanillic and ferulic acids, have been reported to have some of these properties [2].

There are extensive studies describing techniques for determination of phenolic acids in food, including UV spectrophotometry [3], gas chromatography [4], thin-layer chromatography [5] and high-performance liquid chromatography [6,7]. The last, carried out in reversed-phase mode in conventional high pressure liquid chromatographic system (RP-HPLC) is the most common technique employed and in which the octadecyl-silane (C_{18}) is used as the stationary phase, and diode array as the detector [8]. Mobile-phase, stationary phase and elu-

*Corresponding author.

tion system (gradient or isocratic mode) conditions were optimized in developing a RP-HPLC method [9]. The elution systems are usually binary with an aqueous acidified polar solvent (the solvent "A"—responsible for the minimization of peak tailing) such as aqueous acetic acid, perchloric acid, trichloroacetic acid, trifluoracetic acid, phosphoric acid, dihydrogen-potassium phosphate or formic acid; and a less polar organic solvent (the solvent "B") such as methanol or acetonitrile [10].

In general, the RP-HPLC determinations are time-consuming and this is attributed largely due to the gradient elution mode. Isocratic elution mode, on the contrary, is advantageous, since it tends to be accomplished in a single chromatographic run, with economy of solvent. Additionally, in the RP-HPLC systems, acetonitrile has been preferentially employed as the solvent "B" due to its low viscosity and excellent mixing capacity with water. Nevertheless, the recent worldwide scarcity of acetonitrile caused dramatic price increases, requiring for minimized usage of this solvent [11]. Moreover analysis in RP-HPLC conventional system requires more time when compared to analysis performed in reversed-phase ultra performance liquid chromatographic (RP-UPLC) system.

Few reports using isocratic RP-HPLC for the determination of phenolic acids in fruits have been published [12,13], and there is just one report describing the development of isocratic method using RP-UPLC system for the simultaneous determination of phenolic acids (gallic, 3,5 dihydroxybenzoic, protocatechuic, chlorogenic, gentisic, 4-hydroxybenzoic, caffeic, vanillic, syringic, 3-hydroxybenzoic, coumaric, sinapic, ferulic, salicylic and trans-cinnamic acids) in beverages using mass-spectrometer as the detector [14]. However to the best of our knowledge, an isocratic RP-UPLC method using diode array as detector for determining phenolic acids in fruits has not been reported.

In this paper our objective was to develop an isocratic method for rapid determination of phenolic acids such as chlorogenic, ferulic, gallic, *p*-coumaric, protocatechuic and vanillic acids using RP-UPLC system and to analyze their concentrations in Brazilian tropical fruits of mangaba and umbu. Differents mobile phases at varying flow-rates were used and evaluated to optimize analytical conditions. The method was validated by evaluating selectivity, linear range and limits of detection, quantification and recovery of the phenolic acids.

2. Experimental

2.1. Fruits

Ripe fruits used in this study were purchased from the Central Market in the city of Aracaju, Sergipe state, Brazil. Two tropical fruits, largely cultivated in this region, were selected: mangaba (*Harcornia speciosa* Gomes) and umbu (*Spondia stuberosa* Arruda Camara).

2.2. Standards, Solvents and Reagents

All standards of phenolic acids as well as the uracil standard were purchased from Sigma-Aldrich. (St Louis, MO). Phenolic acids standards were: chlorogenic, ferulic, gallic, *p*-coumaric, protocatechuic and vanillic acid. Reagents (hydrochloric acid, dihydrogen-potassium phosphate, trichloroacetic acid and trifluoroacetic acid) and solvents (acetonitrile, acetone and methanol) were of HPLC grade and these were purchased from Vetec Ltd (Duque de Caxias, Brazil). Dihydrogen-potassium phosphate was used at 5 $mmoL \cdot L^{-1}$, and both trichloroacetic and trifluoroacetic acids were used at $1\% v \cdot v^{-1}$ concentration. All standards were dissolved in methanol. The LC grade water was obtained from milliQ water purification system (Millipore, Bedford, MA, USA).

2.3. UPLC Analysis

Analyses were performed using a Prominence Ultra Performance Liquid Chromatographic system (Shimadzu Inc., Japan) which comprised of a binary pump (model LC-20AD), diode array detector (model SPD-M20A), auto-sampler (model SIL-20A HT), oven (model CTO-20A), controller (model CBM-20A), degasser (model DGU-20A3). The LC-Solution software version 1.2 was used to control the auto-sampler, detector, data acquisition and run settings. A C_{18} reverse-phase column, (Shimpack, Shimadzu Inc. Japan; 50 mm × 3.0 mm, 2.2 μm particle size) was employed for all analysis at a temperature of 40˚C. Peak areas were determined at 254 nm for the gallic, vanillic and protocatechuic acid and 310 nm for the chlorogenic, ferulic and *p*-coumaric acids.

2.4. Parameters Optimization

A three-factor full factorial design (FFD) was used to describe the relationship between the response function and the process variables, as well as to determine the conditions that influenced the retention time. The independent variables studied were: solvent "A" (dihydrogen-potassium phosphate, trichloroacetic acid and trifluoroacetic acid); flow rates of mobile phase ($X1$: 0.4; 0.5 and 0.6 $mL \cdot min^{-1}$) and solvent "B" ($X2$: 8%, 10% e 12%$v \cdot v^{-1}$ of acetonitrile) while response variables were retention time of the six phenolic acids (gallic, chlorogenic, protocatechuic, *p*-coumaric, vanillic and ferulic: $Y1$, $Y2$, $Y3$, $Y4$, $Y5$, $Y6$, respectively). Twenty-seven randomized experiments were performed, based on FFD parameters as cited in **Table 1**. Solvent "A" solutions were prepared using deionized water and pH was adjusted to 2.4.

Table 1. Conditions of full factorial design of 3^3 detailing independent variables, their codes and values.

Independent variable	Unit	Symbol	Coded levels		
			−1	0	+1
Flow rate	$mL{\cdot}min^{-1}$	*X1*	0.4	0.5	0.6
Solvent A	-	-	TCA	TFA	KH_2PO_4
Solvent B	$\%v{\cdot}v^{-1}$	*X2*	8	10	12

2.5. Data Interpretation and Statistical Analysis

Results of statistical factorial design experiments were computed using *Statistica* software version 7.0 (Microsoft, Tulsa-Oklahoma, USA). Statistical models generated were re-parameterized considering only statistically significant effects ($p < 0.05$). The influence of the three independent variables, were primarily based on the statistical analysis of retention time (*Y*) as response-variable at ($p < 0.05$). Additionally, parameters such as number of efficient theoretical plates (N_{ef}) [15] resolution between critical adjacent peaks (R_S) and minimal usage of acetonitrile were also considered in optimizing the analytical conditions.

2.6. Sample Preparation and Clean-Up

For sample preparation, pulp weight of about 2 g was used. The SPE cartridges used for clean-up of samples were a Strata C_{18}-E cartridge 500 mg-6 mL and a Strata NH_2 cartridge 500 mg - 6 mL (Phenomenex™, Torrance-California, USA). Two sets of samples were prepared. The first one, non-fortified sample—without any addition of phenolic acids, while the second set was fortified (addition of 0.66 μg of each standard of phenolic acids before elution). The cartridges were conditioned with 10 mL of methanol, followed by 10 mL of distilled water. Both sets of samples were directly loaded onto the cartridges, washed and followed by elution. Both cartridges were eluted with 2 mL of 0.1 $mol{\cdot}L^{-1}$ of HCl, and then continuously eluted with 10 mL of methanol. Additionally, another sequence of solvents was used being 10 mL of methanol and 2 mL of acetone. These eluates were collected in a flask and then evaporated under vacuum in a rotary evaporator at 40°C to dryness. Each dried extract of sample was re-suspended in 2 mL of methanol.

2.7. Identification and Quantification of Phenolic Acids

A volume of 2 μL of the extract prepared as described in item 2.6 was injected in UPLC System (item 2.3). The identification of phenolic acids in the samples was carried out by comparing their retention times and the spectra obtained from those of the standards injected at identical analytical conditions. The quantification of all phenolic acids was done based on the peak area of the compounds related to the calibration curves prepared from injecting the standard solutions.

2.8. Method Validation

2.8.1. Selectivity, Sensitivity and Linearity

Linear least-squares regression was employed to calculate the slope, intercept and correlation coefficient. The calibration graphs were constructed by plotting each chromatographic peak area against its corresponding concentration. Selectivity was evaluated by comparing the chromatograms of sample extracts to those of standards and possible interferences were investigated considering shapes of peaks, retention time and absorption spectra. Additionally, blank chromatogram at selected wavelengths (254 nm for gallic, vanillic and protocatechuic acid and 310 nm for chlorogenic, *p*-coumaric and ferulic acid), was used to confirm nonexistence of these compounds or any other interferences. The method sensitivity was determined by the angular coefficients (slope) of the calibration graphs for each phenolic acid. The linearity of the detector responses for the studied compounds was represented by the linear regression coefficient (r^2).

2.8.2. Limit of Detection (LOD) and Limit of Quantification (LOQ)

Peaks which produced signals between 3 and 10 times the amplitude of noise were selected and LOD calculated by using the equation: $LOD = 2 \cdot \left(C.H.N^{-1}\right)$ where, *C* is the amount of analyte which produced such signal; *H* is the height of the peak; and *N* is the amplitude of noise. The quantification limits (LOQ) were stated as the 3 times the limit of detection ($LOQ = 3 \times LOD$) and these were also determined experimentally by using varying concentrations of phenolic acids [16].

2.8.3. Recovery

The extracts were spiked with known amounts of each standard in triplicate and analyzed. The recovery was determined as the quotient of the determined concentration of phenolic acids (after chromatographic analysis) per concentration added of individual standards. Results were expressed in percentage.

3. Results and Discussion

The variables such as flow rate (*X1*) and the usage of solvent B (*X2*) were selected based on preliminary studies, which indicated that flow rates higher than 0.6 $mL{\cdot}min^{-1}$ implied in lower resolution between all contiguous pairs of peaks ($R_s < 1.3$). On the other hand, flow rates lower than 0.4 $mL{\cdot}min^{-1}$ extended total retention time to more than 15 minutes which was an undesirable feature in this work as it aimed for shorter retention times. Furthermore based on preliminary studies, concentrations of solvent B (acetonitrile) lower than 8% also extended total retention time to more than 15 minutes, and values higher than 12% decreased resolution to undesirable values ($R_s < 1.3$). The effect of independent variables (flow rate and percentage of solvent B) was evaluated by retention time data for all six phenolic acids at a significant level of $p < 0.05$ and statistical data analysis results are presented in **Table 2**.

3.1. Optimal Analytical Conditions

The statistical models reported the influence of the variables on response-factor by the second-order polynomial regression equation. For retention time (Y), effect of flow rate ($X1$) and percentage of solvent "B" ($X2$) were significant ($p < 0.05$) in first-order linear effect and second-order quadratic effect (**Table 2**).

Based on the conditions predicted by the statistical model, **Figure 1** shows flow rate and solvent "B" (acetonitrile) range that promoted lower retention times for chlorogenic acid. Dark-green colour indicates the optimal region representing flow rate of 0.6 $mL \cdot min^{-1}$ and the usage of 12% acetonitrile. Further experiments were performed to determine resolution, using 8%, 10% and 12% of acetonitrile.

Differences among three solvents "A" (KH_2PO_4, TCA and TFA) as well as, interactive effects of all independent variables were not statistically significant ($p < 0.05$). Moreover the present study reveals that the retention time was not affected by using different solvents "A" (**Tables 3 - 5**).

Since the retention time response was almost similar for the three solvents A (KH_2PO_4, TCA and TFA), KH_2PO_4 was selected as solvent A over other acids tried in this study.

Table 2. Polynomial equations and statistical parameters (p < 0.05) calculated from full factorial experimental design experiments.

Dependent variables Y (retention time)	r^2 (%)	Parameterized models
Gallic acid ($Y1$)	99.8	$Y1 = 0.849 - 0.381X1 - 0.065X1^2 - 0.131X2 - 0.016X2^2$
Vanillic acid ($Y2$)	99.8	$Y2 = 1.340 - 0.610X1 - 0.105X1^2 - 0.407X2 - 0.039X2^2$
Chlorogenic acid ($Y3$)	99.9	$Y3 = 2.249 - 0.959X1 - 0.135X1^2 - 1.527X2 - 0.209X2^2$
Protocatechuic acid ($Y4$)	99.7	$Y4 = 2.866 - 1.283X1 - 0.209X1^2 - 1.403X2 - 0.137X2^2$
p-coumaric acid ($Y5$)	99.9	$Y5 = 5.430 - 2.393X1 - 0.402X1^2 - 3.286X2 - 0.351X2$
Ferulic acid ($Y6$)	99.9	$Y6 = 7.949 - 3.488X1 - 0.651X1^2 - 5.591X2 - 0.674X2^2$

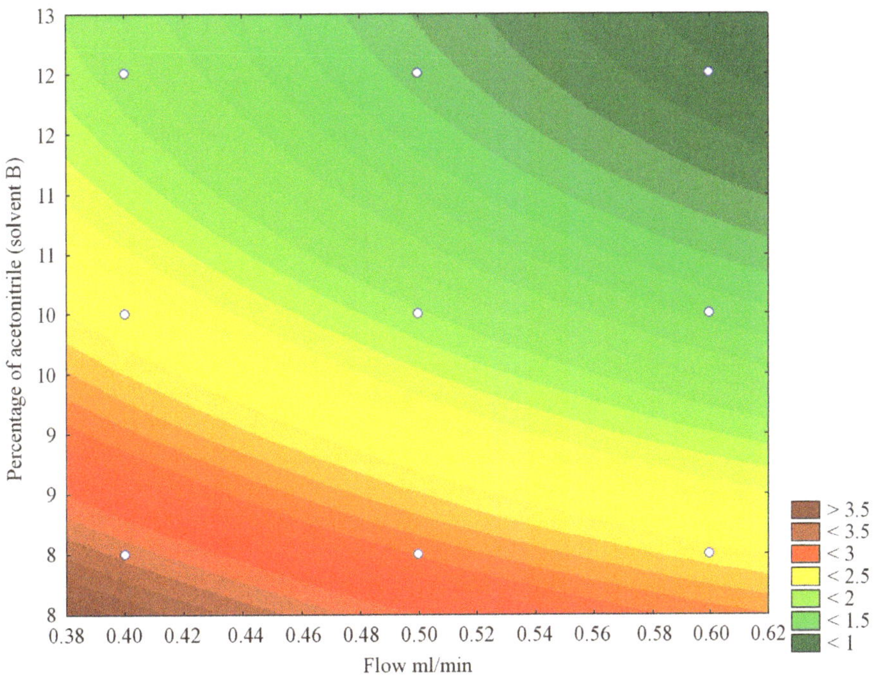

Figure 1. Contour surface plot showing the combined effect of flow rate ($X1$) and percentage of acetonitrile ($X2$) on the retention time of chlorogenic acid ($Y2$).

Table 3. Retention time data of phenolic acids for three solvents "A" (KH_2PO_4, TCA and TFA) and solvent "B" (8% acetonitrile) at different flow rates.

Solvent A	Flow Rate ($mL \cdot min^{-1}$)	Retention time (min)					
		Gallic	Vanillic	Chlorogenic	Protocatechuic	*p*-coumaric	Ferulic
KH_2PO_4	0.4	0.377	1.146	3.150	3.654	7.986	12.505
TCA	0.4	0.365	1.108	2.784	3.511	7.644	11.893
TFA	0.4	0.373	1.121	2.883	3.525	7.693	12.024
KH_2PO_4	0.5	0.309	0.931	2.584	2.973	6.482	10.179
TCA	0.5	0.296	0.892	2.248	2.826	6.180	9.627
TFA	0.5	0.301	0.903	2.350	2.848	6.235	9.739
KH_2PO_4	0.6	0.260	0.781	2.177	2.500	5.521	9.096
TCA	0.6	0.253	0.754	1.910	2.389	5.221	8.175
TFA	0.6	0.256	0.761	1.989	2.404	5.252	8.207

Table 4. Retention time data of phenolic acids for three solvents "A" (KH_2PO_4, TCA and TFA) and solvent "B" (10% acetonitrile) at different flow rates.

Solvent A	Flow Rate ($mL \cdot min^{-1}$)	Retention time (min)					
		Gallic	Vanillic	Chlorogenic	Protocatechuic	*p*-coumaric	Ferulic
KH_2PO_4	0.4	0.286	0.857	1.871	2.615	5.553	8.321
TCA	0.4	0.281	0.830	1.671	2.493	5.268	7.861
TFA	0.4	0.286	0.841	1.758	2.531	5.343	7.954
KH_2PO_4	0.5	0.231	0.696	1.615	2.172	4.541	6.800
TCA	0.5	0.226	0.670	1.361	2.020	4.278	6.368
TFA	0.5	0.232	0.681	1.437	2.047	4.325	6.442
KH_2PO_4	0.6	0.199	0.589	1.310	1.792	3.816	5.740
TCA	0.6	0.191	0.565	1.152	1.713	3.639	5.433
TFA	0.6	0.195	0.574	1.222	1.732	3.672	5.524

Table 5. Retention time data of phenolic acids for three solvents "A" (KH_2PO_4, TCA and TFA) and solvent "B" (12% acetonitrile) at different flow rates.

Solvent A	Flow rate ($mL \cdot min^{-1}$)	Retention time (min)					
		Gallic	Vanillic	Chlorogenic	Protocatechuic	*p*-coumaric	Ferulic
KH_2PO_4	0.4	0.235	0.664	1.186	1.940	3.954	5.671
TCA	0.4	0.231	0.643	1.069	1.867	3.796	5.419
TFA	0.4	0.232	0.644	1.113	1.864	3.785	5.424
KH_2PO_4	0.5	0.190	0.537	0.963	1.568	3.199	4.604
TCA	0.5	0.187	0.521	0.872	1.509	3.063	4.385
TFA	0.5	0.188	0.522	0.904	1.509	3.066	4.397
KH_2PO_4	0.6	0.169	0.454	0.815	1.322	2.702	3.888
TCA	0.6	0.157	0.442	0.742	1.279	2.598	3.724
TFA	0.6	0.159	0.441	0.767	1.278	2.601	3.742

3.2. Usage of Lower Concentration of Acetonitrile

The lower concentration of 8% acetonitrile promoted minimal satisfactory resolution (R_S) for a critical pair of peaks (R_S = 1.3 for the pair chlorogenic/protocatechuic acid) and high N_{ef} for the analytes (**Tables 6** and 7).

Acetonitrile as solvent "B" (*X*2) had a greater influence on retention time of all phenolic acids. Increases in less-polar component of mobile phase in RP-HPLC/UPLC systems, tend to dislocate the analyte toward mobile phase, resulting in decreased retention time of compounds. There were sharp and well separated peaks for all phenolic acids even when acetonitrile concentration was 8%. **Figure 2** shows these chromatograms along with those obtained with the usage of 12% acetonitrile. It

Table 6. Number of efficient theoretical plates (N_{ef}) of phenolic acids in different combinations of flow rate and percentage of solvent B (acetonitrile) using KH_2PO_4 as the solvent A.

Phenolic acid	N_{ef} values for different flow rates (mL·min^{-1}) and solvent B concentration (%v·v^{-1}) hyphenated below								
	0.4 - 8	0.4 - 10	0.4 - 12	0.5 - 8	0.5 - 10	0.5 - 12	0.6 - 8	0.6 - 10	0.6 - 12
Gallic	671.40	658.30	864.70	919.30	1082.70	587.70	500.10	383.60	219.00
Vanillic	431.80	317.78	295.59	640.91	398.85	232.51	356.61	337.73	337.91
Chlorogenic	1450.35	837.55	611.65	1612.46	1317.12	746.34	997.16	742.51	577.98
Protocatechuic	1952.21	1642.43	1625.03	2515.63	1838.06	1579.79	1752.14	1371.49	1521.92
p-coumaric	2612.27	2706.73	2880.27	3465.58	3268.80	2420.29	2728.26	2238.18	2947.87
Ferullic	3214.07	3119.78	3018.74	3437.05	3352.06	3104.90	3825.81	3503.52	3556.88

Table 7. Resolution of contiguous peaks of phenolic acids in different combinations of flow rate and percentage of solvent B (acetonitrile) using KH_2PO_4 as the solvent A.

Contiguous peaks	Resolution of contiguous peaks for different flow rates (mL·min^{-1}) and solvent B concentration (%v·v^{-1}) hyphenated below								
	0.4 - 8	0.4 - 10	0.4 - 12	0.5 - 8	0.5 - 10	0.5 - 12	0.6 - 8	0.6 - 10	0.6 - 12
Gallic and Vanillic acid	>4.0	3.4	3.4	>4.0	>4.0	2.9	3.3	3.1	1.1
Vanillic and Chlorogenic acid	>4.0	>4.0	3.0	>4.0	>4.0	3.0	>4.0	>4.0	3.1
Chlorogenic and Protocatechuic acid	1.5	2.9	3.9	1.6	2.9	4.0	1.3	2.5	3.7

is noteworthy from the figures that gallic, vanilic and protocatechuic acids (peaks 1, 2 and 4, respectively) are larger (**Figure 2(a)**) when measured at 254 nm while chlorogenic, ferulic and *p*-coumaric acids (peaks 3, 5 and 6, respectively) possess larger areas at 310 nm measurement (**Figure 2(b)**). Similar behavior was observed in the flow rate variable. Flow rate of about 0.6 mL·min^{-1} promoted decrease in retention times while values in the flow rate such as 0.4 mL·min^{-1} resulted in higher retention times of phenolic acids. Based on these results the conditions for the identification of all six phenolic acids were optimized: Usage of KH_2PO_4 5 mmol·L^{-1} as solvent A; flow rate of 0.6 mL·min^{-1}, 8% acetonitrile as usage of solvent B.

Amakura *et al* [12] used 15% to 20% of acetonitrile as solvent B for the separation of five phenolic acids (gallic, chlorogenic, caffeic, ellagic and ferulic acids) in grape, apple, pomegranate and prune fruit juices. However, in this work in order to reduce the cost of using acetonitrile, and at the same time get satisfactory results, acetonitrile concentration was reduced from 12% to 8%v·v^{-1}, and even at this concentration, the peaks were distinct and quite well-separated.

3.3. Clean-Up

All phenolic acids except gallic acid, presented recovery percentage higher or equal to 82.3% with Strata C_{18}-E cartridge (data not shown). A Strata NH_2 cartridge was employed to select gallic acid after elution of sample with Strata C_{18}-E cartridge and, as a result, best recovery (89.9%) of gallic acid was observed after using this procedure. Gallic acid recovery using cartridge containing amino group (Strata NH_2), was 89.9% and 70.0% for the mangaba pulp with and without the usage of acetone in extraction step, respectively, and for the umbu pulp it was 78.9% and 50.2%, respectively (**Figure 3(a)**). These results are in agreement with those obtained by Amakura *et al* [12] who utilized combined Sep-Pak Plus C_{18} and Bond Elute (NH_2) cartridges. Phenolic acids investigated in current study belong to two main subgroups: hydroxybenzoic acids (gallic, protocatechuic and vanillic acids); and hydroxycinnamic acids or its esters (ferulic, *p*-coumaric and chlorogenic acids) [17]. Cartridges containing linked amino groups are best suited to separate polar compounds in organic mixtures and the gallic acid molecule is the most polar compound among all the hydroxybenzoic acids tested in current work.

3.4. UPLC Method Validation

The standardized method by optimizing analytical conditions was evaluated for the determination of 6 phenolic acids in mangaba and umbu fruit pulps. The chromatograms were evaluated for selectivity, linearity, limits of quantification and recovery of each of the phenolic acids under study. The selectivity was assessed by comparing the chromatograms of the extracts from fruit pulps to those of the standards phenolic acids. Peak shapes, retention time and spectral purity of the peaks was considered for evaluation of possible interferences. No interfering peaks were found in the blank chromatograms at wave-

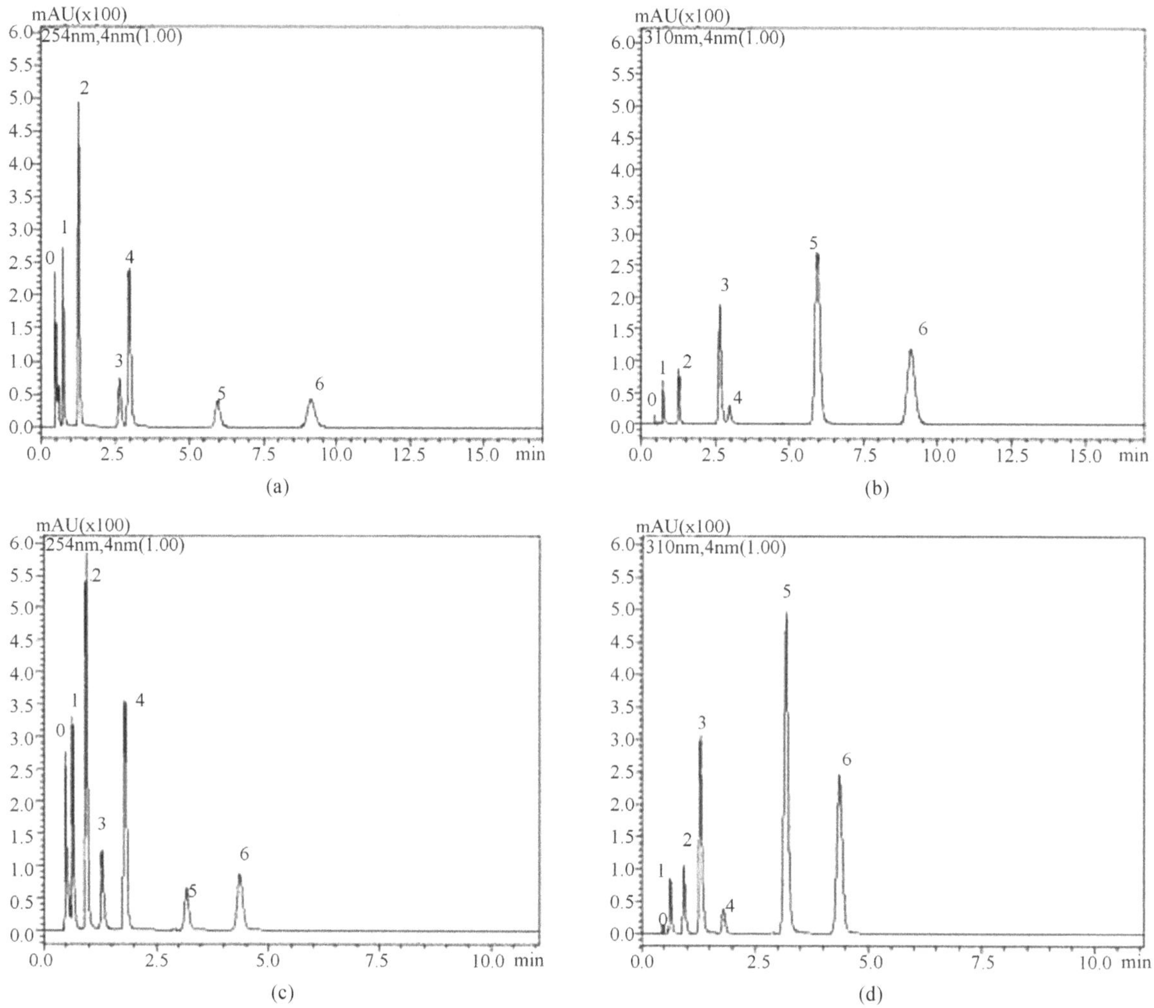

Figure 2. Chromatograms demonstrating the effect of acetonitrile concentration on phenolic acids retention times, at 0.6 $mL{\cdot}min^{-1}$ flow rate: 8% acetonitrile at 254 nm (a) and 310 nm (b) wave lengths; 12% acetonitrile at 254 nm (c) and 310 nm (d) wavelengths. The peak numbers: 0—uracil, 1—gallic acid, 2—vanillic acid, 3—chlorogenic acid, 4—protocatechuic acid, 5—ferulic acid and 6—p-coumaric acid.

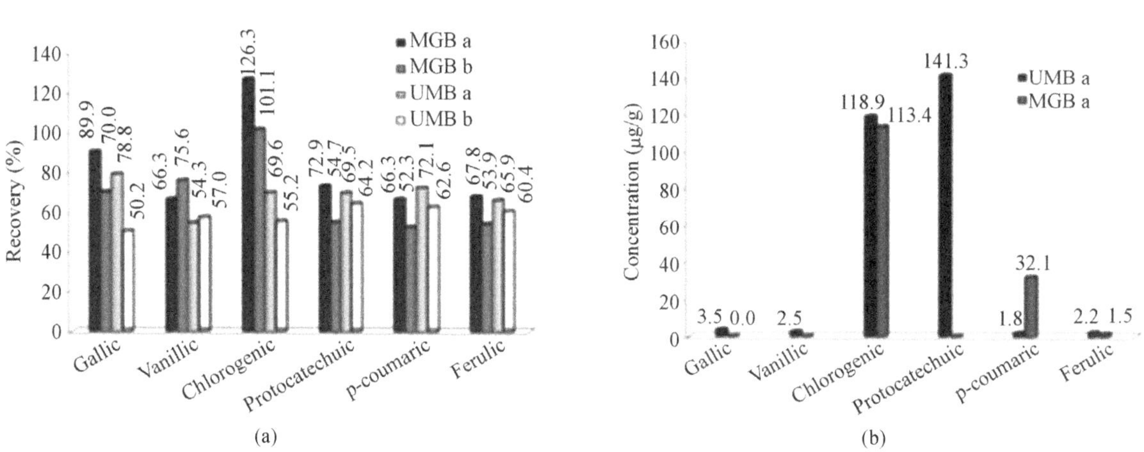

Figure 3. (a) Recovery percentage of phenolic acids in fortified pulps of mangaba and umbu with and without usage of acetone at the extraction step, respectively. (b) Concentration of phenolic acids in non-fortified pulps of mangaba and umbu using optimal extraction condictions. Codes MGBa and MGBb represent mangaba fruit pulp extracted with and without acetone, respectively; UMBa and UMBb represent umbu fruit pulp extracted with and without acetone, respectively.

lengths used for measurement of phenolic compounds.

3.4.1. Linearity

The data on linearity and sensitivity for phenolic acids are presented in **Table 8**. Linearity did not differ statistically ($p < 0.05$) for gallic, protocatechuic and chlorogenicacids in which correlation coefficients were higher than 0.997 for all phenolic acids. The *p*-coumaric acid curve showed highest value of sensitivity among all phenolic acids (**Table 8**). The concentration of all standards varied in the range 0.5 - 2.8 μg·mL^{-1} and good linearity was achieved for all phenolic acids in this range.

3.4.2. Limits of Detection (LOD) and Quantification (LOQ)

Detection limits for various phenolic acids varied from 14 to 94 ng·mL^{-1} (**Table 8**). These values are in accordance, and comparable with the results obtained by Shahrzad and Bitsch [13] and Amakura *et al* [12] who developed an isocratic method to quantify five phenolic acids (gallic, chlorogenic, caffeic, ellagic and ferulic acid) out of which three phenolic acids being gallic, chlorogenic and ferulic acids were subject of study in this work. However, they did not analyze protocatechuic, *p*-coumaric and vanillic acids in fruit juices (apple, pomegranate and prune). The results obtained for the quantification limits show that the optimized method is sensitive enough for determination of the phenolic acids in mangaba and umbu fruit pulps.

3.4.3. Recovery

Addition of acetone in extraction step enhanced the recovery of chlorogenic acid in both fruit pulps (**Figure 4**). This augmentation is thought to be due to the affinity of carbonyl group of acetone with the carboxyl group from chlorogenic acid. When the acetone was added after elution with methanol, chlorogenic acid recovery increased from 101.1% to 126.3% for mangaba, and from 55.2% to 69.6% for umbu (**Figure 3(a)**). Percentages of recovery for various phenolic acids ranged from 66.3% to 126.3% for mangaba pulp when it was extracted with 2 solvents (methanol and acetone) and relatively lower values (52.3% to 101.1%) were obtained when pulp was extracted only with methanol. For the umbu extract recoveries ranged from 65.9% to 78.8% when methanol and acetone was employed in extraction and from 54.2% to 64.2% with methanol only (**Figure 3(a)**).

The presence of different forms of phenolic acids requires suitable extraction conditions, and among all other aspects, the most important is being the polarity of extracting solvents. As a consequence, combined solvents are more effective for screening different classes of phenolic acids than the usage of only one solvent alone [18,19].

Figure 3(b) shows the concentrations of phenolic acids in non-fortified pulps of both fruits obtained under optimized extraction conditions, **Figure 4** shows the chromatograms for phenolic acids measured at different wave lengths (254 nm for gallic, vanillic and protocatechuic acids and 310 nm for chlorogenic, *p*-coumaric and ferulic acids). Higher concentrations of protocatechuic and chlorogenic acid were obtained for umbu (141.3 and 118.9 μg·g^{-1} of pulp, respectively) and higher concentrations of chlorogenic and *p*-coumaric acids concentrations for mangaba (113.4 and 32.1 μg·g^{-1}, respectively). Gallic, vanillic and protochatecuic acids were not detected in mangaba pulp and ferulic acid was present in low concentration (1.5 μg·g^{-1}). In umbu pulp, gallic, vanillic, *p*-coumaric and ferulic acids were found in lower concentrations being 3.5, 2.5, 1.8 and 2.2 μg·g^{-1}, respectively.

These experimental data indicate that the phenolic acid standards behaved in a similar way as to the sample extracts. Based on these results on various analytical parameters studied in the method development, the optimized extraction (usage of methanol-acetone) and analytical conditions at 0.6 mL·min^{-1} flow rate and 8% of acetonitrile as solvent B could be considered suitable and adequate for the analysis of six phenolic acids in mangaba and umbu tropical fruit juices.

Table 8. Sensitivity, linearity, limits of detection and quantification of phenolic acids under optimized analytical conditions.

Phenolic acid	Sensitivity* (angular coefficient α)	r^2 of Linearity*	Limit of detection (μg·mL^{-1})	Limit of quantification (μg·mL^{-1})	
				Calculated values (3 × LOD)	Experimental values
Gallic	3349674	0.998	0.014	0.042	0.039
Vanillic	7012870	0.997	0.029	0.087	0.100
Protocatechuic	13707820	0.998	0.068	0.204	0.209
Chlorogenic	5877090	0.997	0.023	0.069	0.070
p-coumaric	4351783	0.997	0.094	0.282	0.277
Ferulic	8426684	0.999	0.023	0.069	0.081

*Concentration range of all standards: 0.5 - 2.8 μg·mL^{-1}.

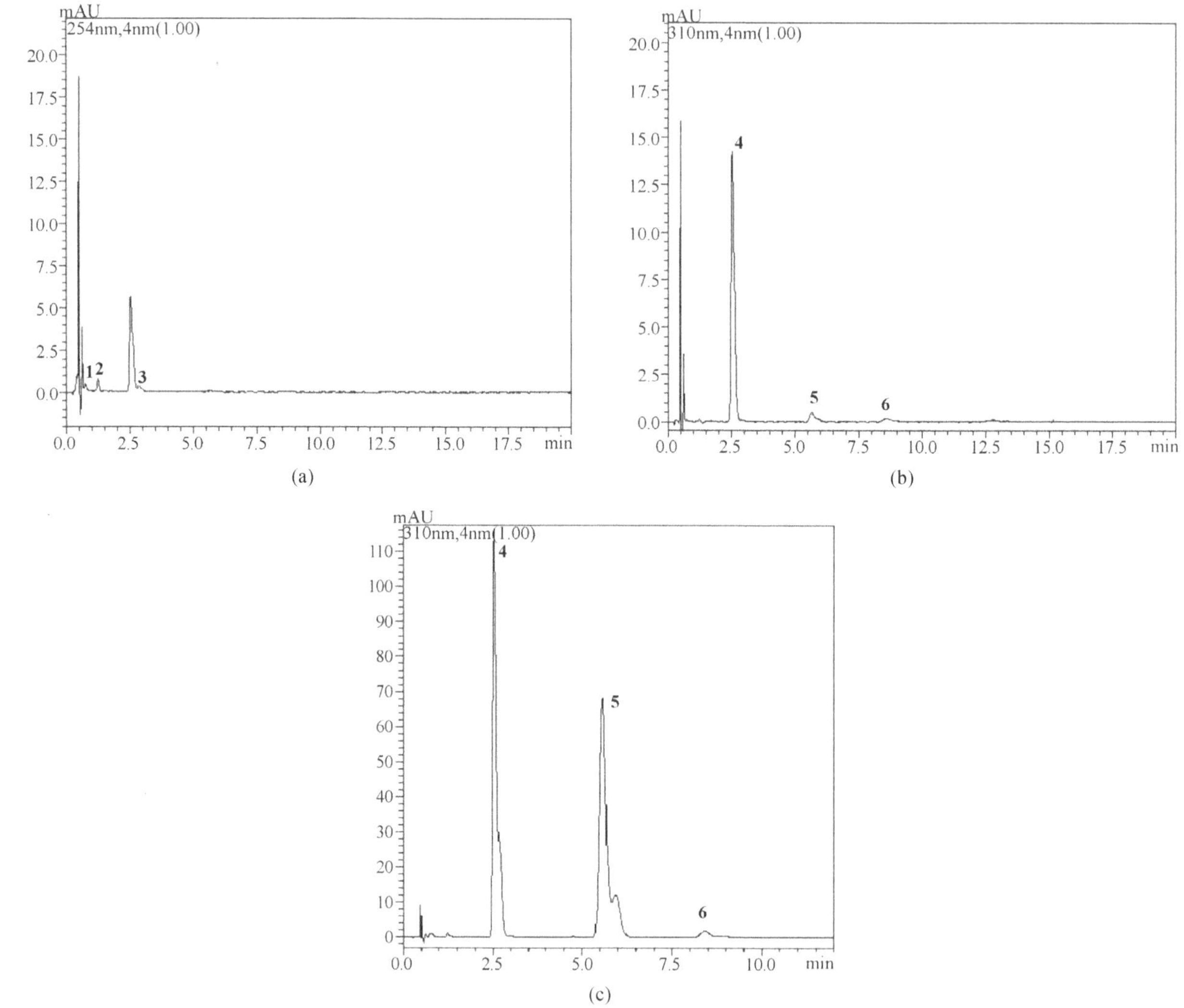

Figure 4. Chromatograms showing separation of gallic, vanillic and protocatechuic acids (1, 2 and 3, respectively) and chlorogenic, *p*-coumaric and ferulic acids (4, 5 and 6, respectively) in samples of umbu (figure a measured at 254 nm and figure b measured at 310 nm) and mangaba (figure c measured at 310 nm) fruit pulps.

4. Conclusion

A rapid UPLC method was developed for separation, identification and quantification of six phenolic acids viz. gallic, chlorogenic, protocatechuic, *p*-coumaric, vanillic and ferulic acids in tropical fruits of mangaba and umbu. The experimental results indicate that mangaba pulp contains higher concentrations of chlorogenic, and *p*-coumaric acids which were 113.4 and 32.1 $\mu g \cdot g^{-1}$, respectively while chlorogenic acid concentration in umbu pulp was equivalent, being 118.9 $\mu g \cdot g^{-1}$. However, protocatechuic acid concentration was higher (141.3 $\mu g \cdot g^{-1}$) in umbu pulp.

5. Acknowledgements

All authors thank the financial support provided from the project, National Institute of Science and Technology (INCT) for Tropical Fruits from CNPq, Brazil offered for promoting research work undertaken in this study. The co-authors (Edelvio and Rita) thank CAPES (Ministry of Education, Brazil) for providing fellowship to them.

REFERENCES

[1] WHO—World Health Organization, "Diet, Nutrition and the Prevention of Chronic Diseases: Report of a Joint WHO/FAO Expert Consultation," WHO Reports, Rome, 2003. http://www.who.int/en/

[2] M. G. L. Hertog, P. C. H. Hollman and B. Van de Putte, "Content of Potentially Anticarcinogenic Flavonoids of Tea Infusions, Wines and Fruit Juices," *Journal of the Agriculture and Food Chemistry*, Vol. 41, No. 8, 1993, pp. 1242-1246.

[3] B. Lapornik, M. Prosek and A. G. Wondra, "Comparison of Extracts Prepared from Plant By-Products Using Dif-

ferent Solvents and Extraction Time," *Journal of Food Engineering*, Vol. 71, No. 2, 2005, pp. 214-222.

[4] T. Y. Chu, C. H. Chang, Y. C. Liao and Y. C. Chen, "Microwave-accelerated Derivatization Processes for the Determination of Phenolic Acids by Gas Chromatography-Mass Spectrometry," *Talanta*, Vol. 54, No. 6, 2001, pp. 1163-1171.

[5] M. Nardini and A. Ghisell, "Determination of Free and Bound Phenolic Acids in Beer," *Food Chemistry*, Vol. 84, No. 1, 2004, pp. 137-143.

[6] R. J. Robbins and S. R. Bean, "Development of a Quantitative High-Performance Liquid Chromatography-Photodiode Array Detection Measurement System for Phenolic Acids," *Journal of Chromatography A*, Vol. 1038, No. 1-2, 2004, pp. 97-105.

[7] R. Tsao and R. Yang, "Optimization of a New Mobile Phase to Know the Complex and Real Polyphenolic Composition: Towards a Total Phenolic Index Using High-Performance Liquid Chromatography," *Journal of Chromatography A*, Vol. 1018, No. 1, 2003, pp. 29-40.

[8] R. J. Robbins, "Phenolic Acids in Foods: An Overview of Analytical Methodology," *Journal of the Agriculture and Food Chemistry*, Vol. 51, No. 10, 2003, pp. 2866-2887.

[9] H. S. Lee, "HPLC Analysis of Phenolic Compounds," In: L. M. L. Nollet, Ed., *Food Analysis by HPLC*, 2nd Edition, Marcel Dekker, Inc., New York, 2000.

[10] H. M. Merken and G. R. Beecher, "Measurement of Food Flavonoids by High Performance Liquid Chromatography: A Review," *Journal of the Agriculture and Food Chemistry*, Vol. 48, No. 3, 2000, pp. 577-599.

[11] J. G. Bordonaba, P. Crespo and L. A. Terry, "A New Acetonitrile-Free Mobile Phase for HPLC-DAD Determination of Individual Anthocyanins in Blackcurrant and Strawberry Fruits: A Comparison and Validation Study," *Food Chemistry*, Vol. 129, No. 3, 2011, pp. 1265-1273.

[12] Y. Amakura, M. Okada, S. Tsuji and Y. Tonogai, "Determination of Phenolic Acids in Fruit Juices by Isocratic Column Liquid Chromatography," *Journal of Chromatography A*, Vol. 891, No. 1, 2000, pp. 183-188.

[13] S. Shahrzad and I. Bitsch, "Determination of Some Pharmacologically Active Phenolic Acids in Juices by High-Performance Liquid Chromatography," *Journal of Chromatography A*, Vol. 741, No. 2, 1996, pp. 223-231.

[14] J. Gruz, O. Novak and M. Strnad, "Rapid Analysis of Phenolic Acids in Beverages by UPLC-MS/MS," *Food Chemistry*, Vol. 111, No. 3, 2008, pp. 789-794.

[15] A. J. P. Martin and R. L. M. Synge, "A New Form of Chromatogram Employing Two Liquid Phases," *Biochemical Journal*, Vol. 35, No. 12, 1941, pp. 1358-1368.

[16] S. M. Muñoz, M. Montserrat, B. Olga and J. Guasch, "Determination of Some Flavan-3-ols and Anthocyanins in Red Grape Seed and Skin Extracts by HPLC-DAD: Validation Study and Response Comparison of Different Standards," *Analytical Chimica Acta*, Vol. 628, No. 1, 2008, pp. 104-110.

[17] L. Bravo, "Polyphenols: Chemistry, Dietary Sources, Metabolism and Nutritional Significance," *Nutrition Review*, Vol. 56, No. 11, 1998, pp. 317-333.

[18] M. Markom, M. Hasan, R. W. D. Wan, H. Singh and J. M. Jahim, "Extraction of Hydrostable Tannins from *Phyllanthus niruri* Linn.: Effect of Solvents and Extraction Methods," *Separation and Purification Technology*, Vol. 52, No. 3, 2007, pp. 487-496.

[19] K. A. Ross, T. Beta and S. D. Arntfield, "A Comparative Study on the Phenolic Acids Identified and Quantified in Dry Beans Using HPLC as Affected by Different Extraction and Hydrolysis Methods," *Food Chemistry*, Vol. 113, No. 1, 2009, pp. 336-344.

11

Volatile Components in the Flower, Pedicellate Ovary and Aqueous Residue of *Cymbidium* sp. (ORCHIDACEAE)

Víctor García Gaytán[1], María de las Nieves Rodríguez Mendoza[1], Marcos Soto Hernández[2], Libia I. Trejo-Téllez[1], Martha E. Pedraza Santos[3], Guadalupe Valdovinos Ponce[4]

[1]Postgrado en Edafología, Colegio de Postgraduados, Texcoco, México; [2]Postgrado en Botánica, Colegio de Postgraduados, Texcoco, México; [3]Facultad de Agrobiología "Presidente Juárez", Uruapan, México; [4]Postgrado en Fitosanidad, Colegio de Postgraduados, Texcoco, México.

ABSTRACT

Cymbidium sp. has several ornamental, medicinal and cosmetic uses. Essential oils were obtained through extraction by hydrodistillation of flower and pedicellate ovary. The chromatographic profile was obtained by thin layer chromatography (TLC), combined with gas chromatography-mass spectrometry (GC/MS), for the identification and relative quantification of various substances in three samples from the orchid (flower, pedicellate ovary and aqueous residue). With TLC, five compounds were identified in the flower, with the most abundant being terpineol, linalool and zingiberene, and five in the pedicellate ovary, with borneol, cineole, and *β*-bisabolene being in the greatest concentration, while for the aqueous residue, the most abundant was geraniol. Using GC/MS, 25 volatile components associated with the flower were identified, of which the most abundant were linalool (22.92%), 4-methyl-phenol (15.07%) and *p*-Menth-1-en-8-ol (12.32%). In pedicellate ovaries, 13 components were identified, with the most abundant being 4-methyl-phenol (31.24%), bicycle [4.4.0] dec-1-ene, 2-isopropyl-5-methyl-9-methylene-(17.74%) and 2-hexanone (10.24%), while in the aqueous residue 18 components were found, with the most abundant being 4-methyl-phenol (18.71%) and 2- cyclohexen-1-ol (14.60%).

Keywords: Thin Layer Chromatography; Gas Chromatography; Orchidaceae; Hydrodistillation

1. Introduction

Plants synthesize and emit chemical compounds that are not only beneficial to humans, but also play a key role in the survival and reproduction of these organisms, such as defense against herbivores and pathogens and mating attraction for pollination [1-3]. In orchids, the phytochemical compounds in roots, tubers, rhizomes, pseudobulbs and leaves have been used to treat a range of diseases [4-7]. The family Orchidaceae is the largest within the angiosperms, consisting of more than 35,000 species and about 750 genera. The family Orchidaceae, whose structural and anatomical characteristics are very diverse, is widely found in temperate and tropical regions, but not in Antarctica [8,9]. The genus *Cymbidium* provides an important commercial ornamental species; it is one of the most popular and sought-after orchids internationally due to its beautiful flowers. It has also been used in traditional medicine for the treatment of paralysis, boils and fever. Compounds with antibacterial and cytotoxic activity, such as phenanthrenes, have been isolated from the fresh roots of *Cymbidium* Great Flower Marie Laurencin. In *Cymbidium goeringii*, compounds have been isolated with hypotensive activity and diuretic activity (peptidoglycan monomer) [10,11]. The study and evaluation of the volatile compounds emitted by orchids and their possible medical applications is a broad and emerging research field [12,13]. Against this background, in this study, the volatile components in the flower, ovary and aqueous residue of *Cymbidium* sp. were isolated, identified and quantified in order to contribute to the chemical knowledge of this genus of orchid.

2. Materials and Methods

The work was carried out in the Phytochemistry Laboratory at the Montecillo campus of the Colegio de Postgraduados. Fresh flowers of *Cymbidium* sp. were removed from the flower spike, and ovaries were separated from the flowers and cut into small pieces. The total amount of fresh flowers and ovary obtained was 549 g and 71.22 g, respectively. Both were weighed using an OHAUS electronic digital scale (Model CT1200-S).

2.1. Analysis of the Flower, Pedicellate Ovary and Aqueous Residue by Thin Layer Chromatography (TLC)

Hydrodistillation was performed for three hours to extract the compounds in the flowers and pedicellate ovaries; 30 ml of each distillate were collected, which made three extractions with 30 ml of dichloromethane (Cl_2CH_2) each. The organic extract was dried with anhydrous sodium sulfate, the solvent was evaporated using a rotary evaporator and the oil yield was obtained by weight differences. The compounds were stored in vials until analysis. The chromatographic profile was obtained by thin layer chromatography (TLC), using a silica gel 60 F_{254} plate with aluminum base (10 × 10 cm). From the samples dissolved in dichloromethane, 10 - 20 μL were taken and placed in a timely manner on the plate. This was developed in a system of toluene ethyl acetate (93:7 v:v) with prior chromatographic chamber saturation. Developing was done with the reagent: 2% vanillin in EtOH, 10% sulfuric acid in EtOH, at 110°C for 5 min [14].

2.2. Analysis of the Flower, Pedicellate Ovary and Aqueous Residue by Gas Chromatography-Mass Spectrometry (GC/MS)

Quantitative analysis was performed by gas chromatography-mass spectrometry (GC/MS) with a gas chromatograph (HP-6890) coupled with a mass detector (HP-5973). The separation was performed using an HP-5 ms capillary column (5% phenyl methyl siloxane, 30 m × 250 μm × 0.25 μm). The operating conditions of the chromatograph were: initial temperature of 60°C for 2 min, thereafter increasing 4°C min^{-1}; final temperature was 250°C, injection port at 250°C; the split 10:1; helium gas flow at 1.5 mL·min^{-1}, quadrupole at 230°C and 1388 mv voltage. An aliquot of the oil was diluted in 1 ml of dichloromethane, with 1 μL of the resulting solution being injected. Identification and quantification was based on retention times and rates, as well as data from the National Institute of Standards and Technology (NIST) mass spectra library.

3. Results and Discussion

3.1. Characterization of the Components in the Flower, Pedicellate Ovary and Aqueous Residue by TLC

Using the TLC technique, five components were identified in the essential oil of the flower of *Cymbidium* sp. five in the ovary oil and two in the aqueous residue (**Figure 1**). The components that stand out in the flower are terpineol with a retention factor (Rf) value of 0.14 and brown coloring, linalool with an Rf of 0.38 and blue coloring, zingiberene with an Rf of 0.52 and blue coloring and the presence of sesquiterpenoids with an Rf of 0.6 and violet coloring. In the ovary the components that stand out are: borneol with an Rf of 0.18 and brown coloring, cineole with an Rf of 0.4 and blue coloring, *β*-bisabolene with an Rf of 0.53 and blue coloring and the presence of other terpenes with an Rf of 0.58 and violet coloring. The TLC data contrast with the results of [15] who, in analyzing the floral components of two species of *Cymbidium* (*C. faberi* and *C. virescens*) by GC/MS, obtained 0.14% *α*-terpineol, 0.21% linalool, 0.65% zingeberene, 0.30% 1,8-cineole and 0.31% *β*-bisabolene.

3.2. Identification and Quantification of the Substances in the Flower, Pedicellate Ovary and Aqueous Residue by GC/MS

The GC/MS technique enabled detecting 21 components in the floral structure of *Cymbidium* sp. of which 19 were identified by comparing mass spectra and retention times (**Figure 2**). **Table 1** lists the components identified from the chromatogram of the extracts, of which the most abundant are: linalool (22.92%) with a retention time (Rt) of 10.30, 4-methylphenol (15.07%) with an Rt of 16.07, p-Menth-1-en-8-ol (12.32%) with an Rt of 12.19 and camphor (5.48%) with an Rt of 10.14.

In flowers of the orchid *Dendrophylax lindenii* (Lind-

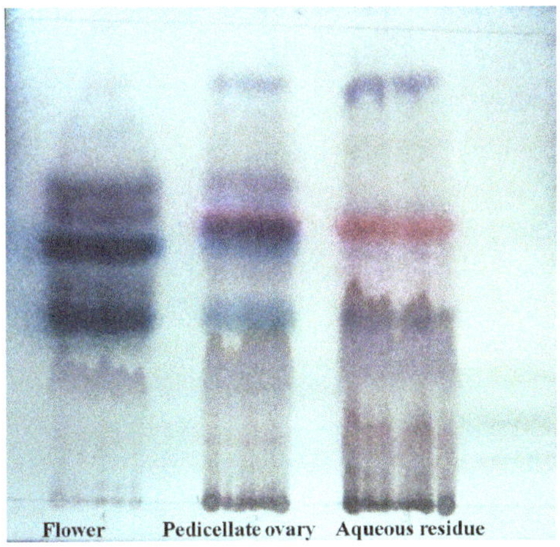

Figure 1. Analysis of the flower, ovary and aqueous residueof *Cymbidium* sp. by thin layer chromatography (TLC).

Table 1. Components in the floral structure by GC/MS in *Cymbidium* sp.

Peak number	Retention time	Compound name	% area
1	4.34	2-Hexanone	1.55
2	7.77	1-Hexanol	2.11
3	9.07	Linalool oxide	2.19
4	9.43	cis-Linalool oxide	1.73
5	10.14	canphor	5.48
6	10.30	Linalool	22.92
7	10.44	1-Octanol	0.69
8	10.75	Bicycle [2.2.1] heptan-3-one, 6,6-dimethyl-2-methylene	0.44
9	10.9	(1R)-(+)-Norinone	0.79
10	11.1	L-4-terpineol	3.78
11	11.47	Myrtenal	0.83
12	11.72	L-Pinocarveol	4.77
13	11.89	unidentified	7.35
14	11.98	Cryptone	1.33
15	12.19	p-ment-1-en-8-ol	12.32
16	12.26	L-Borneol	4.64
17	12.41	cis-Verbenone	0.48
18	12.46	unidentified	1.59
19	12.68	D-Carvone	0.79
20	13.25	Myrtenol	3.67
21	13.48	2,4-Decadienal, (E,E)-	1.72
22	13.66	cis-Carveol	1.20
23	13.75	2,6-Octadien-1-ol, 3, 7-dimethyl-	2.64
24	16.07	Phenol, 4-methyl-	15.07
25	17.58	alpha-Bisabolool	2.07

ley) Bentham through analysis by GC/MS, a low percentage of linalool (5%) was found [16]. By using GC/MS to analyze the volatile components in the flowers of two species of *Cymbidium* (*C. faberi* and *C. virescens*) in Japan, linalool was identified in only *C. faberi* and at a very low concentration (0.21%) [15]. Linalool is a highly sought-after aromatic compound in the fragrance industry. This monoterpene alcohol chemical has several biological activities, among which are antimicrobial, anti-inflammatory, antineoplastic and anti-oxidant properties. Additionally, several studies have confirmed the positive effects of this compound in the central nervous system; it also plays an important role in nature during the pollination of plant species to thereby ensure their reproduction and survival [17]. In the case of the ovary extract and the aqueous residue, the most abundant components were 4-methylphenol (31.24 and 18.71%, respectively) (**Tables 2** and **3**) with Rt values of 16.07 and 16.08 (**Figures 3** and **4**).

Phenolic compounds have been used as natural antioxidants and have become increasingly important for humans because of their medicinal benefits; moreover, their use as an additive in food preservation is also of great interest due to their reported antimicrobial activity [18-20].

Other compounds identified were bisabolol at a concentration of 2.07% and benzaldehyde at 1.49% in floral organ and aqueous residue, respectively. The percentages of these compounds are higher than those reported in flowers of *C. faberi* and *C. virescens* by [15], with percentages of 0.6% and 0.2% respectively for bisabolol and benzaldehyde. A phytochemical study of *Cymbidium aloifolium* with a extract of 75% methanol found a high content of compounds such as flavonoids, reducing sugars,

Table 2. Components in the pedicellate ovary by GC/MS in *Cymbidium* sp.

Peak number	Retention time	Compound name	% area
1	4.33	2-Hexanone	10.24
2	4.68	3-Pentanol, 3-methyl-	1.41
3	5.69	3-Hexanol	1.48
4	6.01	Hexane,1-methoxy-	2.78
5	6.14	Cyclopentanol, 1-methyl-	3.04
6	7.62	Cyclopentanol, 3-methyl-	1.20
7	9.32	2-Cyclohexen-1-ol	1.72
8	10.28	Linalool	5.46
9	12.17	p-menth-1-en-8-ol	4.62
10	13.26	cis-Geraniol	1.38
11	13.48	unidentified	2.43
12	13.75	trans-Geraniol	4.18
13	14.69	unidentified	1.64
14	15.01	1-Nonadecene	1.69
15	15.55	Cembrane	1.32
16	15.75	unidentified	1.72
17	16.07	Phenol, 4-methyl-	31.24
18	16.18	unidentified	4.50
19	17.21	Bicycle [4.4.0] dec-1-ene, 2-isopropyl-5-methyl-9-methylene-	17.74

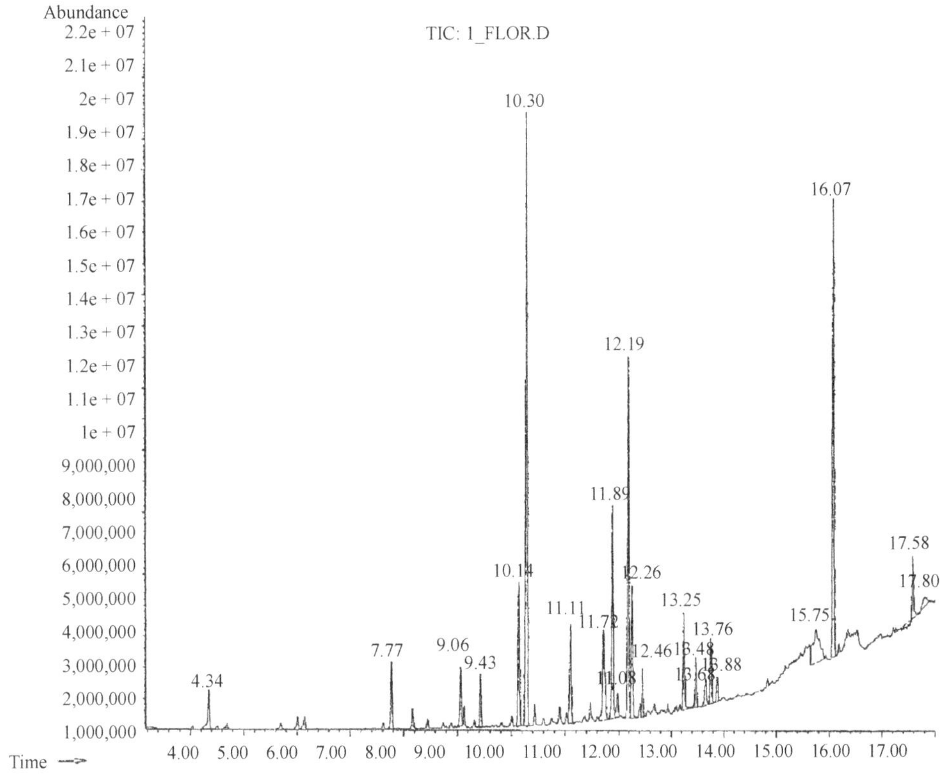

Figure 2. Identification and quantification of the substances in the flower by GC/MS in *Cymbidium* sp.

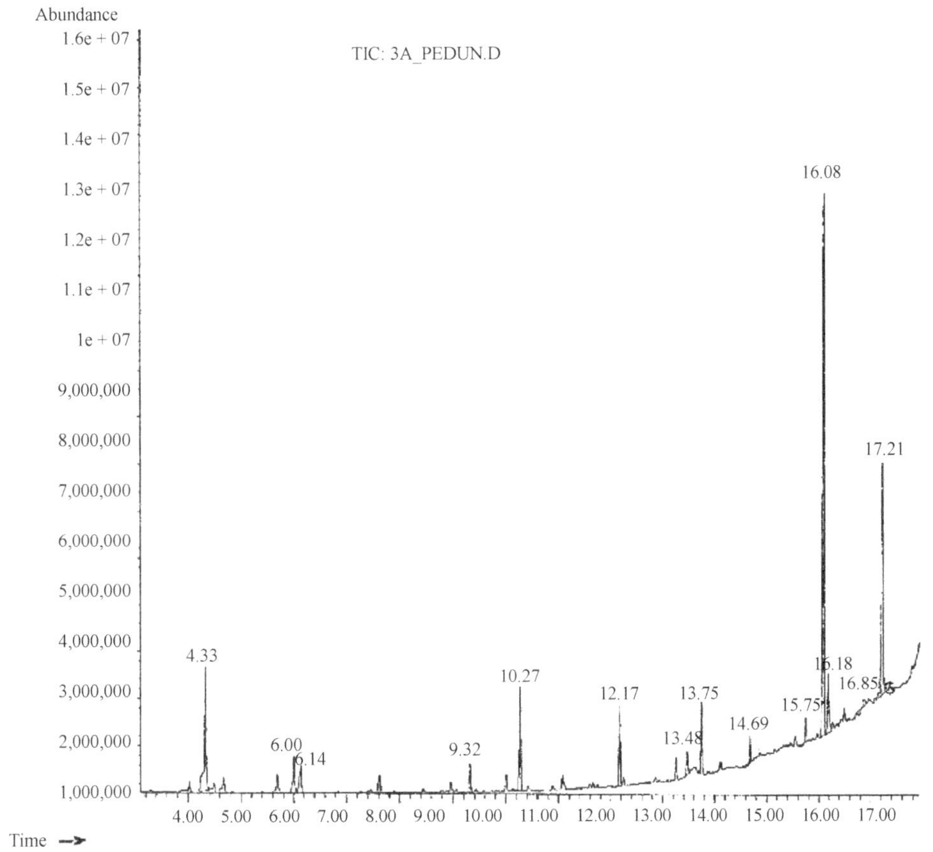

Figure 3. Identification and quantification of the substances in the ovary by GC/MS in *Cymbidium* sp.

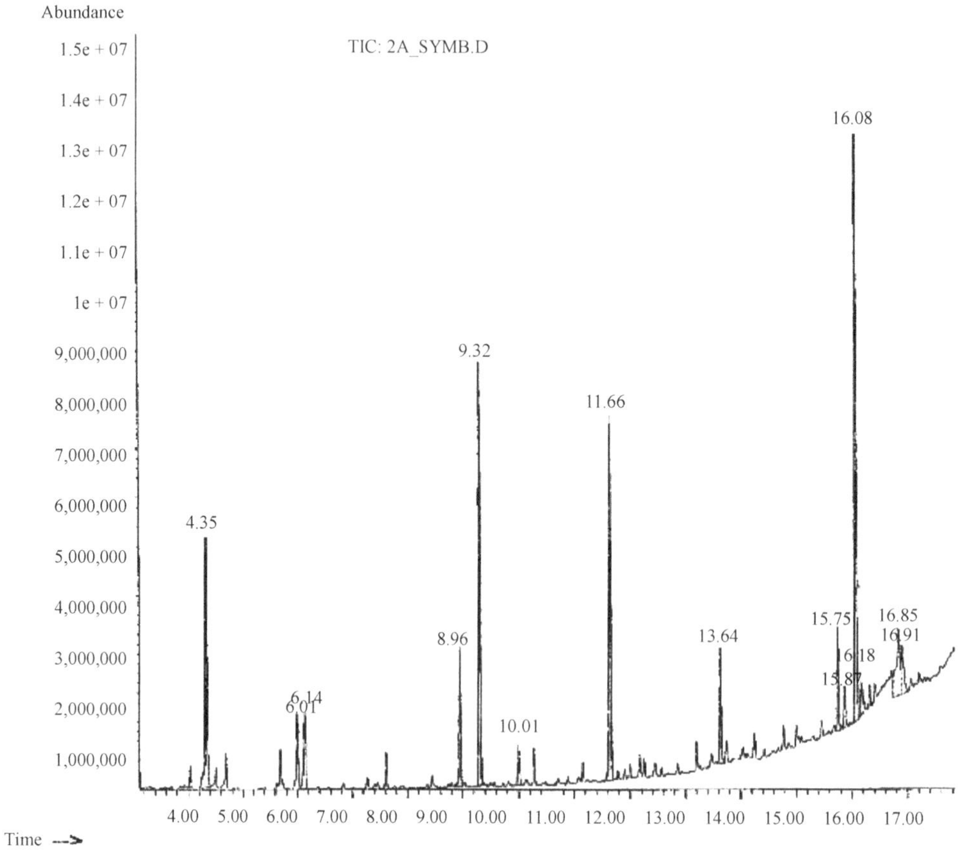

Figure 4. Identification and quantification of the substances in aqueous residue by GC/MS in *Cymbidium* sp.

Table 3. Components in the aqueous residue by GC/MS in *Cymbidium* sp.

Peak number	Retención time	Compound name	% Area
1	4.35	2-Hexanone	9.40
2	2.69	3-Methyl-3-pentanol	1.50
3	5.69	3-Hexanol	1.45
4	6.01	2-Pentanol, 4-methyl-	2.72
5	6.14	Cyclopentanol, 1-methyl-	3.25
6	6.61	Cyclopentanol, 3-methyl-	1.27
7	8.86	2-Cyclohexen-1-one	5.12
8	9.33	2-Cyclohexen-1-ol	14.60
9	10.01	Benzaldehyde	1.49
10	10.28	Linalool	1.25
11	11.66	2-Chlorocyclohexanol	12.65
12	12.17	L-α-Terpineol	0.95
13	13.64	Hexanoic acid	4.19
14	13.75	2,6-Octadien-1-ol, 3,7-dimethyl-	0.70
15	16.08	Phenol, 4-methyl-	18.71

Continued

16	16.18	unidentified	2.80
17	16.42	1-Hexacosene	1.83
18	16.71	E,E,Z-1,3,12-Nonadecatriene-5,14-diol	6.81
19	16.85	(R)-(-)-14-Methyl-8-hexadecin-1-ol	6.81
20	16.91	Octacosanol	3.39

cyanogenic glycosides, terpenes and tannins; however, each of the components were not characterized [21]. Phytochemical studies on leaves of 142 species in 72 genera of the family Orchidaceae, including *Cymbidium*, [22] showed that concentrations of volatile components in a species varies considerably from one geographic region to another; the most common compounds were flavone C-glycosides (53%) and flavonoids (37%); however, the compounds were not uniform and their distribution showed a strong correlation with the geographical location of the plant.

4. Conclusion

In the floral organs of the orchid *Cymbidium* sp., we identified a larger number of terpenes such as linalool

(22.29%), as well as 4-methyl-phenol and *p*-Menth-1-en-8-ol, than in the ovary and aqueous residue where the component with the greatest concentration in both cases was 4-methyl-phenol (31.24% and 18.71%, respectively), which provides evidence of their biological properties and potential in cosmetics and perfumery. The TLC and GC/MS methods of analysis enabled identifying the compounds in the floral organs, ovary and aqueous residue of *Cymbidium* sp. that had not been carried out in previous studies.

5. Acknowledgments

Victor García Gaytan wishes to thank the LPI-4 Priority Line of Research (Agribusiness, Agro-ecotourism and Landscape Architecture), Mexico's National Science and Technology Council (CONACYT) and the Phytochemistry Laboratory at the Montecillo Campus of the Colegio de Postgraduados for their support and sponsorship of this research. Also thanks to M.Sc Manolo Muñiz for his technical assistance in the operation of the GC/EM analysis.

REFERENCES

[1] A. Jurgens and A. M. Viljoen, "Chemical Diversity and Biological Functions of Plant Volatiles," *South African Journal of Botany*, Vol. 76, No. 4, 2010, pp. 607-611.

[2] F. P. Schiestl, "The Evolution of Floral Scent and Insect Chemical Communication," *Ecology Letters*, Vol. 13, No. 5, 2010, pp. 643-56.

[3] C. Soler, M. Hossaert-McKey, B. Buatois, J. M. Bessiére, B. Schatz and M. Proffit, "Geographic Variation of Floral Scent in a Highly Specialized Pollination Mutualism," *Phytochemistry*, Vol. 72, No. 1, 2011, pp. 74-81.

[4] M. Cai, Y. Zhou, S. Gesang, C. Bianba and L. S. Ding, "Chemical Fingerprint Analysis of Rhizomes of *Gymnadenia conopsea* by HPLC-DAD-MSn," *Journal of Chromatography B*, Vol. 844, No. 2, 2006, pp. 301-307.

[5] C. F. Allen, M. E'dwards, L. M. Williamson, W. Kitson-Piggott, H. U. Wagner, B. Camara and C. J. Hospedales, "Sexually Transmitted Infection Service Use and Risk Factors for HIV Infection among Female Sex Workers in Georgetown, Guyana," *Journal of Acquired Immune Deficiency Syndromes*, Vol. 43, No. 1, 2006, pp. 96-101.

[6] M. Adams, W. Alther, M. Kessler, M. Kluge and M. Hamburger, "Malaria in the Renaissance: Remedies from European Herbals from the 16th and 17th Century," *Journal of Ethnopharmacology*, Vol. 133, No. 2, 2011, pp. 278-88.

[7] M. M. Hossain, "Therapeutic Orchids: Traditional Uses and Recent Advances—An Overview," *Fitoterapia*, Vol. 82, No. 2, 2011, pp. 102-40.

[8] N. W. Albert, S. Arathoon, V. E. Collette, K. E. Schwinn, P. E. Jameson, D. H. Lewis, H. Zhang and K. M. Davies, "Activation of Anthocyanin Synthesis in *Cymbidium* Orchids: Variability between Known Regulators," *Plant Cell Tissue and Organ Culture*, Vol. 100, No. 3, 2010, pp. 355-360.

[9] K. Yoshikawa, T. Ito, K. Iseki, C. Baba, H. Imagawa, Y. Yagi, H. Morita, Y. Asakawa, S. Kawano and T. Hashimoto, "Phenanthrene Derivatives from *Cymbidium* Great Flower Marie Laurencin and Their Biological Activities," *Journal of Natural Products*, Vol. 75, No. 4, 2012, pp. 605-609.

[10] M. M. Hossain, M. Sharma and P. Pathak, "Cost Effective Protocol for *in Vitro* Mass Propagation of *Cymbidium aloifolium* (L.) Sw.—A Medicinally Important Orchid," *Engineering in Life Sciences*, Vol. 9, No. 6, 2009, pp. 444-453.

[11] K. Watanabe, R. Tanaka, H. Sakurai, K. Iguchi, Y. Yamada, C. S. Hsu, C. Sakuma, H. Kikuchi, H. Shibayama and T. Kawai, "Structure of Cymbidine A, a Monomeric Peptidoglycan-Related Compound with Hypotensive and Diuretic Activities, Isolated from a Higher Plant, *Cymbidium goeringii* (Orchidaceae)," *Chemical and Pharmaceutical Bulletin*, Vol. 55, No. 5, 2007, pp. 780-783.

[12] J. Gershenzon and N. Dudareva, "The Function of Terpene Natural Products in the Natural World," *Nature Chemical Biology*, Vol. 3, No. 7, 2007, pp. 408-414.

[13] M. Dunkel, U. Schmidt, S. Struck, L. Berger, B. Gruening, J. Hossbasch, I. S. Jaeger, U. Effmert, B. Piechulla, R. Eriksson, J. Knudsen and R. Preissner, "SuperScent—A Database of Flavors and Scents," *Nucleic Acids Research*, Vol. 37, Suppl. 1, 2009, pp. D291-D294.

[14] H. Wagner and S. Bladt, "Plant Drug Analysis. A Thin Layer Chromatography Atlas," Springer Verlag, Berlín, 1996.

[15] A. Omata, S. Nakamura, K. Yomogida, K. Moriai, Y. Ichikawa and I. Watanabe, "Volatile Components of To-Yo-Ran Flowers (*Cymbidium-faberi* and *Cymbidium-virescens*)," *Agricultural and Biological Chemistry*, Vol. 54, No. 4, 1990, pp. 1029-1033.

[16] J. J. Sadler, J. M. Smith, L. W. Zettler, H. T. Alborn and L. W. Richardson, "Fragance Composition of *Dendrophylax lindenii* (Orchidaceae) Using a Novel Technique Applied *in Situ*," *Journal of Environmental Sciences*, Vol. 1, No. 2, 2011, pp. 137-141.

[17] G. P. P. Kamatou and A. M. Viljoen, "Linalool—A Review of a Biologically Active Compound of Commercial Importance," *Natural Product Communications*, Vol. 3, No. 7, 2008, pp. 1183-1192.

[18] A. Sousa, I. C. Ferreira, R. Calhelha, P. B. Andrade, P. Valentão, R. Seabra, L. Estevinho, A. Bento and J. A.

Pereira, "Phenolics and Antimicrobial Activity of Traditional Stoned Table Olives '*Alcaparra*'," *Bioorganic & Medicinal Chemistry*, Vol. 14, No. 24, 2006, pp. 8533-8538.

[19] X. Zhu, H. Zhang and R. Lo, "Phenolic Compounds from the Leaf Extract of Artichoke (*Cynara scolymus* L.) and Their Antimicrobial Activities," *Journal of Agricultural and Food Chemistry*, Vol. 52, No. 24, 2004, pp. 7272-7278.

[20] I. Oliveira, A. Sousa, I. C. Ferreira, A. Bento, L. Estevinho and J. A. Pereira, "Total Phenol, Antioxidant Potential and Antimicrobial Activity of Walnut (*Juglans regia* L.) Green Husks," *Food and Chemical Toxicology*, Vol. 46, No. 7, 2008, pp. 2326-2331.

[21] M. Maridass, M. I. Hussain and G. Rajua, "Phytochemical Survey of Orchids in the Tirunelveli Hills of South India," *Ethnobotanical Leaflets*, Vol. 2008, No. 12, 2008, pp. 705-712.

[22] C. A. Williams, "The Leaf Flavonoids of the Leaf the Orchidaceae," *Phytochemistry*, Vol. 18, No. 5, 1970, pp. 803-813.

12

Recent Advances in Molecular Diagnostics and Treatment of Heart Muscle Diseases

—From Biopsy-Focused to Systemic Diagnostics

Dirk Lassner[1*#], Christine Sabine Siegismund[1*], Jenny Stehr[1], Maria Rohde[1], Felicitas Escher[1,2], Carsten Tschöpe[2], Ulrich Michael Gross[1], Uwe Kühl[1,2], Heinz-Peter Schultheiss[2]

[1]Institute of Cardiac Diagnostics and Therapy, Berlin, Germany; [2]Department of Cardiology and Pneumonology, Charité-Universityhospital Berlin, Campus Benjamin Franklin, Berlin, Germany.

ABSTRACT

Cardiomyopathies are acute or chronic disorders of myocardium. Diagnostic characterization of disease entities demands endomyocardial biopsy analyses with histological, immunohistochemical and molecular biological tissue evaluations to establish a final diagnosis. Only such biopsy-based information allows so far a specific treatment of distinct cardiomyopathy subgroups. In order to reduce sampling error, tissue specimens have to be obtained and analyzed to get clinically relevant data for specific treatment options. Specific gene expression and microRNA (miRNAs) profiles as well as genetic markers add additional valuable information which not only reduce the sampling error but also improve patient management. Advantages of such biomarkers result from their general expression within the entire altered myocardium. Thus, obtained information does not depend on small tissue areas reached by biopsy. This very fact allows prediction of a myocardial infection even in virus-negative areas adjoining positive biopsy specimen. The combination of multiple deregulated miRNAs or genes into one disease specific diagnostic profile demands the integration of new profiling technologies in the routine workflows of cardiological laboratories. In future, multiplex approaches allowing rapid and absolutely reliable identification of inflammatory or virally-induced myocardial diseases will replace single-plex methods such as direct detection of viral genomes in one single biopsy. miRNAs are stable biomarkers which are not only detectable in tissue samples but also in body fluids. Consequently, the determination of distinct miRNA patterns in e.g. peripheral blood samples will provide a systemic diagnostic approach for the characterization of distinct cardiomyopathies by means of non-invasive methods. This will reduce the number of undiagnosed patients who have to undergo endomyocardial biopsy for final confirmation of their myocardial complaints. The resulting molecular diagnostics will pave the way from biopsy focused interpretation to systemic analysis of cardiomyopathies. To reach this goal, the set-up of modern diagnostics harks back to the broad portfolio of high-end analytical techniques and tools.

Keywords: Biomarker; Cardiomyopathy; Endomyocardial Biopsy; Microrna; Systemic Diagnostics

1. Introduction

Cardiovascular diseases constitute the most common cause of death in Western European Countries. According to surveys of the European Society of Cardiology (ESC) 12 million European patients are suffering from heart failure, including two million with non-ischemic dilated cardiomyopathy (DCM). Although myocarditis and inflammatory cardiomyopathy may result from virtually any infectious agent or be a part of an underlying inflammatory or autoimmune process, viral infections and the antiviral immune responses are considered the most frequent causes [1-3].

Myocarditis and inflammatory cardiomyopathy have a wide ranging clinical presentation, contributing to the difficulties in clinical diagnosis and classification. In acute disease, the clinical presentation can range from asymptomatic ECG or echocardiographic abnormalities, to infarct-like presentation, to symptoms of cardiac dysfunction, arrhythmias or heart failure and to hemodynamic collapse. In contrast, the presenting symptoms in sub-acute myocarditis and inflammatory cardiomyopathy

*The first 2 authors contributed equally to this work.
#Corresponding author.

are more subtle and insidious, often with dilated cardiomyopathy, heart failure and arrhythmias. The symptoms are often compatible with moderate ventricular dysfunction, such as fatigue and dyspnea. Under specific circumstances acquired diseases may resemble hypertrophic (HCM), restrictive (RCM) or arrhythmic right ventricular cardiomyopathy (ARVC), clinically.

The spontaneous clinical course of acute and chronic disease is unpredictable and depends on the present activity and cause of the ongoing disease (viral infection, inflammation, auto-antibodies, and toxic injury), the extent of the already developed irreversible myocardial damage, and the genetic predisposition of the individual patient. Finally, acquired cardiomyopathies can also occur in patients with pre-existing other heart diseases and thereby occasionally cause additional diagnostic problems.

Since pathophysiological changes of acquired infectious and non-infectious heart muscle diseases occur at a cellular and subcellular level, imaging techniques such as CMR or echocardiography can provide noninvasive tissue characterization but fail in revealing the true underlying causes especially in chronic disease. Confirmation of a specific pathogen or inflammation requires a direct examination of the affected myocardial tissue by endomyocardial biopsy (EMB).

2. Endomyocardial Biopsy

Invasive removal of tissue samples by EMBs is always necessary when an exact diagnosis cannot be obtained by other clinical methods and is influencing the following treatment.

This presupposes, in addition to comprehensive information and clear question to the pathologist that the testing laboratory can perform a full diagnosis on the current state of knowledge for the clinically suspected disease. An incomplete diagnosis may provide an incomplete picture of the disease and thus leads to misinterpretations and possibly incorrect treatment decisions.

A basic prerequisite for a clinically relevant biopsy diagnostic is the removal of a sufficient number (more than 10 biopsies) high quality and sufficiently large (2 - 3 mm^2) tissue samples from different areas of the myocardium. To avoid squeezing artifacts and retain sterility a needle is taken to remove the biopsy from the biopsy forceps. If the sample predominantly contains thrombus, fat or connective tissue, additional samples need to be collected. Apart from specific conditions such as ARVD/C, in which the characteristic tissue changes mainly occur in certain myocardium areas of the right ventricle, the tissue diagnostic for most questions seems to have a similar sensitivity for left or right ventricular biopsy. To what extent a targeted biopsy of abnormal diagnostic areas of the myocardium, e.g. regional wall motion abnormality, improves the sensitivity of biopsy diagnostic is the subject of ongoing investigations.

The highly complex staged diagnostics of endomyocardial biopsies should be performed only in specialized and certified laboratories with standardized protocols under constant entrainment of internal control samples in all work areas. The constant participation in inter-laboratory surveys for the laboratory testing methods and regular inspections of laboratories by independent monitoring bodies ensure the quality and thus the significance of tissue diagnostics.

The fixation of the biopsies for histological investigations is directly performed in buffered 4% - 10% formaldehyde at room temperature. For immunohistological inflammation diagnostics and molecular studies a fixation in RNAlater (from Ambion, USA) is necessary. RNAlater is particularly suitable for the preservation of unstable RNA at room temperature.

For a meaningful tissue diagnosis the myocardium should be optically controlled during sectioning to contain an evaluable proportion of myocardium exceeding 80% and the tissue samples need to be analyzed completely.

3. Histological Analysis

The histological examination of paraffin sections by different staining protocols (H&E, EvG, PAS, Azan) detects myocardial cell death, scars, fibrosis, disarrays, cardiomyocytes changes, pathological vascular conditions, granulomas, and inflammatory cell differentiation. Storage disorders such as amyloidosis, iron deposits, glycogen and others can be excluded or specified by additional staining, e.g. immunohistochemical differentiation of amyloid subtypes and optionally electron microscopic analyzes (**Table 1**). The EMB diagnosis of myocarditis was based on histomorphological criteria according to the Dallas classification [4]. Active myocarditis is defined by simultaneous detection of inflammatory infiltrates and present lysis of cardiomyocytes. Active myocarditis or storage diseases are found in 5% of examined endomyocardial biopsies in histology.

4. Immunohistochemical Examination on Intramyocardial Inflammation

Immunohistochemical diagnostics are based on application of specific primary antibodies on cryo-fixed tissue section and following detection of coupled primary antibody by secondary antibody. The secondary antibody is conjugated with an enzyme complex producing a precipitating coloured complex by use of staining solution.

Coloured immunospots are counted digitally by application of established digital imaging analysis software

Table 1. Overview on current biopsy diagnostic parameters and predictive value for organ-specific and systemic diagnosis.

Section of routine biopsy diagnostics	Diagnostic analysis of	Invasive Procedure		Non-invasive Procedure	Therapeutic Options
		Focal detection (biopsy related)	**Organ-specific parameter** (detection in neighbouring myocardial tissue)	**Systemic parameter** (detection in peripheral blood)	
Histology					
H&E					
Elastica-van-Giesson (EvG)	Active myocarditis, storage diseases, giant cell myocarditis, cardiac sarcoidosis	Yes	No	No	Immunosuppression
PAS					(Virus-negative)
Azan					
Congored	Amyloidosis	Yes	Yes?	Yes?	Transplantation
v. Kossa	Deposits of Ca^+	Yes	?	No	
Alcianblue	Acid mucopolysaccharides, odema	Yes	?	No	
Stain on iron	Deposits of Fe+	Yes	?	No	
SiriusRed	Myocardial fibrosis	Yes	Yes?	No	
Immunohistochemistry					
CD3	Intramyocardial inflammation	Yes	No	No	
*LFA*1	Intramyocardial inflammation	Yes	No	No	Immunosuppression
Mac-1	Intramyocardial inflammation	Yes	No	No	(Virus-negative)
Perforin	Intramyocardial inflammation	Yes	No	No	
HLA-1	Tissue activation	Yes	Yes?	No	
CD54	Tissue activation	Yes	Yes?	No	
CD106	Vascular activation	Yes	?	No	
CD31	Microvessel density	Yes	?	No	
*Collagen*1/*Collagen* 3	Myocardial fibrosis	Yes	?	No	
desmosomal proteins for ARVD diagnostics	Suspect on ARVD	Yes	Yes	No	
Molecular Virology/ Microbiology					
Adenovirus	Viral cardiomyopathy	Yes	No	No	Interferon-Beta
Coxsackievirus	Viral cardiomyopathy	Yes	No	No	Interferon-Beta
*Erythrovirus (Parvovirus B*19) *and viral replication*	Vascular cardiomyopathy	Yes	Yes?	No	Telbivudine
Human Herpesvirus 6 (*HHV*6)	Immunosuppressive situation	Yes	No	No	
chromosomally integrated Human Herpesvirus 6 (*ciHHV*6)	Immunosuppressive situation	Yes	Yes	No	Valganciclovir
Epstein-Barr-Virus (EBV)	Immunosuppressive situation	Yes	No	No	
Borrelia burgendorferi	Microbial cardiomyopathy	Yes	No	No	Antibiotics
Hepatitis B and C	Viral cardiomyopathy?	Yes	No	No	Antiviral

Continued

CMV	Viral cardiomyopathy?	Yes	No	No	Valganciclovir
Other		Yes	No	No	
Immunology					
Immunofluorescence on autoantibodies	Autoimmune cardiomyopathy	No	No	Yes	Immunoabsorption
Measurement of cardiac autoantibodies (ELISA)	Autoimmune cardiomyopathy	No	No	Yes	Immunoabsorption
Measurement of chemokines and cytokines in sera/plasma	Systemic inflammation	No	No	Yes	Interferon-Beta
Novel biomarkers					
Gene expression profiling	Inflammatory and virally-induced heart muscle diseases	**Yes**	**Yes**	**Yes?***	Silencing RNAs
MicroRNA expression profiling		**Yes**	**Yes**	**Yes***	Antagomirs
Genetic analysis	Genetically based cardiomyopathies	**Yes**	**Yes**	**Yes***	

?—no sufficient data available, Yes?—detection in other biopsies possible, not sure, Yes*—possible detection of specific parameters in body fluids which represent biopsy-located situation.

for calculating area fractions, numbers of immunospots and area of myocardial tissue [5]. The digital imaging system consists of a microscopic unit, a digital camera and supporting analyzing software. Immunohistochemical analysis of intramyocardial inflammation and activation of different tissue fractions (myocytes, capillary) in further biopsies supports the results achieved by histology.

Immunohistochemical analyses are carried out on frozen sections (two EMB) in order to allow detection of elevated inflammatory cell subsets including non-paraffin staining antibodies, e.g. CD3, CD11a (LFA-1), CD11b (MAC-1), CD45R0 (memory or activated lymphocytes), Perforin-positive cytotoxic lymphocytes and increased expression of adhesion molecules CD54 (ICAM), CD106 (VCAM) and HLA-1 as marker for tissue activation.

Perforin-positive, cytotoxic cells were detected in biopsies during acute and chronic phases of infection. Increased number of perforin-positive cells in myocardial tissue is associated with more lesions of cardiomyocytes and poor clinical prognosis of affected patients [6].

Whereas the number of cellular infiltrates (lymphocytes, macrophages) is varying from biopsy to biopsy, the expression of adhesion molecules is globally indicating the tissue activation. Introduction of HLA-1 and CD54 in routine biopsy diagnostics improves prediction of inflammatory processes in myocardium without onsite identification of infiltrative cellular foci [5]. Patients with detectable intramyocardial inflammation are classified as borderline myocarditis due to immunohistochemistry [4]. Inflammatory processes were diagnosed in more than 40% of all bioptated patients with high predictive value for clinical course of these patients [7].

Novel diagnostic markers for cardiomyopathies are the microvessel density (MVD) and the quantitative evaluation of fibrosis. MVD in endomyocardial biopsies is a critical marker in the development of heart muscle diseases. Diagnostics of MVD is performed by the immunohistochemistry of endothelial surface marker CD31. Chronic inflammation or viral infections of myocardium are reducing MVD resulting in malfunction of myocardium and inducing atrophy or hypertrophy of cardiomyocytes.

Myocardial fibrosis detected by CMR is an independent and incremental predictor of mortality and sudden cardiac death in DCM patients [8]. In biopsies the quantitative measurement of intramyocardial fibrosis is performed by immunohistochemistry with Collagen 1 and 3 specific antibodies or staining with Sirius Red and subsequent digital imaging analysis. Correlation of fibrotic areas in myocardial tissue for the risk stratification of acquired cardiomyopathies requires further investigation.

5. Molecular Virology for Detection of Myocardial Infections

Microbial genomes are determined, quantified and sequenced using PCR-based methods (**Table 1**). Nested-PCR protocols consisting of two sequentially performed PCR assays, where the amplicon of first assay is the template for second reaction. This procedure is highly sensitive and enables us to detect very low copy numbers of viral genomes (**Figure 1**). Amplified PCR reactions are separated on agarose gel electrophoresis in ethidium

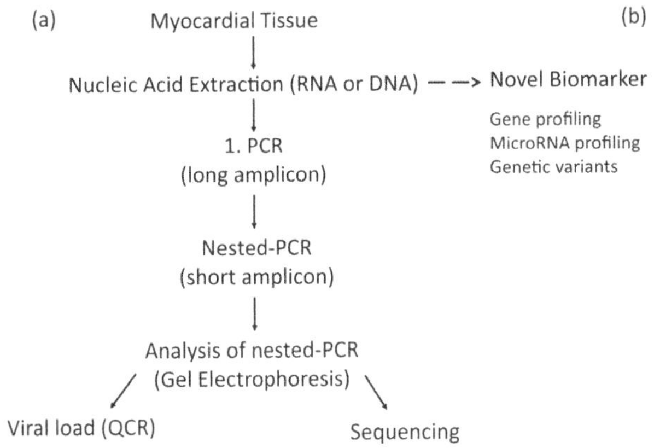

Figure 1. Flowchart of routine diagnostic procedure for detection of cardiotropic viruses (a) and future application of novel biomarker (b) for detection of viral infections, inflammatory diseases or genetic variants predicting development of cardiomyopathies.

bromide containing buffer. This allows the subsequent visualization of generated PCR product by UV fluorescence. All positive PCR reactions are sequenced as quality control, checked by manual alignment and online with NCBI database for confirmation of corresponding virus strains and/or estimation of specific virus subtypes or variants [9].

Depending on the two types of viral nucleic acids the isolation of DNA and RNA is performed in separate extraction procedures. The current viral load monitoring for an effective treatment is performed by quantitative PCR. The transcriptional activity of virus in myocardial tissue or peripheral blood cells will be routinely determined for the two most frequent cardiotropic viruses Erythrovirus and HHV-6 by nested-RT-PCR and QPCR (**Figure 1**). Patients with active erythroviral infection reported a higher frequency of chest pain, systolic left ventricular ejection fraction and accompanied increase of left ventricular end-diastolic diameter. Because these differences were not associated with myocardial inflammation viral replication seems to be an independent prognostic marker [10].

Human herpesvirus 6 A and B (HHV-6A/B) are possible pathogenetic causes of myocarditis and idiopathic-cardiomyopathy subsets [11]. The prevalence of chromosomally integrated HHV-6 (ciHHV-6) is approximately 0.8% of HHV-6-positive endomyocardial biopsies. Identifying individuals with ciHHV-6 is important because the complete HHV-6 genome is present in every cell in their body [12] and a permanent reactivation of this virus in all tissues is assumed. Detection of ciHHV-6 is only possible by QPCR. Persistent high viral loads of HHV-6 genomes in blood cells or tissues are excluding active infection and confirm ciHHV-6 presence. Elimination of chromosomally integrated virus is impossible but transcriptional activity of ciHHV-6 is reduced under treatment with antiviral drug. Monitoring of viral RNA load is the best indicator for an effective therapy or reactivation of ciHHV-6.

The use of these sensitive techniques to detect the most common cardiotropic viruses (such as enteroviruses, adenoviruses, erythrovirus, human herpesvirus 6, Epstein-Barr virus and in the Far East also hepatitis C) reveals a virus infection in about 50% - 60% of patients who are biopsied under the clinical suspicion of myocarditis or DCMI. The clinical significance is clearly demonstrated only for some cardiotropic viruses [13-15].

Current molecular biological workflow for detection of myocardial infection is a combination of qualitative (nPCR) and quantitative detection of viral genomes by nucleic acid isolation systems, thermocycling, gel electrophoresis and real-time QPCR, and subsequent sequencing of generated PCR amplicons by DNA sequencing devices (**Table 2**).

6. Indirect Immunofluorescence Testing of Autoimmunity

Autoimmune cardiomyopathy is an immune-mediated chronic inflammation of the myocardial tissue. Generation of autoantibodies are induced by infection of heart muscle by viruses or by intramyocardial inflammation. The indirect immunofluorescence test is the analytical method of choice for simultaneous screening on different autoantibodies reacting against myocardial tissue. Indirect immunofluorescence assays based on monkey heart tissue slides as a widely similar reference to human tissue. Reaction of patient sera with structural components of presented heart tissue gives first indication of autoantigenic origin (basal membrane, mitochondria and intercalating discs). Cell lines with corresponding virus infection, cells or animal models with overexpression of desired autoantigen were generated. These cultured cells or animal tissue slides were placed as test agents on microscopic slides. Reaction of patient sera with presented reference material corresponds to existing autoantibodies against individual autoantigens [16,17]. The immunological analysis is made by fluorescence microscopy (**Table 2**).

7. Immunoassays

The immune system protects the body from infection by creating and maintaining barriers that prevent bacteria and viruses from entering the body. If a pathogen penetrates these barriers, and gets into the body, the innate immune system is equipped with specialized cells that detect, and often eliminate, the invader before it is able to reproduce, potentially causing serious injury to the host. The innate immune system protects the host by estab-

Table 2. Analytical methods in current and future diagnostics of cardiomyopathies.

Cardiac disease	Subdivision	Standard Biopsy Diagnostics			Immunology, indirect immunofluorescence	New Biomarker Biopsy	Biopsy or PBMC	Novel Biomarkers in PBMC and Serum	Therapeutic Options
		Histology	Immunhisto-chemistry	Molecular Biology		Gene + miRNA Profiling	Genetics	mRNA + miRNA Profiling	
Dilated Cardiomyopathy		+				+	+	+	Heart failure therapy
Dilated Cardiomyopathy with inflammation		+	+			+	-	+	Immune suppression
Autoimmune Cardiomyopathy		+	+		+		-		
Acute Myocarditis		+	+			+	-	+	Immune suppression
Borderline Myocarditis		+	+			+	-	+	Immune suppression
Idiopathic Giant Cell Myocarditis		+				+	-	+	Immune suppression
	Adenovirus			+	+	+	-	+	Antiviral therapy
	Human Herpesvirus 6		+	+	+	+	-	+	Antiviral therapy
Virus-induced Cardiomyopathy	Erythrovirus (Parvovirus B19)			+	+	+	-	+	Antiviral therapy
	Coxsackievirus persistent			+	+	+	-	+	Antiviral therapy
	Coxsackievirus spontaneously eliminated			+		+	-	+	No therapy necessary
Analytical Methods		Light and Fluorescence Microscopy and digital imaging analysis		nPCR + QPCR + Sequencing	Fluorescence Microscopy, Multiplex-ELISA	QPCR Array	Sequencing, SNP array, NGS	QPCR Array	

Grey-currently applied methods in routine diagnostics of endoymocardial biopsies, NGS-next-generation-sequencing, nPCR-nested-PCR, QPCR-quantitative PCR, PBMC-peripheral blood mononuclear cells, SNP-single nucleotide polymorphism

lishing humoral, chemical and cellular barriers to infection. Inflammatory response is produced by chemical factors including specialized chemical mediators, called cytokines. Cytokine levels in peripheral blood correspond to systemic situation in patients initiated by various, but often global factors like infections or inflammatory processes. Cytokines or chemokines are estimated in blood serum for characterization of present immune response. High level of interferon-beta in serum results in spontaneous elimination of enterovirus from myocardial tissue and is associated with reduced long term mortality in cardiac patients [15].

Immunoassays on cardiac autoantibodies [17] or cellular hormones like adiponectin [18] are good tools for monitoring clinical cause of cardiomyopathies and support biopsy diagnostics. Limitations are the availability of corresponding assays for multiple cardiac autoantigens. For the quantitative determination of autoantibodies or antibodies against infectious agents, cells, tissue sections or purified, biochemically characterized substances are used as antigen substrates.

Introduction of multiplexing technologies allows the simultaneous measurement of up to 500 analytes in a single patient sample. Multiplex-ELISA is using fluorescent microparticles (beads) with immobilized antibodies against different cytokines, chemokines or other antigens. The incubation of these beads with patients serum results in binding of specific antigens and subsequent detection by a second antibody which is labeled with a second dye. This approach allows also the measurement of multiple cardiac autoantibodies in one assay. Final analysis of this microparticle based test is performed by dual-color flow cytometry, discriminating different colors and sizes of applied beads. Novel set-ups are using final fluorescence microscopic analysis, placing beads by sedimentation or magnetic placement at the bottom of reaction vessels for digital imaging analysis [16,19].

Immunological set-up in cardiology consists of manual or automated performance of ELISA detection of antigens of interest and a full-automated analysis by flow cytometric or microscopic principle (**Table 2**).

8. Profiling Technologies for miRNAs or Gene Expression—The Way to Organ-Specific and Systemic Diagnosis

Positive diagnostic findings in examined biopsies are confirming the myocardial condition, but negative results will not exclude infections or intramyocardial inflammation. Increasing number of biopsies will reduce biopsy-depending sampling error, but could never overcome it completely. This fact is indicating the need of global and stable biomarkers to identify disease entities in the effected organ preventing and overcoming sampling error (**Figure 2**).

First investigations indicate that deregulation of myocardial gene expression is not limited to focal biopsy area, and therefore allows the identification of specific disease situations of the examined patient analyzing neighboring biopsies [20]. It becomes difficult to diagnose complex diseases by one parameter. Current diagnostic biomarker panels are consisting of a few deregulated genes. Viral infections or massive infiltration of inflammatory cells in myocardium could be diagnosed in neighboring tissue samples without histological confirmation of these agents. We identified a gene pattern differentiating virally-induced or inflammatory cardiomyopathies like active myocarditis or giant cell myocarditis from inflammation-free or non-infected myocardium. Such disease specific profiles will change during effective treatment and thereby could also be applied for therapy monitoring.

miRNAs are small non-coding regulatory molecules (17 – 24 nt) and were recently identified as important regulators of genetic expression in myocardial tissue [10, 21]. They modulate gene expression by enhanced degradation of protein-coding mRNAs or sequestration from the translational apparatus. The role of miRNAs in physiologic and pathologic processes and the capability to correlate expression changes with disease states highlights their value as novel molecular biomarkers [22]. Currently available miRNA profiles allow the identification of preceding cardiotropic infection even in PCR-negative biopsies [10]. The clinical course of this patient is predictable at the point of primary biopsy and a disease-directed therapy could be initiated immediately to prevent myocardial injuries.

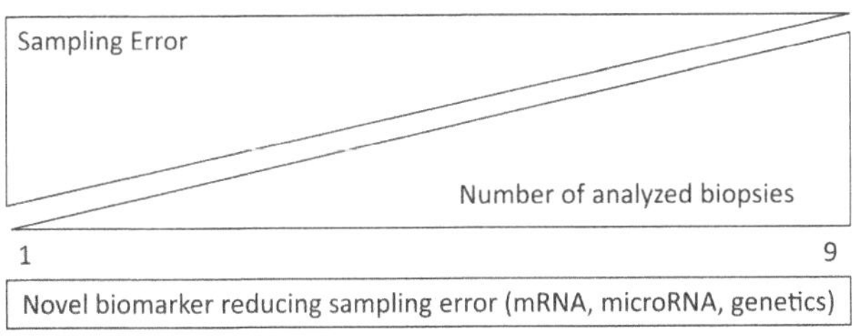

Figure 2. Biopsy-independent novel biomarkers for reduction of sampling error at a limited number of biopsies.

miRNAs have been demonstrated to play crucial roles in the etiology of a variety of common human diseases. They are involved in intra- and intercellular communication. miRNAs were detected not only in human tissue but also in serum or plasma influencing systemic response on organ-specific alterations. These findings emphasize the potential of circulating miRNAs as stable blood-based markers for the detection of cardiomyopathies [23]. The investigation of miRNA expression profiles in body fluids opens a completely new diagnostic approach to detect and characterize focal diseases by systemic markers with an easily taken and abundant sample. The detection of bladder cancer by determination of miRNA profiles in urine as an example for the development of non-invasive tests to detect human diseases is a tremendous benefit to both patients and healthcare systems [24,25].

Although the incidence of annual new cases suffering of cardiomyopathies is more than 25,000, only about 5000 patients are bioptated per year in Germany. Application of miRNA profiling in blood samples is an option to identify more patients with severe forms of cardiomyopathies before visiting a catheter lab. This blood based miRNA screening could reduce the disproportion of under diagnosed cardiac patients by strong indication for an endomyocardial biopsy to diagnose the origin of heart muscle disease.

Quantitative PCR for gene expression or miRNAs is the method of choice for profiling assays by high sensitivity and broad dynamic range of detection. Low density QPCR arrays or microplate based set-ups are convenient for diagnostic use of expression panels for up to 50 analytical parameters (**Table 2**) [26].

9. Proof of Disease-Related Gene Mutations for Cardiomyopathies

Current guidelines recommend genetic screening (evidence level A) for ARVD/C, HCM and DCM with conduction abnormalities or extra cardiac manifestations [27]. Although for many cardiomyopathies a genetic predisposition is shown, the genetic analysis is costly and time consuming yet therapeutically relevant for only a few diseases. Next generation sequencing (NGS) technologies will enable each lab to perform genetic testing for a low price but the interpretation of genetic data is still requesting great expertise. Now NGS is entering the diagnostic level. Advantage of de novo sequencing of all cardiac genes is the identification of temporary unknown mutations in patients with expressed heart failure problems.

Nevertheless recent studies identified a massive overpresentation of previously described cardiomyopathy-related genetic variants in a healthy control cohort with genotype prevalence thousand fold higher than in diseased patients [28]. This finding makes genetic testing without prior clinic-diagnostically definition questionable. Only an endomyocardial biopsy supported genetic study could identify genetic markers which are associated with occurrence of genetic cardiomyopathies excluding myocardial inflammation or microbial infections as reason for preceding heart muscle disease.

Drawback of all genetic testing is the trust on imaging techniques as primary cardiological diagnostics. A differential diagnosis of infiltrative processes of myocardium or wall hypertrophy in acute inflammatory diseases caused by oedema is requested besides a first documented myocardial wall hypertrophy or restrictive filling pattern in addition to the usually genetically, primary hypertrophic and restrictive cardiomyopathies. In biopsies of patients suspicious for HCM or ARVD were often found storage diseases, massive infiltration of inflammatory cells or viral genomes undetectable by echocardiography, CMR or CT. By additional laboratory and cardiac enzyme tests or tissue biopsies, it is usually possible to make a specific diagnosis. Morbus Fabry documented by measuring the alpha-galactosidase activity of leukocytes is well treated by enzyme replacement therapy. In the case of histologically diagnosed amyloidosis immunohistological classification of the corresponding subtype is required, for example to distinguish an age-related from a genetically determined ATTR form that requires family counseling [29].

In general, genetic testing is unable to detect viral infections or inflammation in myocardium. Nevertheless the enormous power of NGS to read out millions of sequences fragments in a very short period [30] will allow the parallel identification of pathological mutations in genomic DNA and the included microbial genomes in the same patient sample.

Routineously applied methods for confirmation of existing disease-relevant mutations are direct sequencing of small genomic regions harbouring the single nucleotide polymorphism (SNP) or PCR based detection of individual mutations, if the distance between neighbouring mutated site is too far for sequencing. Diagnostic sets of up to 50 SNP PCRs could be placed on low density arrays and allow the simultaneous analysis of multiple gene aberations in on assay [31] (**Table 2**).

10. Diagnostic-Based Treatment Decisions

Since a significant proportion of acute myocardial disease heals spontaneously and associated arrhythmias may regress completely, mechanical circulatory support and implantation of permanent pacemakers or automatic defibrillators (ICD's) should be only used if it is not possible to stabilize the patients by a symptomatic therapy. These measures are often the only therapeutic option for genetic cardiomyopathies.

Develop patients a rapidly progressive pump dysfunction despite optimal medical treatment, the prognosis can often be improved by rapid and sustained cause-oriented treatment. In acute heart failure, biopsy-based information is mandatory to recognize virus negative inflammation that requires immediate and tailored immunosuppression in order to reduce early mortality and improve long-term outcome. An immediate biopsy diagnostics with an immediately following initiation of therapy is particularly indicated in the eosinophil, the necrotic and giant cell myocarditis, which often respond well to early immunosuppressive treatment while untreated a very high mortality occurs (**Table 2**) [32,33].

In chronic inflammation, virus-negative patients with post-infectious or auto-immune inflammatory processes responded well to immunosuppression in early clinical trials. Virus-positive patients do not improve or even deteriorate upon anti-inflammatory treatment [34].

Clinical investigation in the last ten years revealed at least one therapeutic option for the four most important cardiotropic viruses (adenovirus, enterovirus, erythrovirus, HHV-6) with different grades of efficiency (**Table 1**). Myocardial injury is also occurring under optimal heart failure therapy progressive due to persistence of cardiotropic viruses. For anti-viral treatment currently two approaches are existing: immunomodulation supporting intrinsic immune response or antiviral drugs affecting life cycle of corresponding virus. Patients with chronic heart failure due to persistent enterovirus and adenovirus infections of the myocardium responded well to a six months' interferon-beta (IFN-β) course, but have untreated a really poor prognosis [15,35]. Complete elimination of viral genome was proven by follow-up biopsies taken three month after termination of the antiviral therapy [15]. Virus clearance was paralleled by an improvement of mean left ventricular function, a decrease in ventricular size, and amelioration of heart failure symptoms. Other viruses, e.g. parvovirus B19 or human herpesvirus 6, respond less well upon IFN-β treatment with respect to virus clearance. Despite incomplete virus clearance, these patients improved clinically following reduction of virus load and of endothelial function [36].

Erythrovirus (Parvovirus B19) is the most frequent cardiotropic virus. In 10% of all erythrovirus DNA positive patients an active replication is detectable by measurement of viral RNA [10]. Clinical symptoms of these patients are improving shortly after initialization of antiviral treatment by replication inhibitor Telbivudine (**Table 1**).

Valcytc is an effective antiviral drug for treatment of cytomegalovirus (CMV) infections in immunocompromised transplant patients. CMV is a member of herpesvirus family. Less than 1% of all bioptated patients are harboring ciHHV-6. This life-long infection is characterized by permanent generation of virus and resulting in a clinical situation similar to immunosuppression [10]. The treatment with Valcyte is not influencing the load of viral DNA in these patients but is reducing or eliminating the viral RNA hindering de novo synthesis of virus particles. Patient's general condition is improving by overcoming the immunocompromised situation. Valcyte should be an effective drug for conventional HHV-6 too, but corresponding trials are still missing due to low viral titers in myocardium and few patients with suspected HHV-6 myocarditis.

The efficiency of antiviral or immunosuppressive therapy may be reviewed only by a control biopsy. The absence of inflammatory markers, elimination of treated virus or at minimum a reduction of viral load in myocardium is a good predictor of successful treatment. Imaging techniques and physical activity testing are confirming clinical improvement.

Modern therapeutic options are based on inhibition of single gene expression by silencing RNAs or oligonucleotide-based therapies [37]. Blocking of one specific miRNA or miRNA groups is more effective by regulation of whole gene cascades or multiple gene families [38]. Current application of silencing mRNAs, antisense oligonucleotides or miRNA blocking molecules (antagomirs) are subject of intensive research but not available yet for routine drug treatment.

Novel biomarkers in myocardial tissue or peripheral blood for therapy monitoring will change their profile during therapy indicating the change of molecular marker on focal or systemic level.

11. Conclusions

General guidelines for diagnosis, treatment and corresponding quality analyses are missing for various, especially acquired cardiomyopathies [39]. The unequivocal diagnosis and thereupon based specific treatment in these patients is often only possible by direct examination of the myocardial tissue or special supplementary examinations, since the pathophysiological changes in these diseases occur at the cellular level. In this regard, therapy guiding diagnostics must be standardized and controlled.

In order to recognize whether an infectious disease may hide behind uncharacteristic cardiac symptoms and consequently proof a myocarditis unambiguously poses an important clinical challenge. A reliable diagnostic differentiation between virus-positive or virus-negative myocarditis and thereby indication of the infectious state needs to be accomplished to allow successful administration of a pathogen-tailored treatment strategy. This approach cannot be reached by routine clinical diagnosis but demands biopsy-based information with complete histological, immunohistochemical, and molecular work-up of tissue samples. Although EMB diagnostics are revealing only an extract of heart's current situation, this is still the gold standard for final confirmation of causative reasons of cardiomyopathies. This fact is indicating the need of global organ-specific or systemic markers which are disclosing the condition of the entire heart and therefore overcoming the sampling error.

Beside genetic markers for a limited subgroup of cardiomyopathies, novel biomarkers as miRNA and gene expression profiling are introduced in molecular examination of EMB presenting a global picture of heart muscle and overcoming the limitation of biopsy-focused diagnostics. These profiling approaches will improve the diagnosis of heart muscle disease and the monitoring of initiated therapeutic approaches in treatment of different forms of virally-induced or inflammatory cardiomyopathies by minimizing the sampling error. Measurement of disease-related miRNA profiles in peripheral blood could facilitate identification of currently undiagnosed cardiac patients by strong indication for endomyocardial biopsy examination.

12. Acknowledgements

Development of some diagnostic procedures and treatment strategies were supported by grants of the German Research Foundation (DFG), Transregional Collaborative Research Centre "Inflammatory Cardiomyopathy-Molecular Pathogenesis and Therapy" (SFB TR19, Z1), and two grants of the Federal Ministry of Education and Research (BMBF, Germany) for KMU innovative program (No.616 0315296, 0316141A).

REFERENCES

[1] P. P. Liu and H.-P. Schultheiss, "Myocarditis," In: Baunwalded, Ed., *Heart Disease*, 8th Edition, W. B. Saunders Co., Philadelphia, 2008, pp. 1775-1792.

[2] H.-P. Schultheiss, U. Kühl and L. T. Cooper, "The Management of Myocarditis," *European Heart Journal*, Vol. 33, No. 21, 2011, pp. 2616-2625.

[3] L. T. Cooper, K. L. Baughman, A. M. Feldman, *et al.*, "The Role of Endomyocardial Biopsy in the Management

of Cardiovascular Disease," *European Heart Journal*, Vol. 28, 2007, pp. 3076-3093.

[4] H. T. Aretz, "Myocarditis: The Dallas Criteria," *Human Pathology*, Vol. 18, No. 6, 1987, pp. 619-624.

[5] M. Noutsias, M. Pauschinger, K. Ostermann, F. Escher, J. H. Blohm, H.-P. Schultheiss and U. Kühl, "Digital Image Analysis System for the Quantification of Infiltrates and Cell Adhesion Molecules in Inflammatory Cardiomyopathy," *Medical Science Monitor*, Vol. 8, No. 5, 2002, pp. 59-71.

[6] J. C. Silverio, L. M. de-Oliveira-Pinto, A. A. da Silva, G. M. de Oliveira and J. Lannes-Vieira, "Perforin-Expressing Cytotoxic Cells Contribute to Chronic Cardiomyopathy in Trypanosomacruzi Infection," *International Journal of Experimental Pathology*, Vol. 91, No. 1, 2010, pp. 72-86.

[7] I. Kindermann, M. Kindermann, R. Kandolf, K. Klingel, B. Bültmann, T. Müller, A. Lindinger and M. Böhm, "Predictors of Outcome in Patients with Suspected Myocarditis," *Circulation*, Vol. 118, No. 6, 2008,pp. 639-648.

[8] A. Gulati, A. Jabbour, T. F. Ismail, K. Guha, J. Khwaja, S. Raza, K. Morarji, T. D. Brown, N. A. Ismail, M. R. Dweck, E. Di Pietro, M. Roughton, R. Wage, Y. Daryani, R. O'Hanlon, M. N. Sheppard, F. Alpendurada, A. R. Lyon, S. A. Cook, M. R. Cowie, R. G. Assomull, D. J. Pennell and S. K. Prasad, "Association of Fibrosis with Mortality and Sudden Cardiac Death in Patients with Nonischemic Dilated Cardiomyopathy," *JAMA*, Vol. 309, No. 9, 2013, pp. 896-908.

[9] U. Kühl, D. Lassner, M. Pauschinger, U. M. Gross, B. Seeberg, M. Noutsias, W. Poller and H.-P. Schultheiss, "Prevalence of Erythrovirus Genotypes in the Myocardium of Patients with Dilated Cardiomyopathy," *Journal of Medical Virology*, Vol. 80, No. 8, 2008, pp. 1243-1251.

[10] U. Kühl, M. Rohde, D. Lassner, U. M. Gross, F. Escher and H. P. Schultheiss, "miRNA as Activity Markers in Parvo B19 Associated Heart Disease," *Herz*, Vol. 37, No. 6, 2012, pp. 637-643.

[11] U. Kuhl, M. Pauschinger, M. Noutsias, *et al.*, "High Prevalence of Viral Genomes and Multiple Viral Infections in the Myocardium of Adults with 'Idiopathic' Left Ventricular Dysfunction," *Circulation*, Vol. 111, 2005, pp. 887-893.

[12] P. E. Pellett, D. V. Ablashi, P. F. Ambros, *et al.*, "Chromosomally Integrated Human Herpesvirus 6: Questions and Answers," *Reviews in Medical Virology*, Vol. 22, No. 3, 2012, pp. 144-155.

[13] C. T. Bock, K. Klingel and R. Kandolf, "Human Parvovirus B19-Associated Myocarditis," *The New England Journal of Medicine*, Vol. 362, No. 13, 2010, pp. 1248-1249.

[14] A. L. Caforio, F. Calabrese, A. Angelini, *et al.*, "A Prospective Study of Biopsy-Proven Myocarditis: Prognostic Relevance of Clinical and Aetiopathogenetic Features at Diagnosis," *European Heart Journal*, Vol. 28, No. 11, 2007, pp. 1326-1333.

[15] U. Kühl, D. Lassner, J. von Schlippenbach, W. Poller and H. P. Schultheiss "Interferon-Beta Improves Survival in Enterovirus-Associated Cardiomyopathy," *Journal of the American College of Cardiology*, Vol. 60, No. 14, 2012, pp. 1295-1296.

[16] A. Willitzki, R. Hiemann, V. Peters, U. Sack, P. Schierack, S. Rödiger, U. Anderer, K. Conrad, D. P. Bogdanos, D. Reinhold and D. Roggenbuck, "New Platform Technology for Comprehensive Serological Diagnostics of Autoimmune Diseases," *Clinical and Developmental Immunology*, Vol. 2012, 2012, Article ID: 284740.

[17] N. Deubner, D. Berliner, A. Schlipp, G. Gelbrich, A. L. Caforio, S. B. Felix, M. Fu, H. Katus, C. E. Angermann, M. J. Lohse, G. Ertl, S. Störk, R. Jahns, Etiology, Titre-Course and Survival-Study Group, "Cardiac Beta1-Adrenoceptor Autoantibodies in Human Heart Disease: Rationale and Design of the Etiology, Titre-Course, and Survival (ETiCS) Study," *European Journal of Heart Failure*, Vol. 12, No. 7, 2010, pp. 753-762.

[18] C. Skurk, F. Wittchen, L. Suckau, H. Witt, M. Noutsias, H. Fechner, H.-P. Schultheiss and W. Poller, "Description of a Local Cardiac Adiponectin System and Its Deregulation in Dilated Cardiomyopathy," *European Heart Journal*, Vol. 29, No. 9, 2008, pp. 1168-1180.

[19] K. L. Kellar, R. R. Kalwar, K. A. Dubois, D. Crouse, W. D. Chafin and B. E. Kane, "Multiplexed Fluorescent Bead-Based Immunoassays for Quantitation of Human Cytokines in Serum and Culture Supernatants," *Cytometry*, Vol. 45, No. 1, 2001, pp. 27-36.

[20] F. Wittchen, L. Suckau, H. Witt,C. Skurk, D. Lassner, H. Fechner, I. Sipo, U. Ungethüm, P. Ruiz, M. Pauschinger, C. Tschope, U. Rauch U, Kühl, H.-P. Schultheiss and W. Poller, "Genomic Expression Profiling of Human Inflammatory Cardiomyopathy (DCMi) Suggests Novel Therapeutic Targets," *European Heart Journal*, Vol. 85, No. 3, 2007, pp. 257-271.

[21] S. V. Naga Prasad, Z. H. Duan, M. K. Gupta, *et al.*, "Unique microRNA Profile in End-Stage Heart Failure Indicates Alterations in Specific Cardiovascular Signaling Networks," *The Journal of Biological Chemistry*, Vol. 284, 2009, pp. 27487-27499.

[22] D. Baek, J. Villen, C. Shin, F. D. Camargo, S. P. Gygi and D. P. Bartel, "The Impact of microRNAs on Protein Output," *Nature*, Vol. 4355, No. 7209, 2008, pp. 64-71.

[23] P. S. Mitchell, R. K. Parkin, *et al.*, "Circulating microRNAs as Stable Blood-Based Markers for Cancer Detection," *PNAS*, Vol. 105, 2008, pp. 10513-10518.

[24] G. Wang, E. S. Chan, B. C. Kwan, P. K. Li, S. K. Yip, C. C. Szeto and C. F. Ng, "Expression of miRNAs in the Urine of Patients with Bladder Cancer," *Clinical Genitourinary Cancer*, Vol. 10, No. 2, 2012, pp. 106-113.

[25] P. Puerta-Gil, R. Garcia-Baquero, A. Y. Jia, S. Ocana, M. Alvarez-Mugica, J. L. Alvarez-Ossorio, C. Cordon-Cardo, F. Cava and M. Sanchez-Carbayo, "miR-143, miR-222, and miR-452 Are Useful as Tumor Stratification and Noninvasive Diagnostic Biomarkers for Bladder Cancer," *American Journal of Pathology*, Vol. 180, No. 5, 2012, pp. 1808-1815.

[26] E. V. Grigorenko, E. Ortenberg, J. Hurley, A. Bond and K. Munnelly, "miRNA Profiling on High-Throughput OpenArray™ System," *Methods in Molecular Biology*, Vol. 676, 2011, pp. 101-110.

[27] R. E. Hershberger, J. Lindenfeld, L. Mestroni, C. E. Seidman, M. R. Taylor and J. A. Towbin, "Genetic Evaluation of Cardiomyopathy: A Heart Failure Society of America Practice Guideline," *Journal of Cardiac Failure*, Vol. 15, No. 2, 2009, pp. 83-97.

[28] C. Andreasen, J. B. Nielsen, L. Refsgaard, A. G. Holst, A. H. Christensen, L. Andreasen, A. Sajadieh, S. Haunsø, J. H. Svendsen and M. S. Olesen, "New Population-Based Exome Data Are Questioning the Pathogenicity of Previously Cardiomyopathy-Associated Genetic Variants," *European Journal of Human Genetics*, 2013.

[29] D. Lassner, M. Rohde, U. M. Gross, F. Escher, H. P. Schultheiss, R. P. Linke and U. Kühl, "Classification of Four Chemically Different Amyloid Types in Routine Endomyocardial Biopsies by Advanced Immunohistochemistry," *Amyloid*, Vol. 18, Suppl. 1, 2011, pp. 76-78.

[30] D. A. Wheeler, M. Srinivasan, M. Egholm, *et al.*, "The Complete Genome of an Individual by Massively Parallel DNA Sequencing," *Nature*, Vol. 452, No. 7189. 2008, pp. 872-876.

[31] K. Hopp, K. Weber, D. Bellissimo, S. T. Johnson and B. Pietz, "High-Throughput Red Blood Cell Antigen Genotyping Using a Nanofluidic Real-Time Polymerase Chain Reaction Platform," *Transfusion*, Vol. 50, No. 1, 2010, pp. 40-46.

[32] L. J. Cooper, J. M. Hare, H. D. Tazelaar, *et al.*, "Usefulness of Immunosupp. Ression for Giant Cell Myocarditis," *American Journal of Cardiology*, Vol. 102, 2008, pp. 1535-1539.

[33] A. Frustaci, C. Chimenti, F. Calabrese, et al., "Immunosupp. Ressive Therapy for Active Lymphocytic Myocarditis: Virological and Immunologic Profile of Responders versus Nonresponders," *Circulation*, Vol. 107, 2003, pp. 857-863.

[34] A. Frustaci, M. A. Russo and C. Chimenti, "Randomized Study on the Efficacy of Immunosupp. Ressive Therapy in Patients with Virus-Negative Inflammatory Cardiomyopathy: The TIMIC Study," *European Heart Journal*, Vol. 30, No. 16, 2009, pp. 1995-2002.

[35] U. Kuhl, M. Pauschinger, P. L. Schwimmbeck, B. Seeberg, C. Lober and M. Noutsias, "Interferon-Beta Treatment Eliminates Cardiotropic Viruses and Improves Left Ventricular Function in Patients with Myocardial Persistence of Viral Genomes and Left Ventricular Dysfunction," *Circulation*, Vol. 107, No. 22, 2003, pp. 2793-2798.

[36] C. Schmidt-Lucke, F. Spillmann, T. Bock, U. Kühl, S. Van Linthout, H.-P. Schultheiss and C. Tschöpe "Interferon Beta Modulates Endothelial Damage in Patients with Cardiac Persistence of Human Parvovirus b19 Infection," *The Journal of Infectious Diseases*, Vol. 201 No. 6, 2010, pp. 936-945.

[37] T. M. Wheeler, A. J. Leger, S. K. Pandey, A. R. MacLeod, M. Nakamori, S. H. Cheng, B. M. Wentworth, C. F. Bennett and C. A. Thornton, "Targeting Nuclear RNA for *in Vivo* Correction of Myotonic Dystrophy," *Nature*, Vol. 488, No. 7409, 2012, pp. 111-115.

[38] T. G. Hullinger, R. L. Montgomery, A. G. Seto, B. A. Dickinson, *et al.*, "Inhibition of miR-15 Protects against Cardiac Ischemic Injury," *Circulation Research*, Vol. 110, No. 1, 2012, pp. 71-81.

[39] L. T. Cooper, K. L. Baughman, A. M. Feldman, A. Frustaci, M. Jessup, U. Kuhl, G. N. Levine, J. Narula, R. C. Starling, J. Towbin and R. Virmani, American Heart Association, American College of Cardiology, European Society of Cardiology, Heart Failure Society of America and Heart Failure Association of the European Society of Cardiology, "The Role of Endomyocardial Biopsy in the Management of Cardiovascular Disease: A Scientific Statement from the American Heart Association, the American College of Cardiology, and the European Society of Cardiology. Endorsed by the Heart Failure Society of America and the Heart Failure Association of the European Society of Cardiology," *Journal of the American College of Cardiology*, Vol. 50, No. 19, 2007, pp. 1914-1931.

Abbreviations

ciHHV-6 = chromosomally integrated human herpesvirus 6,
CMR = cardiac magnetic resonance,
DCM = Dilated Cardiomyopathy,
EMB = endomyocardial biopsy,
ESC = European Society of Cardiology,
miRNA = microRNA,
MVD = microvessel density,
NGS = next generation sequencing,
PBMC = peripheral blood mononuclear cells,
PCR = polymerase chain reaction,
QPCR = quantitative PCR,
SNP = single nucleotide polymorphism.

NAA Algerian Laboratory Evaluation Processed by WEPAL and IAEA during 2011-2012

Lylia Hamidatou, Hocine Slamene, Tarik Akhal, Samir Begaa, Mohammed Messaoudi, Zouranen Boussaad

Neutron Activation Analysis Department, Nuclear Research Centre of Birine, Djelfa, Algeria.

ABSTRACT

Participation in inter-comparison programs is an important process to enhance the accuracy and precision of the analytical techniques used in the laboratories. At Es-Salam Research Reactor, our laboratory uses two functional analytical techniques such as INAA and k_0-INAA in the research projects and for the completion of the service works. Recently, our laboratory has participated in two inter-comparison tests organized by IAEA within the framework of the AFRA project to assess the analytical performance of eighteen (18) African analytical laboratories participating in the RAF/4/022 project, Enhancement of Research Reactor Utilization and Safety by taking part in analytical proficiency testing IAEA in conjunction with WEPAL, the Wageningen Evaluating Programs for Analytical Laboratories. The Proficiency Testing tests related to the determination of major, minor and trace elements in materials of the International Soil and Plant Analytical Exchange (Wepal codes ISE, IPE), were held in the fourth quarter of 2011 and the first quarter of 2012, respectively. Concentrations of thirty (30) elements such as Al, As, Ba, Br, Ca, Cd, Ce, Co, Cr, Cs, Fe, Ga, K, La, Mg, Mo, Mn, Na, Nd, Rb, Sb, Sc, Sr, Th, Ti, U, V, W, Zn, Zr were determined in ISE and IPE samples. All results were evaluated and reported in this paper.

Keywords: Laboratory; Evaluation; NAA Method; Performance; Accreditation; Proficiency Testing

1. Introduction

The Algerian research reactor (Es-Salam) is a 15 MW heavy water reactor type, operating since 1992. Instrumental neutron activation analysis (INAA) was introduced in 1993 at the laboratory of the neutron activation analysis department, in Nuclear Research Centre of Birine, Algeria. In our research reactor, the k_0-method was successfully developed using the HØGDAHL convention and WESTCOTT formalism [1,3]. In 2012, Hamidatou L et all reported in this paper "k_0-NAA quality assessment in an Algerian laboratory by analysis of SMELS and four IAEA reference materials using Es-Salam Research reactor" the internal quality control of the k_0-NAA technique [4]. The concept of QC/QA, internal and external validation is considered as an advanced stage in the life cycle of an analytical method. This technique was used for practical applications in several areas of life such as: nutrition, health related to environmental field, mining and industrial fields [5-7].

Recently, there is a growing interest in the use of the NAA method for trace element determination in the industrial sector. The richness of the service analytics in our NAA laboratory has been in the diversity of solid samples. The dominant method used has been neutron activation analysis at Es-Salam research reactor during 20 years of its activity. The viewpoint is that the neutron activation analysis remains the best tool to determine the contents of trace elements in all types of samples [8,9].

At Es-salam research reactor, the INAA and k_0-NAA techniques are fully applied by researchers. In addition, the development of other analytical techniques based on the neutron activation is also an achievement such as the cyclic delayed neutron counting technique and the application of RNAA to the proportioning of iodine in food salt [10,11]. The NAA laboratory work is to make an effort to connect the unique features of NAA activities in a strategic way for the national goals. A reactor centre, whether it is in a university or other institution, is a substantial investment in the human and financial resources of a developing nation.

The current status of the NAA method, the analytical procedures and the nuclear activities of our laboratory are presented in the chapter "Concepts, Instrumentation and Techniques of Neutron Activation Analysis" newly published in the book: Imaging and Radioanalytical Techniques in Interdisciplinary Research—Fundamentals and Cutting-edge Applications [9].

A regular independent assessment of the technical performance of a laboratory is necessary to assure the validity of measurements or tests, and should be part of an overall quality strategy. The main objective of a PT scheme is to help the participant to assess the accuracy [12] of its measurements. In addition, participation in an appropriate PT scheme is required for laboratories seeking recognition of their competence through accreditation against the standard ISO/IEC 17025 [13] or ISO 15189 [14]. In some sectors participation in specific schemes can be mandatory [15].

PT schemes are operated for the benefit of the laboratory participants. However, other parties also have a legitimate interest in PT schemes. These include customers of analytical laboratory services, accreditation bodies, regulatory authorities and other end-users of the laboratory results. The authors, Isabel Taverniers and all reported in the publication "Trends in quality in the analytical laboratory: Analytical method validation and quality assurance", the different approaches to validation, gives attention to the different characteristics of method performance with the concepts of single-laboratory, or in house validation, inter-laboratory or collaborative study [16]. The basic use of PT is to assess the performance for the conduct of specific measurements and calibrations.

This paper focused our experience on the inter-laboratory exercises undertaking for the evaluation of the analytical laboratory competency during the last decade in particular for the two new Proficiency Testing rounds organized by the IAEA under the RAF/4/022 project [17] in conjunction with WEPAL, the Wageningen Evaluating Programs for Analytical Laboratories (www.wepal.nl) [18].

2. Synopsis Competency Tests Participation of the NAA-CRNB Laboratory

Our NAA research laboratory was participated in several inter-comparison tests organized in the frame work of AFRA Research Reactor Projects. In the first participations, the work of analysis is based on the inter-comparison results of African laboratories. The last PT testing participation was to evaluate our research laboratory among a great number reached 500 laboratories on an international scale. This contribution was carried out in the PT program launched by WEPAl organisation. In this context, a brief description was presented for all participation.

2.1. The AFRA IV-7 Proficiency Test 2003 Using CMIP-P1 Rock Materials

Our first participation using NAA laboratory on the proficiency test was carried out in the frame work of the AFRAIV-7 project named CMIP-AFRA-P1 Test 2003. It was requested from the laboratories to determine the trace and minor elements in geological materials. The proficiency test study was organized by the CSIR-NML in collaboration with IAEA and NECSA for the laboratories of the Research Reactor Centres in Africa. About 11 laboratories from 5 countries were participated in this study for the determination of 16 trace and minor elements (Au, Ca, Cd, Co, Cr, Cs, Fe, K, Mn, Na, Ni, Pb, Rb, Th, U, Zn) in a rock-type test material. The analytical methods used in this study are INAA, EDXRF, GFAAS, IC and ICP-MS. This work aimed to illustrate the comparability of results produced by the participating laboratories [19]. The results of the proficiency testing exercises were evaluated by Experts to the IAEA and a follow-up Inter-comparison feedback Meeting was held in Algeria during 29 Nov-3 Dec 2003 [20].

2.2. The AFRA IV-7 Proficiency Test 2005 Using Medicinal Plants and Aquatic-Biomonitors Materials

Two years later, AFRA IV-7 (2005) was the second Proficiency Test within the AFRA IV-7 Research Reactor Project for Socio-economic development. Thirteen laboratories from eight countries were participated in this study. The laboratories were requested to analysis 31 elements in five biological materials that were distributed by the project. One of the distributed samples was an IAEA reference material (IAEA-155, Whey powder) to check the results of the nuclear analytical technique applied for all five materials. The focus of this project was on the evaluation of foodstuffs for their nutritive values, the determination of the element composition of medicinal plants and the analysis of aquatic-biomonitors for environmental monitoring, using nuclear and related analytical techniques, with neutron activation analysis as the common technique [21]. The results of the proficiency testing exercises were evaluated by Experts to the IAEA and a follow-up Inter-comparison feedback Meeting was held in Libya during 2005 [22].

2.3. The AFRA IV-12 Proficiency Test 2007 Using Foodstuff, Medicinal Plants and Aquatic Biomonitors Materials

The AFRAIV-12 Inter-laboratory Analysis Test 2007 was the third inter-laboratory analysis test within that project. The laboratories were asked to analyze about 43 essential and toxic elements using nuclear and related analytical techniques, with neutron activation analysis as the common technique. Five materials were distributed to the participants representing foodstuff, medicinal plants and aquatic biomonitors according to the focus of the RAF/4/020 AFRA IV-12 (2006-2010) project [23].

For the first time, Es-Salam research reactor of Algeria had the opportunity to take part with two techniques INAA and k_0-INAA for which the k_0-NAA technique was newly developed (2004) in our department. Indeed, two sets of independent results obtained by INAA and k_0-INAA techniques were sent for evaluation. Each technique was used as an analytical tool regarded as an independent laboratory while associating a number and a particular code in event the Algerian laboratories with CRNB/INAA and CRNB/k_0-NAA, respectively. The participating laboratories met to discuss their results in the presence of the experts of IAEA. All results obtained by the African analytical laboratories were evaluated by expects of the IAEA and the meeting for was held in Kenitra, Morocco from 14 to 18 July 2008 to discuss the results evaluations [24,25].

2.4. WEPAL PT Scheme 2011.4-2012.1 Rounds in Conjunction with IAEA/AFRA Project

The goal of this paper focuses the assessment of our scientific activities through the PT evaluation of the NAA laboratory. The principal actions carried out within the framework of the RAF/4/022 project: "Enhancement of Research Reactor Utilization and Safety" led to the evaluation of the African laboratories with the advantage to participate on the PT program of WEPAL. The Proficiency Testing rounds related to the determination of (trace) elements in soil and plant materials were held by WEPAL for which 4 ISE and 4 IPE samples have been used in each round during 2011/4 and 2012/1, respectively. WEPAL, a provider of such Inter-labortaory comparison schemes, is accredited by the Dutch Council for Accreditation, for compliance with the International Standard ISO17043:2010. The activity within IAEA AFRA RAF 4/022 coincided with similar assessments amongst European and Latin American analytical laboratories under RER 1007 and RLA 0037, respectively.

In this study, we aim to evaluate of the competency of our laboratory among the proficiency test enhancement of low and medium power research reactor (RR) utiliztion is often pursued by increasing the neutron activation analysis (NAA) activities. Over the years, the IAEA has stimulated the orientation of NAA groups worldwide on fields of application in which large amounts of samples may exist for analysis. Whereas the markets for NAA laboratories may have been identified, an underestimated problem remains the quality assurance and quality control (QA&QC), which limits tremendously the commercial routine application of this powerful technique.

Indeed, the IAEA facilitated during 2011-2012 laboratories, participating in IAEA/AFRA project RAF/4/022 to assess their analytical performance by taking part in analytical proficiency testing IAEA in conjunction with WEPAL, the Wageningen Evaluating Programs for Analytical Laboratories. Two such rounds have been organized, and the results evaluated by the IAEA supported by Experts in nuclear and non-nuclear analytical techniques. Potential sources of error, technical and organizational, were identified for the individual laboratories, action plans for improvement were drafted and accepted, and the Agency was recommended continuing this approach for proficiency testing and evaluation as all participants highly valued the outcome of it.

The 18 participants were representatives of research reactor (RR) operating facilities and analytical laboratories in 9 African Member States (MSs). Of these, 6 are with operational RRs: Algeria, Egypt, Ghana, Morocco and South Africa; the other 3 MSs were: DRC, Libya and Sudan. Others participants were Nigeria, Cameroon Madagascar, Tanzania and Tunisia. The IAEA was supported in the evaluation and discussion of the results by two Experts.

Two functional analytical techniques applied in the NAA laboratory of Es-Salam research reactor are INAA and k_0-INAA. Our vision is to take benefit of these rounds to assess the competency of our laboratory through the use the instrumental and the k_0-standardized neutron activation analysis. Indeed, two techniques were applied to provide the analytical results. About 31 elements were analyzed in different samples of soil and plant materials of WEPAL. The unique possible approach is to make combination between INAA and k_0-INAA results. It means that the final data sheets provided by our laboratory to WEPAL organism include some results obtained by k_0-INAA technique and other by INAA technique. The results of analysis must be reported according to the specific forms for IPE and ISE reports. During the meeting on Inter-Comparison Feedback of NAA and other Analytical Techniques Proficiency Tests, AIEA Project/ RAF4/022—Enhancing Research Reactor Utilization and Safety, 04-08 June 2012, Tunis, Tunisia, all participating laboratories have been presented their results with the same IAEA proposed form [26].

3. Materials and Methods

3.1. IPE and ISE WEPAL Samples

The global program of WEPAL included six sets of samples: ISE (International Soil-Analytical Exchange), IPE (International Plant-Analytical Exchange), SETOC (International Sediment Exchange for Tests on Organic Contaminants), MARSEP (International Manure and Refuse sample Exchange Program), BIMEP (International Biomass Exchange Program) and Q (Quasimeme Laboratory Performance Studies organic contaminants, metals, nutrients in seawater, sediment and biota).

Each quarter, sample kits are mailed to participating laboratories and data are analyzed to evaluate performance. Each mailing and subsequent data analysis is referred to as a round. The reference samples included in a kit are designated to cover a variety of analytical methods used by laboratories.

In the case of the AFRA IV-12-RAF4/022 project within the financial support of IAEA a total of sixteen materials were received from WEPAl organism during 2011-4 and 2012-1 periods. The materials representing different types of soil and plant named ISE and IPE. In each round the samples ISE and IPE are labelled from one to four. **Table 1** illustrates the analyzed WEPAL samples covering plants and soils. Before receiving the WEPAL materials, all samples used in these rounds of the proficiency test have passed the homogeneity test [27-30].

3.2. Instrumental Neutron Activation Analysis Method, INAA

The unknown sample is irradiated together with a calibrator containing a known amount of the element(s) of interest. The calibrator is measured under the same conditions as the sample (sample-to-detector distance, equivalent sample size and if possible equivalent in composition). From comparison of the net peak areas in the two measured spectra the mass of the element of interest can be calculated [9]:

$$m_{x(unk)} = m_{x(cal)} \cdot \frac{\left(\dfrac{N_p}{t_m e^{-\lambda t_d} \cdot \left(1-e^{-\lambda t_m}\right)}\right)_{unk}}{\left(\dfrac{N_p}{t_m e^{-\lambda t_d} \cdot \left(1-e^{-\lambda t_m}\right)}\right)_{cal}} \quad (1)$$

in which $m_{x(unk)}$, $m_{x(cal)}$ mass of the element of interest, in unknown sample and calibrator, respectively; in g unit. In this procedure many of the experimental parameters—such as neutron fluence rate, cross section and photopeak efficiency cancel out at the calculation of the mass and the remaining parameters are all known. This calibration procedure is used if the highest degree of accuracy is required. The relative calibration on basis of element calibrators is not immediately suitable for laboratories aiming at the full multi-element powers of INAA. It takes considerable effort to prepare multi-element calibrators for all 70 elements measurable via NAA with adequate degree of accuracy in a volume closely matching the size and the shape of the samples.

3.3. The k_0-Standardization Neutron Activation Analysis, k_0-INAA

The k_0-based neutron activation analysis (k_0-NAA) technique, developed in 1970s, is being increasingly used for multielement analysis in a variety of matrices using reactor neutrons. In the k_0-INAA the evaluation of the analytical result is based on the so-called k_0-factors that are associated with each gamma-line in the gamma-spectrum of the activated sample. These factors replace nuclear constants, such as cross sections and gamma-emission probabilities, and are determined in specialized NAA laboratories. This technique has been reported to be flexible with respect to changes in irradiation and measuring conditions, to be simpler than the relative comparator technique in terms of experiments but involves more complex formulae and calculations, and to eliminate the need for using multi-element standards. The concentration of an element can be determined as:

$$\rho_x(\text{ppm}) = \frac{\left[\dfrac{C/t_m}{SDCW}\right]_x}{\left[\dfrac{C/t_m}{SDCW}\right]_{Au}} \cdot \frac{1}{k_{0,Au(x)}} \cdot \frac{G_{th,Au} f + G_{epi,Au} Q_{0,Au(\alpha)}}{G_{th,x} f + G_{epi,x} Q_{0,x(\alpha)}} \cdot \frac{\epsilon_{p,Au}}{\epsilon_{p,x}} \times 10^6 \quad (2)$$

where the indices x and Au refer to the sample and the monitor, respectively; WAu and Wx represent the mass of the gold monitor and the sample (in g); Np is the measured peak area, corrected for dead time and true coincidence; S, D, C are the saturation, decay and counting factors, respectively; tm is the measuring time; Gth and Gepi are the self shielding correction for the thermal and

Table 1. List of plant and soil samples analyzed in this work.

Type of matrials	IPE	ISE
2011.4[a]	Leek (allim porrum), Lucerne (Medicago Savitum), Oil palm leaves (Elaeis Guineensis), Grass gr94 (Poaceae)	sandy soil, calcareous brown soil, clay, river clay
2012.1[b]	Maize (Zea mays), Banana leaves (Musa Sapientum) and Beech Leaf (Fugus Sylvatica).	clay soil, loamy soil, sandy clay soil

[a]Oct-Dec 2011. [b]January-March 2012.

epithermal neutrons, respectively.

4. Experimental Procedure

Triplicates were prepared from each sample of IPE and ISE. To be able to exploit both techniques INAA and k_0-INAA in the same experiment, it is necessary to prepare the CRMs, RMs and gold monitors. In this work, the certified reference materials like CRM-GSV4, CRM-NIST-1573a, CRM-GSD12 and reference material such as RM-AIEA-soil7 were used to determine the elemental concentrations by relative technique. In the other hand, the analytical results obtained of CRMs are the subject of internal quality control procedure.

All samples, monitors and standards were placed in clean cylindrical aluminium vials for long irradiation. The gold monitors Al-0.1% Au alloy wire of 0.5 mm diameter provided by IRMM (Institute for Reference Materials and Measurements, Belgium) were prepared for k_0-INAA utilisation. Data acquisition is performed using-Genie-2000 software from Canberra [31]. All spectra are deconvoluted with the commercial software HyperLab 2005 [32] with resulting to files peak table (*.PTF) and a spectrum (*.SPE) which are used as input for KayWare software [33]. This is a specific software package developed for k_0-based NAA at DSM Resolve in the 1990s and at present commercialized by INW Gent and k_0-Ware in the Netherlands.

5. Results and Discussion

The PT program is designated to complete, not replace, a participating laboratory's internal quality control (QC) program. In addition, the PT program may identify analytical problems when the laboratory's internal controls do not submit an application. To get reproducible results, Wepal recommends to the participants the determination of the moisture content in each sample. The procedure is explained in the quarterly IPE-ISE reports of WEPAl for which the materials must be drying during at least 3 hours at 105°C and cooling down in desiccator before weighing. The dry mass factors were determined for all plant and soil WEPAL samples. In our analytical techniques, the moisture content is directly introduced as an input values in the KayWin program.

The results are reported for the RAF/4/022 inter-laboratory proficiency test 2011-2012 and consequently evaluated by WEPAl. The final treatment of results for each type of samples ISE and IPE has to generate some parameter statistics such as: mean, standard deviation, median, MAD and Z-score.

The assessment of the PT rounds participation during 2011-4 and 2012-1 has been evaluated according to the sophisticated statistical methods which are required to obtain meaningful assessments. The model that is chosen calculates population characteristics (mean and standard deviation) from experimental datasets. The model uses an estimate for the probability density function of the measurement process and calculates a best fit based on all observed values. The analytical results of the analyzed materials are presented by three ways per element, sample and participant (laboratory) in the quarterly and annual reports ISE and IPE of WEPAL.

In this study, the evaluated results obtained by our laboratory were carried out by WEPAL for which the proficiency tests of the participant laboratories were processed by the tool of Z-scores. Using this approach, all Z-score values of our laboratory were presented graphically in the **Figures 1-4**. According to the evaluation processed by the PT provider the majority of our results correspond to Z-scores lower than 2.

5.1. PT 2011.4 Round

Elemental concentrations obtained by relative and k_0-NAA methods in eight samples of ISE and IPE materials are expressed by Z-scores accordingly the **Figures 1** and **2**. In this test, 30 elements (Al, As, Ba, Br, Ca, Cd, Ce, Co, Cr, Cs, Fe, Ga, K, La, Mg, Mn, Mo, Na, Nd, Rb, Sb, Sc, Se, Sr, Th, Ti, U, V, W, Zn, Zr) are determined in all WEPAL soil and plant samples.

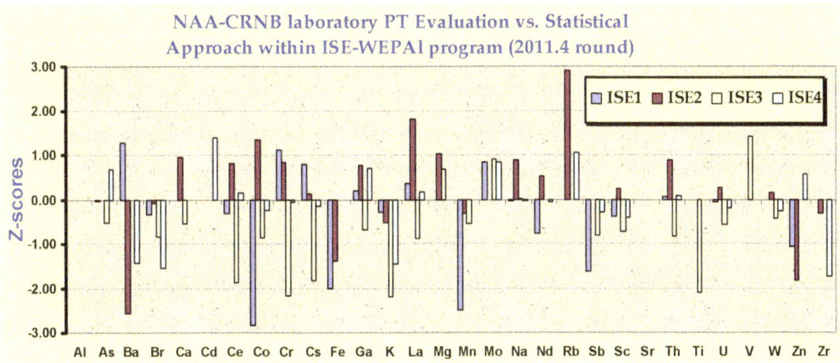

Figure 1. Graphical PT evaluation of the overall z-score results obtained by NAA-CRNB laboratory for 31 elements analyzed in four ISE-WEPAL materials.

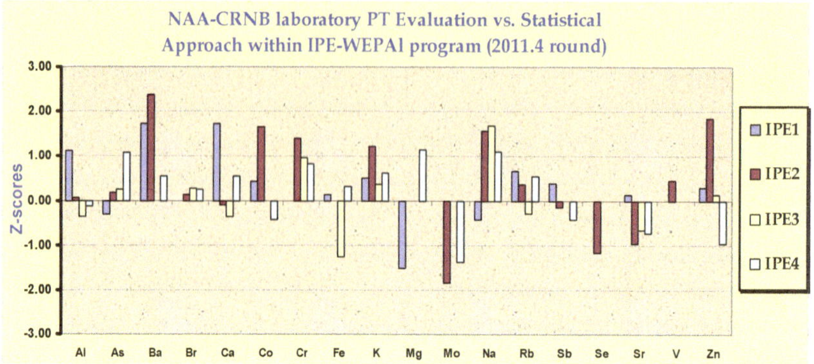

Figure 2. Graphical PT evaluation of the overall z-score results obtained by NAA-CRNB laboratory for 18 elements analyzed in four IPE-WEPAL materials.

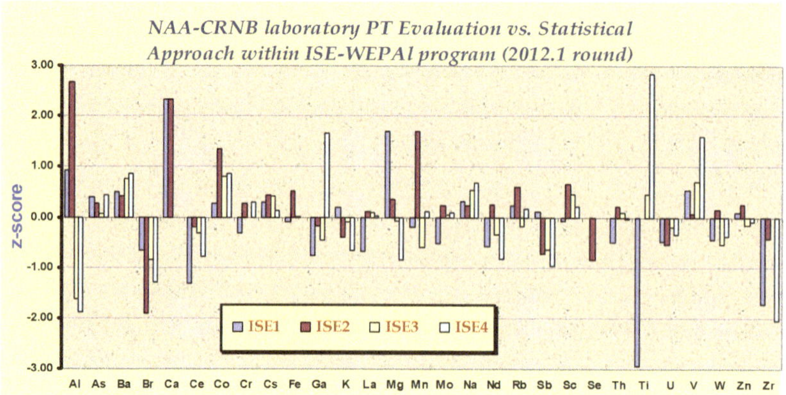

Figure 3. Graphical PT evaluation of the overall z-score results obtained by NAA-CRNB laboratory for 29 elements analyzed in four ISE-WEPAL materials.

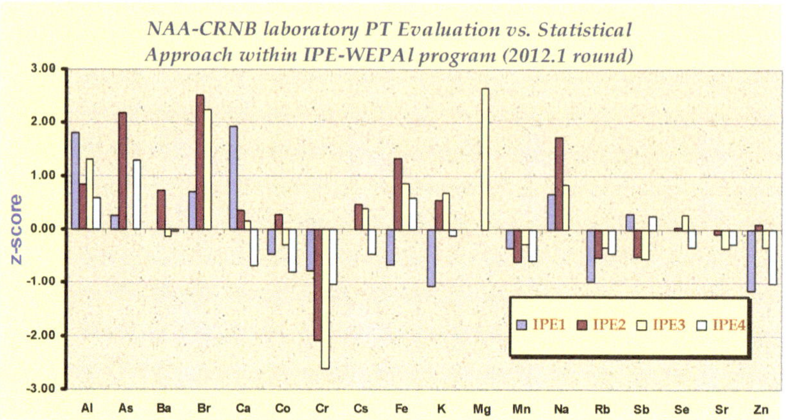

Figure 4. Graphical PT evaluation of the overall z-score results obtained by NAA-CRNB laboratory for 18 elements analyzed in four International Plant Analytical exchange IPE-WEPAL materials.

Two diagrams (**Figures 1** and **2**) illustrate clearly the quality of our results. These allow extracting numerous information noted bellow:

- About 182 concentration values for plant and soil materials are determined by k_0-INAA and INAA methods for the first PT,
- Among the total number of results, 70 and 112 values are reported for IPE and ISE samples, respectively,
- The evaluation of all results revealed that 135 good results having Z-scores lower than 2 from 182 values for which the total success rate is 74.2%.
- In term of biological matrices like IPE, we find 55 good results having a Z-score lower than 2 from 70 data giving about 78.575% of achievement.

- For ISE samples, the same observations are noted. The rate of achievement is 71.43% based on 80 good results from 112 values.

5.2. PT 2012.1 Round

The same interpretations can be made for the second PT 2012.1 round. The comparison between two participating proficiency tests gives a significant improvement of our contribution. The number of accepted values determined by NAA method in the second test is relatively higher than the first. **Figures 3** and **4** presented the distribution of Z-scores versus analyzed elements.

The success rate became more important during our second participation in PT of provider. This indicates that the system calibration like calibration of weighing balance, irradiation channels and HPGe detectors were obtained carefully. In a short-term projection our laboratory seeking accreditation, these measures are an important aspect of the requirements. PT is one of these measures, as well as for example, the use of (certified) reference materials. PT plays a highly valuable role as it provides an objective evidence of the competence of the participant. This evidence can be used to improve the performance of the participant and/or give confidence in the participant's ability to perform a specific measurement (**Table 2**).

During the participation in PT schemes, it is very interesting to assess the quality of obtained results and to make comparison at different stage. In the **Figure 5**, the comparison was presented. The comparison between two PT round 2011.4 and 2012.1 revealed that the number of satisfactory results expressed by accepted Z-scores are very important in each round. In addition, an apparent improvement was observed for the second test. This indicates the role of researcher in the improvement of the analytical process.

6. Conclusions

One of the basic elements in all PT is the evaluation of the performance of each participant. In order to do so, the PT provider has to establish two values, which are used for the performance evaluation. It is clear that the PT plays a highly valuable role as it provides an objective evidence of the competence of the participant. This evidence can be used to improve the performance of the participant and/or give confidence in the participant's ability to perform a specific measurement.

This paper presents the general state of the NAA analytical laboratory with the considerable efforts of NAA group using Es-Salam research reactor facility.

Table 2. Experimental conditions used in this work.

	Short irradiation channel		Long irradiation channel	
Analyzed material	ISE GSD12 Soil7	IPE, GSV4 NIST1573a	ISE, GSD12 Soil7	IPE, GSV4 Lichen336
Neutron spectrum parameters α, f, Φ_{th}	−0.4, 159.42, 2.26 10 10^{+14}		0.027 28.83 4.7 10^{+12} [2]	0.045 20.5 3.76 10^{+13} [1]
Irradiation time	30 sec		4H	8H
Decay time	10 - 20 mn		4 - 6 d 11 - 12 d	6 - 8 d 12 - 16 d
Counting time	500 sec		5400 sec/1st measurement 7200 sec/2nd measurement	
Measured elements	Al, Mg, Mn, Ti, V		As, Ba, Br, Ca, Cd, Ce, Co, Cr, Cs, Fe, Ga, K, La, Mo, Na, Nd, Rb, Sb, Sc, Sr, Th, U, W, Zn, Zr	As, Ba, Br, Ca, Co, Cr, Cs, Fe, K, Mg, Mn, Na, Rb, Sb, Se, Sr, Zn

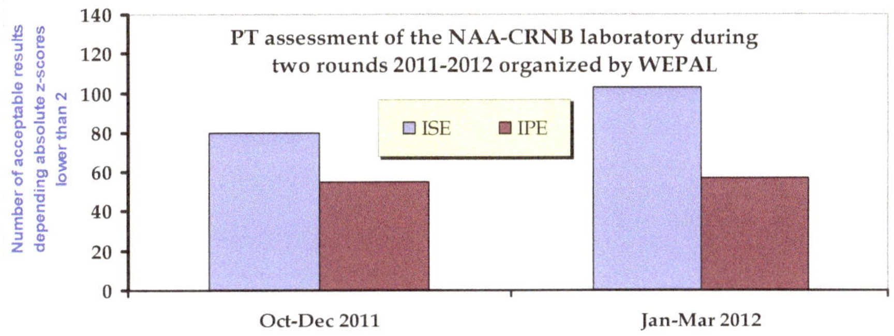

Figure 5. Proficiency test assessment of the NAA laboratory results obtained two rounds 2011.4 and 2012.1.

During the last decade, we acquired a good experiment through our participations in the proficiency tests program and the inter-comparison exercises which were carried out within the framework of the AFRA/RAF projects. In particular, we agreed that the objectives of the IAEA in facilitating these PT rounds under the project RAF/4/022 were achieved. These remains strongly recommend that the IAEA continues to facilitate new PT Rounds with feedback meetings for independent assessment of performance of analytical laboratories. In addition, in the first participation with WEPAl organization we got a clear understanding of the procedures and methodology of PT rounds, and a mechanism for their own evaluation of the outcome thereof.

7. Acknowledgements

This work has been carried out at Es-Salam Research reactor under the framework of AFRA-RAF/4/022 project. We are thankful to the International Atomic Energy Agency for financial support. We also would like to thank Dr Danas Ridikas and Pr Peter Bode for their valuable discussions. Thanks are due to Mhamed Salhi, national coordinator for RAF4/022 project and the NAA group of CRNB and our colleagues from the division of nuclear techniques and applications for their assistance during the experiments.

REFERENCES

[1] L. Alghem, M. Ramdhane, S. Khaled and T. Akhal, "The Development and Application of K_0-Standardization Method of Neutron Activation Analysis at Es-Salam Research Reactor," *Journal of Nuclear Instruments and Methods in Physics Research Section A*, Vol. 556, No. 1, 2006, pp. 386-390.

[2] L. Hamidatou and M. Ramdhane, "Characterization of Neutron Spectrum at Es-Salam Research Reactor Using Høgdahl Convention and Westcott Formalism for the K_0-Based Neutron Activation Analysis," *Journal of Radioanalytical and Nuclear Chemistry*, Vol. 278, No. 3, 2008, pp. 627-630.

[3] L. Hamidatou and H. Benkharfia, "Experimental and MCNP Calculations of Neutron Flux Parameters in Irradiation Channel at Es-Salam Reactor," *Journal of Radioanalytical and Nuclear Chemistry*, Vol. 287, No. 3, 2011, pp. 971-975.

[4] L. A. Hamidatou, S. Dakar and S. Boukari, "K_0-NAA Quality Assessment in an Algerian Laboratory by Analysis of SMELS and Four IAEA Reference Materials Using Es-Salam Research Reactor," *Journal of Nuclear Instruments and Methods in Physics Research Section A*, Vol. 682, 2012, pp. 75-78.

[5] L. A. Hamidatou, S. Khaled, M. Mouzai, B. Zouranen, A. Ararem, A. Alghem and M. Ramdhane, "Instrumental Neutron Activation Analysis of Milk Samples Using the K_0-Standardization Method at Es-Salam Research Reactor," *Physical and Chemical News*, Vol. 45, 2009, pp. 44-47.

[6] L. A. Hamidatou, S. Khaled, T. Akhal and M. Ramdhane, "Determination of Trace Elements in Cigarette Tobacco with the K_0-Based NAA Method Using Es-Salam Research Reactor," *Journal of Radioanalytical and Nuclear Chemistry*, Vol. 281, No. 3, 2009, pp. 535-540.

[7] S. Khaled, M. Mouzai, A. Ararem, L. Hamidatou and B. Zouranen, "Elemental Analysis of Traditional Medicinal Seeds by Instrumental Neutron Activation Analysis," *Journal of Radioanalytical and Nuclear Chemistry*, Vol. 281, No. 1, 2009, pp. 87-90.

[8] L. Hamidatou Alghem, "Mise au Point de la Méthode K_0-NAA en Utilisant la Convention de HϕGDAHL et le Formalisme de WESTCOTT et son Application dans la Nutrition, la Santé et la Minéralogie," Ph.D Thesis, Ferhat Abbes University, 2010.

[9] L. Hamidatou, H. Slamene, T. Akhal and B. Zouranen, "Concepts, Instrumentation and Techniques of Neutron Activation Analysis," In: F. Kharfi, Ed., *Imaging and Radioanalytical Techniques in Interdisciplinary Research—Fundamentals and Cutting Edge Applications*, InTech, Rijeka, 2013, pp. 141-178. http://www.intechopen.com/books/imaging-and-radioanalyal-techniques-in-interdisciplinary-research-fundamentals-and-cutting-edge-applications/concepts-instrumentation-and-techniques-of-neutron-activation-analysis

[10] S. Khaled, S. Belaid, M. Mouzai, L. Alghem, A. Ararem and B. Zouranen, "Development and Application of the Cyclic Delayed Neutron Counting Technique at Es-Salam Research Reactor," *Physical and Chemical News*, Vol. 45, 2009, pp. 39-43.

[11] T. Akhal, M. Mouzai, Y. Sana and M. Ladjel, "Application of RNAA to the Proportioning of Iodine in Food Salt," *The Journal of Physical Chemistry A*, Vol. 45, 2009, pp. 34-38.

[12] ISO 5725-1994, "Accuracy (Trueness and Precision) of Measurement Methods and Results".

[13] ISO/IEC 17025:2005, "General Requirements for the Competence of Testing and Calibration Laboratories".

[14] ISO 15189:2007, "Medical laboratories—Particular Requirements for Quality and Competence".

[15] EURACHEM, "Selection, Use and Interpretation of Proficiency Testing (PT) Schemes," 2nd Edition, 2011.

[16] I. Taverniers, M. De Loose, E. Van Bockstaele, "Trends in Quality in the Analytical Laboratory. II. Analytical Method Validation and Quality Assurance," *TrAC Trends in Analytical Chemistry*, Vol. 23, No. 8, 2004, pp. 535-552.

[17] IAEA/AFRA Project RAF /4/022, Summary Report-Nuclear Sciences and Applications, 31 March 2013.

[18] WEPAL, 2013. www.wepal.nl

[19] A. Barzev and A. Faanhof, "A Proficiency Test Study under the AFRA IV-7 Project Cmip-Afra-P1: Trace and

Minor Elements in Geological Materials. Report to participants," NECSA, 2013.

[20] M. Mouzai and T. Akhal, "Inter-Comparison Feedback Meeting: A Proficiency Test Study under the AFRA IV-7 Project Trace and Minor Elements in Geological Materials," Algiers, 29 November-3 December 2003.

[21] A. Faanhof, "Document RA-07013 (Rev 00) (2006) AFRA IV-7 Proficiency Test 2005: Essential and Toxic Elements in Foodstuffs," NECSA.

[22] M. Mouzai and T. Akhal, "Inter-Comparison Feedback Meeting: A proficiency Test Study under the AFRA IV-7 Project Medicinal Plants and the Analysis of Aquatic-Biomonitors for Environmental Monitoring," Lybia, 2005.

[23] A. Faanhof, "Document RA-07013 (Rev 00) (2008) AFRA IV-12 Inter-Laboratory Analysis Test 2007: Essential and Toxic Elements in Foodstuffs, Medicinal Plants and Aquatic Biomonitors," NECSA.

[24] L. Hamidatou Alghem, S. Khaled and B. Zouranen, "Elemental Determination of Five Materials Representing Foodstuff, Medicinal Plants and Aquatic Biomonitors Using K_0-INAA Method: Meeting on the First Results Inter-Laboratory Competency Test AFRA IV-12 Project, RAF/4/020," Kenitra, 14-18 July 2008.

[25] M. Mouzai and T. Akhal, "Elemental Determination of Five Materials Representing Foodstuff, Medicinal Plants and Aquatic Biomonitors by Relative: Meeting on the First Results Inter-Laboratory Competency Test AFRA IV-12 Project, RAF/4/020," Kenitra, 14-18 July 2008.

[26] L. Hamidatou, T. Akhal, H. Slamene and Z. Boussad, "The Proficiency Testing Tests Related to the Determination of Major, Minor and Trace Elements in Materials of the International Soil and Plant Analytical Exchange Material of WEPAL," *Meeting on Inter-Comparison Feed-Back of NAA and Other Analytical Techniques Proficiency Tests, AIEA Project/RAF*4022—*Enhancing Research Reactor Utilization and Safety*, 4-8 June 2012, Tunis.

[27] Quarterly Report 2011.4, "International Plant-Analytical Exchange," Wageningen University (Environmental Sciences), Wageningen. Prof Testing RvA R 002.

[28] Quarterly Report 2011.4, "International Soil-Analytical Exchange," Wageningen University (Environmental Sciences), Wageningen. Prof Testing RvA R 002.

[29] Quarterly Report 2012.1, "International Plant-Analytical Exchange," Wageningen University (Environmental Sciences), Wageningen. Prof Testing RvA R 002.

[30] Quarterly Report 2012.1, "International Soil-Analytical Exchange," Wageningen University (Environmental Sciences), Wageningen. Prof Testing RvA R 002.

[31] Genie^{TM} 2000 Spectroscopy Software, Version 3.2.1, Canberra, 2009.

[32] HyperLab 2005 System: Installation and Quick Start Guides. HyperLabs Software, Budapest, 2005.

[33] Kayzero for Windows (KayWins) 2005 Software Package for Reactor Neutron Activation Analysis (NAA) Using the K_0-Standardization Method, Version 2, Software Developed by DSM Research, Geleen (NL) for NAA Based on the K_0-Standardization Method Developed at the INW-RUG, Gent (B) and the AEKI, Budapest (H).

14

Development and Validation of an Ion Chromatography Method for the Simultaneous Determination of Seven Food Additives in Cheeses

Marco Iammarino*, **Aurelia Di Taranto**

Istituto Zooprofilattico Sperimentale della Puglia e della Basilicata, Foggia, Italy.

ABSTRACT

Cheeses are characterized by several chemical-physical properties that make it difficult for the microorganisms growth, consequently. The actual European legislation allows the addition of few food additives in this type of food products. In this work, the entire procedure of extraction, purification, chromatographic separations and quali/quantitative determination of seven food additives (sorbic acid, benzoic acid, lactic acid, acetic acid, nitrites, nitrates and phosphates) was developed and applied for the analysis of different types of cheese (mozzarella, cheese spread, semi-hard and hard cheeses). Through validation procedure it was possible to evaluate the most important validation parameters. Extended calibration curves ($r > 0.990$) were obtained for all the analyzed compounds. Recovery values ranged from 72.8% to 98.4% and a good repeatability was obtained, with precision levels in the range of 0.03% - 0.11% ($n = 6$). The potential and feasibility of the method were tested by analysing real samples, such as mozzarella, cheese spread, semi-hard and hard cheeses, confirming that the method is well suited to satisfy the demands for accurate confirmation analyses of seven food additives in cheeses, which is especially valuable in official check analyses and in monitoring schemes.

Keywords: Dairy Products; Cheese; Food Additives; Food Preservatives; Nitrites; Nitrates; Ion Chromatography; Validation

1. Introduction

Dairy products are characterized by several chemical-physical properties that make it difficult for the microorganisms growth. These characteristics are low water activity, low pH, high concentration of sodium chloride, lacking of fermentescible carbohydrates and the possible presence of batteriocins produced by microbial starters and/or of anaerobic condition [1]. These conditions do not make it necessary to the addition of particular food additives, in particular preservatives, in this type of food products; consequently, the actual legislation allows the addition of few food additives in dairy products (**Table 1**). Among these additives, the most used ones are two preservatives (with some restrictions, see **Table 1**): sorbic acid and nitrates, and three acidity regulators: lactic acid, acetic acid and phosphates. Another two important food preservatives, benzoic acid and nitrites, are not admitted [2], but they are considered in this study due to their toxicity and large use in many foodstuffs. In several foodstuffs the contrast action towards different microorganisms, yeast and moulds is exercised through the addition of a mixture of sorbic acid and benzoic acid [3]. The first one cannot be considered as harmful for humans since it can be metabolized like caproic acid [4,5]. For this reason, some limits of use related to sorbic acid are particularly high (up to 2000 mg·kg^{-1} in processed cheeses). As it concerns benzoic acid, it may be considered harmless for humans, at employment doses, because it is not accumulated, but completely eliminated by urine as hippuric acid [6-8]. Nevertheless, high doses of benzoic acid (until to 1000 mg·day^{-1}) may cause oesophagus burning, nausea and headache [9]; moreover, it causes occasional allergic reactions (hives, dermatitis, rhinitis, etc.) in sensitive persons [10]. For these reasons an acceptable daily

*Corresponding author.

Table 1. Food additives analysed and relating legal limits.

Food additive	*E-number*	*Legal limit* (mg·kg^{-1})	*Restrictions*
Sorbic acid—sorbates	E200, 202, 203	1000	Only cheese, prepacked, sliced and cut; layered cheese and cheese with added foods
		quantum satis	Only ripened products surface treatment
		2000	Processed cheese
Acetic acid	E260	*quantum satis*	Only *mozzarella* and whey cheese
Benzoic acid—benzoates	E210, 211, 212, 213	Not admitted	-
Lactic acid	E270	*quantum satis*	Only *mozzarella* and whey cheese
Nitrites	E249, 250	Not admitted	-
Nitrates	E251, 252	150	Only hard, semi-hard and semi-soft cheese
Phosphoric acid—phosphates di-tri and polyphosphates	E338, 339, 340, 341, 450, 451, 452	2000	Only unripened products except *mozzarella*
		20,000	Processed cheese

intake (ADI) for benzoic acid, equal to 0 - 5 mg·kg bw, was established [11] and this food preservative was deleted from the list of authorized food additives in dairy products, since it was considered unnecessary. It is important to underline that the actual legislation indicates that a quantifiable amount of this substance may be detected in certain fermented products resulting from the fermentation process following good manufacturing practice. Consequently, recent studies have deepened the dairy products characteristic and a maximum admissible limit for benzoic acid in cheeses was proposed [12]. Among the most common food preservatives, nitrites and nitrates are well-known due to their large use in meat curing. Notoriously, these food additives are harmful for humans. They may cause methemoglobinemia, which is the pathology due to reaction between nitrates, haemoglobin and myoglobin resulting in the synthesis of methaemoglobin and metmyoglobin [13-15]. Moreover, the reaction between secondary amines and nitrous acid, deriving from nitrates, brings about the synthesis of N-nitroso compounds that are notorious carcinogenic agents [16]. In the actual European Legislation, only nitrates are admitted in dairy products, limited to hard, semi-hard and semi-soft cheese; the limit of use is equal to 150 mg·kg^{-1}. For lactic acid, acetic acid and phosphates, there are no legal limits established by the actual Normative (lactic acid and phosphates are naturally present in dairy products). This is due to not-toxicity of these compounds at the employment doses. However it is compulsory to indicate the addition of these food additives on the product label [17], moreover, high intakes of phosphates may cause significant decreases in some oligoelements absorption [18] and they have been correlated with the bile ducts stones pathogenesis [19]. Actually, an analytical confirmatory method for the simultaneous determination of these food additives in cheeses is not available, consequently, in this work, the entire procedure of extraction, purification, chromatographic separations and quali/determination of seven food additives (sorbic acid, benzoic acid, lactic acid, acetic acid, nitrites, nitrates and phosphates) (**Figure 1**) was developed and applied for the most common types of cheeses (mozzarella cheese, cheese spread, semi-hard and hard cheese). The analytical method was fully validated following an *in-house* validation model according to European Regulations and then evaluating the most important validation parameters.

2. Materials and Methods

2.1. Chemicals

Lactic acid solution (≥85%), acetic acid (≥99.7%), potassium phosphate tribasic (≥98%), nitrite ion standard solution (1000 mg·L^{-1}), nitrate ion standard solution (1000 mg·L^{-1}), sorbic acid (≥99%) and benzoic acid (≥99.5%) were purchased from Sigma-Aldrich (Stenheim, Germany). Sodium carbonate anhydrous (≥99.5%) and sodium hydroxide (50% w/w) were supplied by J. T. Baker (Deventer, Netherlands). All solutions used for ion-exchange chromatography were prepared with ultrapure water (minimal resistance 18.2 MΩ-cm), supplied by Milli-Q RG unit from Millipore (Bedford, MA, USA). Sodium carbonate solutions used as eluents were prepared by dilution in ultrapure water, degassing with nitrogen. The samples used for validation procedure were collected in local stores and they were characterized by the absence of food additives declared on the label. The sample fortified for accuracy tests (precision and recovery) was a semi-hard cheese produced with raw milk of cow origin, characterized by a ripening time of 8 months. The samples used for ruggedness evaluation were 4 mozzarella cheeses (2 from cow milk and 2 from buffalo milk), 4 different brands of cheese spreads and 4 hard cheeses (2 produced with raw milk of sheep origin, 1 with pasteurized milk of cow origin and 1 with raw milk of goat origin). They were characterized by ripening times ranging from 16 to 18 months.

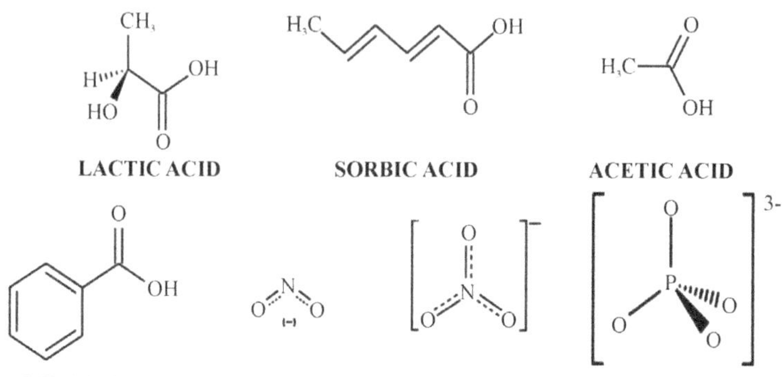

Figure 1. Structures of seven food additives analysed.

2.2. Instrumentation

All chromatographic determinations were performed on a Dionex HPLC system DX500 (Dionex Corporation, Sunnyvale, CA) equipped with an electrochemical detector (model ED40) fit to conductivity mode, a temperature compensated conductivity cell, a GP50 quaternary gradient pump, an injection loop (25 μL) and a Rheodyne injection valve (model RH9125, Cotati, CA, USA). A Dionex anion self-regenerating suppressor (ASRS II, 4 mm) was used for electrochemical suppression at an operating current of 50 mA. The chromatographic separations were accomplished by using an anion-exchange column IonPac® AS9-HC (250 mm × 4 mm i.d., particle size: 9 μm, Dionex Corporation, Sunnyvale, CA) eluted in gradient mode at a flow rate of 1.0 $mL \cdot min^{-1}$. The mobile phase consisted of 0.9 mM Na_2CO_3 (A) and 28.5 mM Na_2CO_3 (B). The experimental separation conditions involved a multilinear gradient operating at room temperature and consisting of a linear gradient from 0.9 mM to 3.7 mM in 5 minutes, from 3.7 mM to 9.2 mM in 1 min, an isocratic step for 19 minutes, then a linear gradient from 9.2 mM to 28.5 mM in 1 minute and 4 minutes at this eluent concentration. The system was then re-equilibrated for 10 min at the initial Na_2CO_3 concentration. The mobile phase reservoir bottles (DX500 2 L bottles, Dionex) were closed and pressurized with pure nitrogen to 0.8 MPa. The system was interfaced via software (PeakNet™, Dionex Corporation, Sunnyvale, CA) to a personal computer for instrumentation control, data acquisition and processing.

2.3. Sample Preparation

A 4-g portion of sample, homogenized by blade ho-mogenizer, was placed in a Falcon tube and mixed with 40 mL of a NaOH 8.5×10^{-3} M solution. This particular solution allows an optimal analytes extraction from matrix, moreover it is essential for the preparation of standard solutions because it brings about an increment of method sensibility (by increasing acids dissociation) and it makes more neutral the solution pH, stabilising the nitrite ion.

The extraction was obtained by placing the tube in ultrasonic bath (Transsonic Digitals, Elma Instruments, Singen, Germany. Ultrasound power: 80%; Heating: 40°C) for 10 minutes and then vortexing (Digital vortex mixer, VWR Int., Milan, Italy) for 1 minute. The sample purification was carried out through a centrifugation (Jouan BR4i centrifuge, Thermo Fisher, Milan, Italy. Speed: 1500 × g, 10 minutes at room temperature) and a filtration of ~5 mL of supernatant with Minisart® GF syringe filters (0.2 μm, Sartorius AG, Goettingen, Germany). A final removing of excess chlorides is necessary. This was obtained through a purification of ~1 mL of filtrate with OnGuard II Ag chromatography filters (Dionex Corporation, Sunnyvale, CA) previously activated with 1 mL of ultrapure water, prior to chromatographic analysis (injection volume: 25 μL).

2.4. Method Validation

The optimized analytical method was submitted to a validation procedure, following an *in-house* validation model developed according to the Regulation No. 882/2004/EC [20] and Decision No. 657/2002/EC [21]. These Regulations describe the analytical parameters to appraise in order to assure the method reliability. These parameters are linearity, detection and quantification limits (LOD and LOQ), selectivity, accuracy (precision and recovery), ruggedness and measurement uncertainty.

Method linearity was verified by injecting 5 standard solutions, obtained by diluting the stock solution (100.0 $mg \cdot L^{-1}$ lactic acid, acetic acid, nitrites, nitrates and phosphates, 1000.0 $mg \cdot L^{-1}$ Benzoic Acid and 2000.0 $mg \cdot L^{-1}$ sorbic acid) with NaOH 8.5×10^{-3} M.

The limit of determination (LOD) and quantification (LOQ) were elaborated according to Miller & Miller [22]: $LOD = 3.3\ s_a/b$ and $LOQ = 10\ s_a/b$, where s_a is the standard deviation of the intercept and b is the slope of the regression line obtained from the calibration curve.

The selectivity is the method capacity to distinguish the analytes from other matrix components. This parameter was investigates through the analyses of twenty samples of cheeses (five mozzarella cheeses, five cheese spreads, five semi-hard cheeses and five hard cheeses) and verifying the absence of interfering peaks in the retention time-window of interest (±2.5% of the retention time of each analyte). It is important to underline that lactic acid and phosphate are substances naturally present in dairy products, consequently, for these additives, the selectivity was evaluated indirectly through accuracy tests.

Method accuracy (precision and recovery) was evaluated by analysing 6 cheese samples spiked with stock solution to obtain final concentrations equal to 20.0 $mg \cdot kg^{-1}$ in lactic acid, 20.0 $mg \cdot kg^{-1}$ in acetic acid, 400.0 $mg \cdot kg^{-1}$ in sorbic acid, 20.0 $mg \cdot kg^{-1}$ in nitrites, 200.0 $mg \cdot kg^{-1}$ in benzoic acid, 20.0 $mg \cdot kg^{-1}$ in nitrates and 20.0 $mg \cdot kg^{-1}$ in phosphates. For lactic acid and phosphates, that are naturally present in dairy products, the concentrations obtained by analysing the "blank" sample (not spiked) were subtracted from those obtained from spiked samples.

Method ruggedness was ascertained under *major changes* conditions toward different types of cheeses (mozzarella, cheese spread, semi-hard and hard cheeses) by using the Youden factorial experimental design [23].

Finally, the measurement uncertainty for each food additive was calculated by using the bottom-up method together with validation data obtained from each step of the analytical procedure [24]. The final values were obtained by the following equation:

$$\bar{u} = \sqrt{\left(\bar{u}(C)\right)^2 + \left(\bar{u}(V_f)\right)^2 + \left(\bar{u}(w)\right)^2}$$

where $\bar{u}$ indicates the relative uncertainty, V_f is the volume of final extract, and w is the sample weight. Four sources of uncertainty were considered for the determination of $\bar{u}(C)$: (a) standards preparation; (b) repeatability; (c) recovery; (d) calibration curve.

3. Results and Discussion

3.1. Method Oprimization

The ion chromatography method for the simultaneous determination of seven food additives in cheeses was optimized starting from a chromatographic separation of organic acids proposed by Dionex Corporation [25]. The optimized gradient elution guaranteed a good resolution of analytes towards endogenous interfering peaks and a good retention time repeatability for each analyte (<0.5%, n = 6).

Different solutions, to use both for standards dilution and for samples extraction, were tested. The most important analytical problem to solve was the nitrites oxidation to nitrates due to the contemporary presence of different acids in the mixture [26]. It is not possible to use buffer solutions, since they cause an increase of interfering signals that compromise method selectivity. In order to basify the solution, NaOH solutions at different concentrations were tested. The best analytes stability (one week) was obtained by using NaOH 8.5×10^{-3} M compared to NaOH 8.5×10^{-2} M, NaOH 8.5×10^{-4} M and ultrapure water. Another analytical problem was the high chlorides concentration of this type of food product that represents an important chromatographic interference for sorbic acid. This inconvenience was solved by using OnGuard II Ag chromatography filters (Dionex Corporation, Sunnyvale, CA) that remove about 90% of total chlorides and make possible the identification of sorbic acid.

3.2. Method Validation

The most important validation parameters appraised through validation procedure are reported in **Table 2**.

All calibration curves showed a determination coefficient (r^2) higher than 0.990. This parameter together with evaluation of intercept (not significantly different from 0), of slope ($t_s/b < 0.22$) and of Mandel's test (linear regression preferable to quadratic) confirmed method linearity for the food additives considered [27]. In **Figure 2** a chromatogram related to an injection of a standard solution is shown.

As it concerns limits of detection, the lower values were obtained for lactic acid (LOD and LOQ equal to 0.6 and 1.9 $mg \cdot kg^{-1}$ respectively) whereas the highest values were registered for sorbic acid (LOD and LOQ equal to 59.4 and 196.0 $mg \cdot kg^{-1}$ respectively). These high values related to sorbic acid do not represent a weakness of this method, since the established legal limits for this food additive in dairy products are relatively high (up to 2000 $mg \cdot kg^{-1}$); subsequently, there is no necessity to reach high method sensibility.

No interfering peaks were observed by analysing twenty samples of cheeses (mozzarella cheeses, cheese spreads, semi-hard cheeses and hard cheeses), so, method selectivity was verified for acetic acid, sorbic acid, benzoic acid, nitrites and nitrates. For lactic acid and phosphates, method selectivity was assured through accuracy tests.

Method accuracy (precision and recovery) was evaluated by injecting 6 spiked samples. Precision was evaluated as CV%, obtaining values in the range 3.0% - 10.9%. Recovery percentages were in the range 72.8% - 98.4%. Since there are no established reference values for the evaluation of accuracy of an analytical method for the determination of food additives in food products, the obtained values were compared with those reported in

Table 2. Analytical method validation parameters.

Food additive	Linearity (r^2)	Specificity	LOD[a] mg·kg^{-1} in the sample	LOQ[b] mg·kg^{-1} in the sample	Recovery (%)	CV%[c] (n = 6)
Lactic acid	0.992	-	0.6	1.9	92.3	9.4
Acetic acid	0.998	No interferences	3.6	11.9	79.6	3.0
Sorbic acid	0.990	No interferences	59.4	196.0	72.8	10.2
Nitrite	0.994	No interferences	2.8	9.3	78.6	10.3
Benzoic acid	0.999	No interferences	16.5	54.5	95.0	7.0
Nitrate	0.997	No interferences	2.1	6.8	88.1	10.9
Phosphate	0.997	-	2.1	7.0	98.4	8.5

[a]Limit of Detection. [b]Limit of Quantification. [c]Coefficient of Variation.

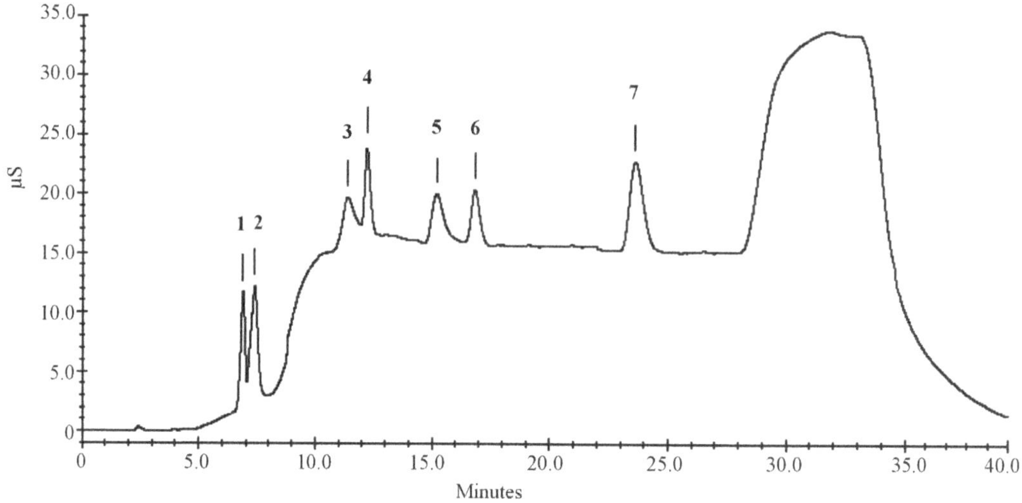

Figure 2. Chromatogram of a standard solution. (1): lactic acid (5.0 mg·L^{-1}), (2): acetic acid (5.0 mg·L^{-1}), (3): sorbic acid (100.0 mg·L^{-1}), (4): nitrites (5.0 mg·L^{-1}), (5): benzoic acid (50.0 mg·L^{-1}), (6): nitrates (5.0 mg·L^{-1}), (7): phosphates (5.0 mg·L^{-1}).

Decision No. 657/2002/EC, resulting substantially conform. In **Figures 3** and **4** comparisons between cheese samples spiked with a standard solution and not-spiked are shown.

Method ruggedness under *major changes* conditions was evaluated by using Youden factorial experimental design. Different sets of real samples: 4 mozzarella cheeses, 4 cheese spreads, 4 semi-hard cheeses and 4 hard cheeses, spiked at the same fortification level used for accuracy tests. The seven factors chosen as variables for Youden test were the matrix and six fictitious factors. The Youden experimental design requires twelve independent experiments: four with validation matrix (semi-hard cheese) and four with each alternative matrices. The results obtained for each alternative matrix, expressed as standard deviation of difference S_{Di} of each food additive, were lower than the estimated method precision (evaluated as twice the repeatability standard deviation). These results confirmed that the matrix variation does not compromise the analytical performances and, consequently, the method is applicable to mozzarella cheese, cheese spread, semi-hard and hard cheese samples.

Measurement uncertainty was calculated for each analyte. The obtained values were in the range 8.6% - 12.7%, confirming method reliability.

3.3. Real Samples Analyses

In order to confirm the reliability of described analytical method, 20 different types of cheese samples (8 mozzarella, 6 cheese spreads, 3 semi-hard cheeses and 3 hard cheeses) were collected in local stores and analysed. No food additives were declared on the label of these samples (except for two mozzarella samples in which citric acid was added). Quantifiable amounts of nitrites, sorbic acid, acetic acid and benzoic acid were not registered. Nitrates quantifiable amounts were detected in two samples: a semi-hard cheese (23.6 mg·kg^{-1}) and a hard cheese sample (11.5 mg·kg^{-1}). The presence of quantifi-

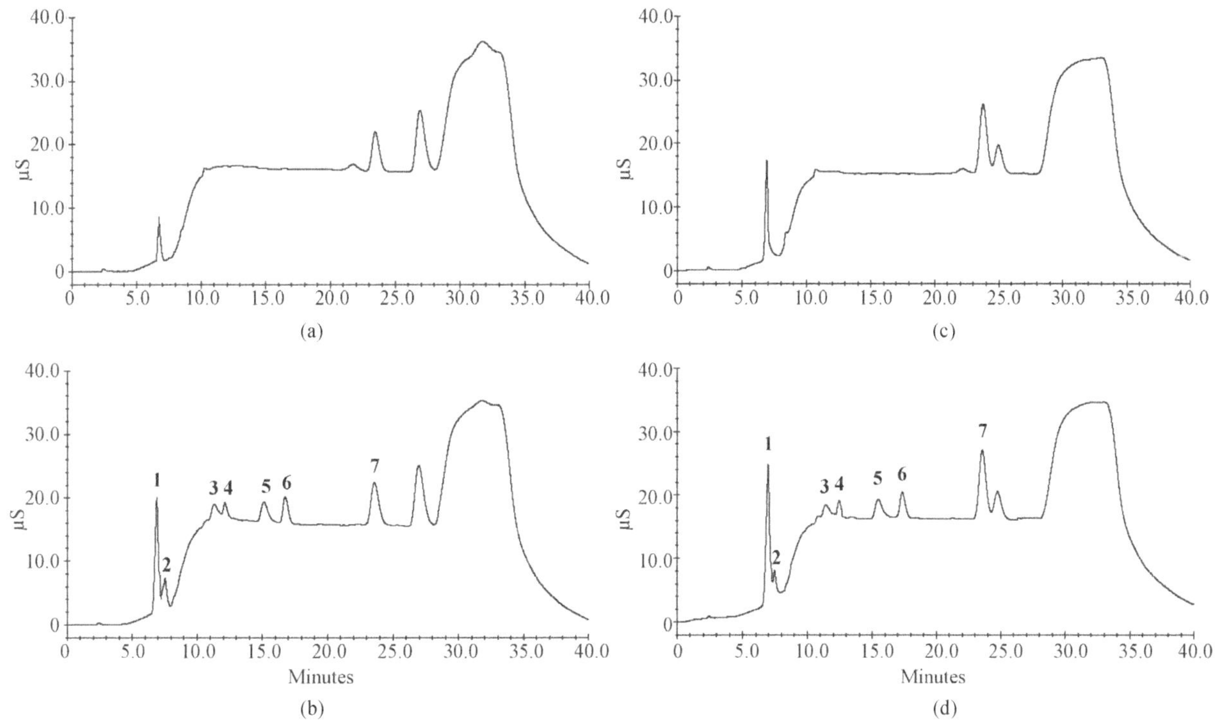

Figure 3. Chromatograms comparison between spiked samples (lactic acid 20.0 mg·kg^{-1}, acetic acid 20.0 mg·kg^{-1}, sorbic acid 400.0 mg·kg^{-1}, nitrites 20.0 mg·kg^{-1}, benzoic acid 200 mg·kg^{-1}, nitrates 20.0 mg·kg^{-1}, phosphates 20.0 mg·kg^{-1}) and not-spiked: Not-spiked Mozzarella cheese (a); Spiked Mozzarella cheese (b); Not-spiked cheese spread (c); Spiked cheese spread (d).

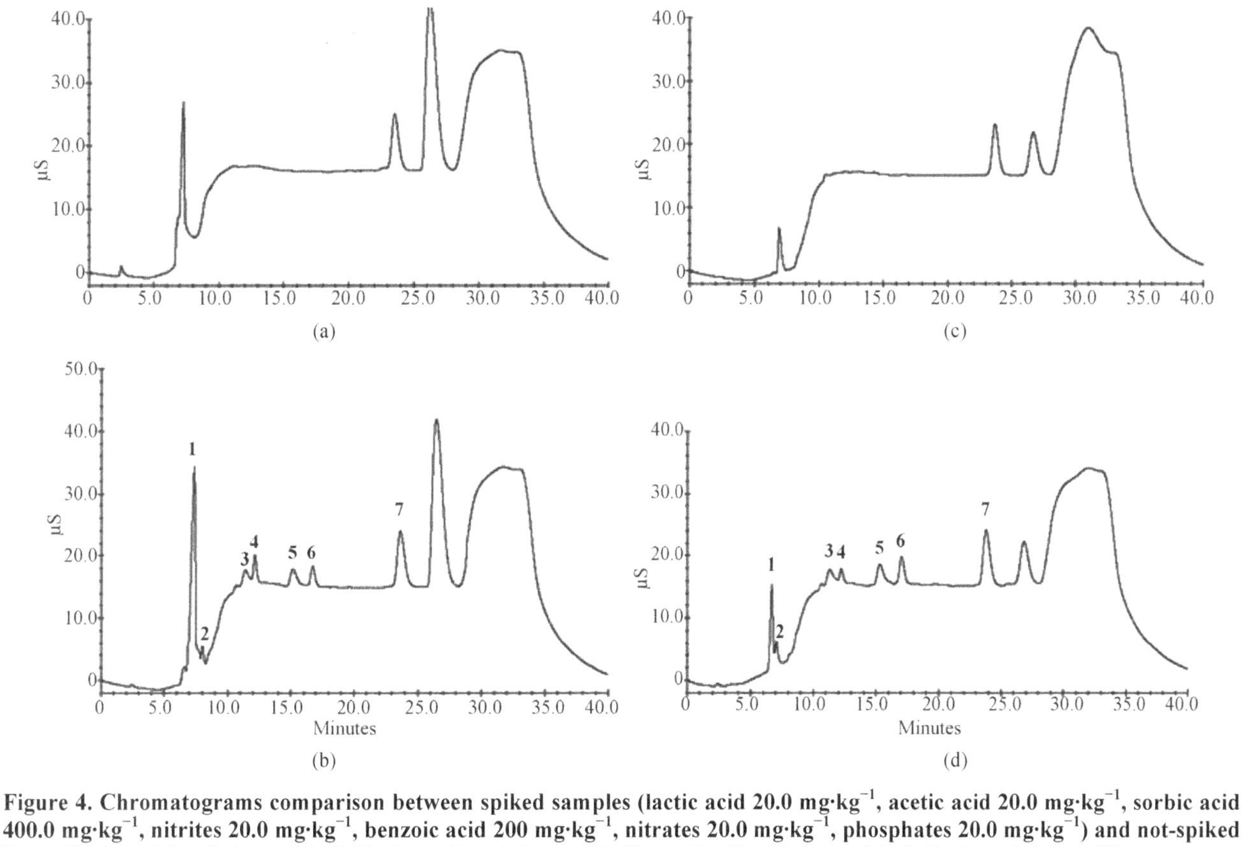

Figure 4. Chromatograms comparison between spiked samples (lactic acid 20.0 mg·kg^{-1}, acetic acid 20.0 mg·kg^{-1}, sorbic acid 400.0 mg·kg^{-1}, nitrites 20.0 mg·kg^{-1}, benzoic acid 200 mg·kg^{-1}, nitrates 20.0 mg·kg^{-1}, phosphates 20.0 mg·kg^{-1}) and not-spiked: Not-spiked semi-hard cheese (a); Spiked semi-hard cheese (b); Not-spiked hard cheese (c); Spiked hard cheese (d).

able residues of nitrates in cheeses was already reported by several authors [28-32]. Lactic acid and phosphates were quantified in the ranges 40.3 - 162.1 mg·kg^{-1} and 17.3 - 71.4 mg·kg^{-1} respectively.

4. Conclusions

In this work, an analytical method for the determination of seven food additives (sorbic acid, benzoic acid, lactic acid, acetic acid, nitrites, nitrates and phosphates) by ion chromatography coupled with suppressed conductivity detection, was developed, validated and applied for the analysis of different types of cheeses (mozzarella cheese, cheese spread, semi-hard and hard cheese).

The optimisation of sample extraction/purification procedure together with the development of a suitable gradient elution allowed an optimal recovery of each analyte, the removal of the most important interfering compounds, a good resolution of analytes towards endogenous interfering peaks and a good retention times repeatability.

Through a validation procedure, carried out by following an *in-house* validation model according to European Regulations, it was possible to evaluate the most important validation parameters. These results, expressed in terms of linearity, detection limits (LOD and LOQ), selectivity, accuracy (precision and recovery), ruggedness and measurement uncertainty, demonstrated that the method is well suited to satisfy the demands for accurate confirmation analyses of seven food additives in cheeses, with a simple and fast sample preparation, a rapid analytical response, which is especially valuable in official check analyses and in monitoring schemes.

REFERENCES

[1] P. F. Fox, P. L. H. Mc Sweeney and C. M. Lynch, "Significance of Non-Starter Lactic Acid Bacteria in Cheddar Cheese," *Australian Journal of Dairy Technology*, Vol. 53, 1998, pp. 83-90.

[2] Commission Regulation, "Regulation (EC) No. 1129/2011 of 11 November 2011 Amending Annex II to Regulation (EC) No 1333/2008 of the European Parliament and of the Council by Establishing a Union List of Food Additives," *Official Journal of the European Communities*, 2011, pp. L295/1-177.

[3] S. Bahruddin, B. Fazlul, I. S. Muhammad, A. Kamaruddzaman and M. T. Khairuddin, "Simultaneous Determination of Preservatives (Benzoic Acid, Sorbic Acid, Methylparaben and Propylparaben) in Foodstuffs Using High-Performance Liquid Chromatography," *Journal of Chromatography A*, Vol. 1073, No. 1-2, 2005, pp. 393-397.

[4] R. F. Witter, E. H. Newcomb and E. Stotz, "The Oxidation of Hexanoic Acid and Derivatives by Liver Tissue *in Vitro*," *The Journal of Biological Chemistry*, Vol. 185, No. 2, 1950, pp. 537-548.

[5] H. J. Deuel, R. Alfin-Slater, C. S. Weil and H. F. Smyth, "Sorbic Acid as a Fungistatic Agent for Foods. I. Harmlessness of Sorbic Acid as a Dietary Component," *Food Research*, Vol. 19, No. 1, 1954, p. 1.

[6] R. J. Feldmann and H. I. Maibach, "Absorption of Some Organic Compounds through the Skin in Man," *Journal of Investigative Dermatology* Vol. 54, 1970, pp. 399-404.

[7] US FDA (United States Food and Drug Administration), "GRAS (Generally Recognized as Safe) Food Ingredients: Benzoic Acid and Sodium Benzoate," Food and Drug Administration, Washington, DC, 1972.

[8] F. Feillet and J. V. Leonard, "Alternative Pathway Therapy for Urea Cycle Disorders," *Journal of Inherited Metabolic Disease*, Vol. 21, No. 1, 1998, pp. 101-111.

[9] WHO (World Health Organization), "Concise International Chemical Assessment Document No. 26. Benzoic Acid and Sodium Benzoate," Geneva, Switzerland, 2000.

[10] M. G. El-Ziney, "CG-MS Analysis of Benzoate and Sorbate in Saudi Dairy and Food Products with Estimation of Daily Exposure," *Journal of Food Technology*, Vol. 7, No. 4, 2009, pp. 127-134.

[11] FAO/WHO (Food and Agriculture Organization of the United Nations/World Health Organization), "Toxicological Evaluation of Certain Food Additives," 67*th Meeting of the* 435 *Joint FAO/WHO Expert Committee on Food Additives* (*JECFA*), 20-29 June 2006, Rome.

[12] M. Iammarino, A. Di Taranto, M. Muscarella and C. Palermo, "Survey of Benzoic Acid in Cheeses: Contribution to the Estimation of an Admissible Maximum Limit," *Food Additives and Contaminants*: *Part B*, Vol. 4, No. 4, 2011, pp. 231-237.

[13] A. M. Fan and V. E. Steinberg, "Health Implications of Nitrate and Nitrite in Drinking Water: An Update on Methemoglobinemia Occurrence and Reproductive and Developmental Toxicity," *Regulatory Toxicology and Pharmacology*, Vol. 23, No. 1, 1996, pp. 35-43.

[14] G. M. McKnight, C. W. Duncan, C. Leifert and M. H. Golden, "Dietary Nitrate in Man: Friend or Foe?" *British Journal of Nutrition*, Vol. 81, No. 5, 1999, pp. 349-358.

[15] O. Matteucci, G. Diletti, V. Principe, E. Di Giannatale, M. M. Marconi and G. Migliorati. "Due casi di Metaemoglobinemia Acuta da Sospetto Avvelenamento da Sodio Nitrito," *Veterinaria Italiana*, Vol. 44, No. 2, 2008, pp. 439-445.

[16] M. C. Archer, "Mechanisms of Action of N-Nitroso Compounds," *Cancer Surveys*, Vol. 8, 1989, pp. 241-250.

[17] Commission Regulation, "Regulation (EC) No. 1169/2011 of 25 October 2011 on the Provision of Food Information to Consumers, Amending Regulations (EC) No 1924/2006 and (EC) No 1925/2006 of the European Parliament and of the Council, and Repealing Commission

Directive 87/250/EEC, Council Directive 90/496/EEC, Commission Directive 1999/10/EC, Directive 2000/13/EC of the European Parliament and of the Council, Commission Directives 2002/67/EC and 2008/5/EC and Commission Regulation (EC) No 608/2004. *Official Journal of the European Communities*, 2011, pp. L304/ 18-63.

[18] M. S. Norma, J. Steinhardt Bour, M. S. Barbara, A. Soullier and M. B. Zemel, "Effect of Level and Form of Phosphorus and Level of Calcium Intake on Zinc, Iron and Copper Bioavailability in Man," *Nutrition Research*, Vol. 4, No. 3, 1984, pp. 371-379.

[19] M. C. Carey, "Pathogenesis of Gallstones," *Recenti Progressi in Medicina*, Vol. 83, No. 7-8, 1992, pp. 379-391.

[20] Commission Regulation, "Regulation (EC) No. 882/2004 of the European Parliament and of the Council of 29 April 2004," *Journal of the European Communities*, 2004, pp. L165,1-141.

[21] Commission Decision, "Decision (EC) No. 657/2002 of 12 August 2002 Implementing Council Directive 96/23 /EC Concerning the Performance of Analytical Methods and the Interpretation of Results," *Journal of the European Communities*, 2002, L221,8-36.

[22] E. J. C. Miller and J. N. Miller, "Statistics for Analytical Chemistry," 3rd Edition, Ellis Horwood PTR Prentice Hall, New York, 1993, p. 115.

[23] W. J. Youden and E. H. Steiner, "Statistical Manual of the AOAC—Association of the Official Analytical Chemists," AOAC-I, Washington DC, 1975.

[24] E. Hund, D. L. Massart and J. Smeyers-Verbeke, "Operational Definitions of Uncertainty" *Trends Analytcal Chemistry*, Vol. 20, No. 8, 2001, pp. 394-406.

[25] Dionex Corporation, "Product Manual for IonPac® AS11-HC IonPac® AG11-HC," Document No. 031333-07, Sunnyvale, 2008.

[26] F. Toldrà, "Handbook of Meat Processing," John Wiley & Sons, Blackwell Publishing, New York, 2010.

[27] L. Brüggemann, W. Quapp and R. Wennrich, "Test for Non-Linearity Concerning Linear Calibrated Chemical Measurements," *Accreditation and Quality Assurance*, Vol. 11, No. 12, 2006, pp. 625-631.

[28] M. Iammarino, A. Di Taranto and M. Cristino, "Endogenous Levels of Nitrites and Nitrates in Wide Consumption Foodstuffs: Results of Five Years of Official Controls and Monitoring," *Food Chemistry*, Vol. 140, No. 15, 2013, pp. 763-771.

[29] M. Iammarino, A. Di Taranto, M. Muscarella and C. Palermo, "Assessment of Natural Levels of Substances with Preservative Effects in Dairy Products," In: H. Castelli and L. du Vale, Eds., *Handbook on Cheese, Production, Chemistry and Sensory Properties*, Nova Publisher Inc., New York, 2013, pp. 559-571.

[30] L. J. Schuddeboom, "Nitrates and Nitrites in Foodstuffs," Council of Europe, Strasbourg, 1993.

[31] A. Topçu, A. A. Topçu, I. Saldamli and M. Yurttagül, "Determination of Nitrate and Nitrite Content of Turkish Cheeses," *African Journal of Biotechnology*, Vol. 5, No. 15, 2006, pp. 1411-1414.

[32] L. Tudor, E. Mitrânescu, L. Tudor and F. Furnaris, "Assessment of Nitrate and Nitrite Content of Romanian Traditional Cheese," Lucrâri Stiintifice Medicinâ Veterinariâ, Timisoara, XL, 2007.

15

Optical Rotatory Dispersion Measurement of D-Glucose with Fixed Polarizer Analyzer Accessory in Conventional Spectrophotometer

Alfons Penzkofer

Faculty of Physics, University of Regensburg, Regensburg, Germany.

ABSTRACT

In the sample compartment of a conventional spectrophotometer, mounting of a polarizer before sample and an analyzer behind sample allows the determination of the optical rotatory dispersion of optical active media by measurement of the transmission ratio of crossed and parallel arranged polarizer and analyzer. A formula for the determination of the angle of rotation is derived from the transmission ratio. The arrangement is applied to determine the molar optical rotation of D-glucose in water in the wavelength range from 220 nm to 820 nm.

Keywords: Optical Activity; Optical Rotatory Dispersion; Specific Optical Rotation; Molar Optical Rotation; Polarimeter; Spectropolarimeter; Grape Sugar; D-Glucose

1. Introduction

Substances with handed (chiral) structure are optically active (circular birefringent) [1-5]. They have different refractive index dispersion, $n(\lambda)$, for right and left circular polarized light, *i.e.* $n_r(\lambda) \neq n_l(\lambda)$ (circular birefringence), and light absorption is different for right and left circular polarized light, *i.e.* $\sigma_r(\lambda) \neq \sigma_l(\lambda)$ or $\varepsilon_r(\lambda) \neq \varepsilon_l(\lambda)$ (circular dichroism, σ is absorption cross-section, $\varepsilon = N_A\sigma/[1000\ln(10)]$ is molar decadic absorptivity, λ is wavelength). Linear polarized light can be thought to be composed of equal amount of right and left circular polarized light. Optical active substances cause rotation of the polarization plane of linear polarized light in their transparency region due to different speed of right and left circular polarized light. In the absorption region linear polarized light becomes additionally elliptically polarized because of different attenuation of right and left circular polarized light.

The rotation of polarization at a fixed wavelength is generally measured with a polarimeter consisting of a monochromatic light source, an entrance polarizer, space for the sample, an exit polarizer (analyzer) on a rotation stage, and a detector [6]. The analyzer is rotated to crossed position (blocking of linear polarized light transmission) in accordance of the rotation of the linear polarization of the optical active sample under investigation. The optical rotatory dispersion (wavelength dependence of optical rotation) is measured with spectropolarimeters consisting of a broadband light source, a spectrometer, and a polarimeter [6].

Here the polarizer and analyzer are oriented perpendicular ($\perp$) and then oriented parallel ($\parallel$). In both cases the transmission through polarizer—sample—analyzer ($T_\perp$ and $T_\parallel$) is measured, and the angle of polarization rotation ϕ is calculated from the transmission ratio $T_\perp/T_\parallel$. A formula for the relation between ϕ and $T_\perp/T_\parallel$ is derived. The polarizer—sample—analyzer arrangement may be illuminated with a monochromatic light source (laser, light emitting diode, lamp with interference filter) for determination of the optical polarization rotation at a fixed wavelength (polarimeter function), or it may be assembled into the sample chamber of a conventional spectrophotometer to determine the wavelength dependence of the optical polarization rotation (spectropolarimeter function).

2. Experimental

The experimental arrangement for the determination of

the optical rotation of polarization at a fixed wavelength is shown in the top part of **Figure 1(a)**. The light source (here a He-Ne laser with λ = 632.8 nm) emits monochromatic light along the z-direction through the polarizer (P1)—sample (S)—analyzer (P2) arrangement. The polarizer orientations and the optical polarization rotations of right (*dextro*) rotatory and left (*laevo*) rotatory samples are illustrated in the bottom part of **Figure 1(a)**.

The light transmission through the P1-S-P2 arrangement is given by

$$T = \frac{S_{out}}{S_{in}} = T_{P1} T_S T_{P2}, \qquad (1a)$$

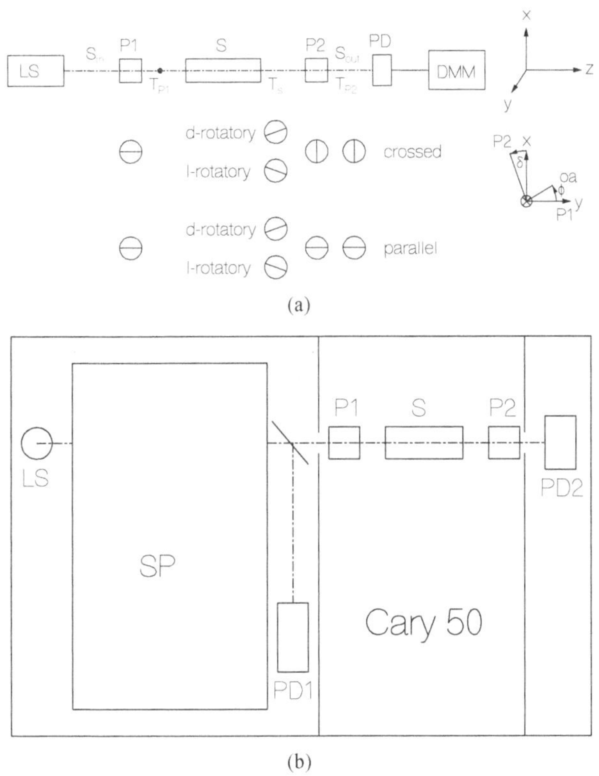

Figure 1. (a) Top part: Experimental setup for single-wavelength optical polarization rotation measurement of an optical active sample. LS: light source (here: He-Ne laser λ = 632.8 nm). P1: Glan polarizer aligned to horizontal polarized light transmission. P2: Glan polarizer operated vertical oriented (crossed to P1) and horizontal oriented (parallel to P1). S: optical active sample. PD: Photo-detector. DMM: Digital multimeter. Bottom part: Polarizer orientations of P1 and P2, polarization plane rotation for d-rotatory and l-rotatory samples, and light polarization direction after P2 (analyzer) are indicated. (b) Experimental setup for optical rotatory dispersion measurement of an optical active sample. In Cary 50 UV-Vis spectrophotometer sample compartment, two Glan-polarizers, P1 and P2, aligned to crossed and parallel orientation, are inserted before and behind the position of sample S. Components of Carry 50 are: LS full-spectrum Xe pulse lamp; SP Cerny-Turner monochromator; PD1 and PD2 silicon diode detectors.

where S_{out} is the transmitted light power, S_{in} is the incident light power, T_{P1} is the transmission through polarizer P1, T_S is the transmission through the optical active sample S, and T_{P2} is the transmission through analyzer P2. The light transmission through the crossed polarizer arrangement with polarization rotation ϕ of the optical active sample is

$$T_{\perp}(\phi) = T(\angle(\text{P2},\text{P1}) = 90^\circ, \phi) = T_{P1} T_S T_{P2,0} \sin^2(\phi). \quad (1b)$$

The light transmission through the parallel polarizer arrangement is

$$T_{\parallel}(\phi) = T(\angle(\text{P2},\text{P1}) = 0^\circ, \phi) = T_{P1} T_S T_{P2,0} \cos^2(\phi), \quad (1c)$$

where $T_{P2,0}$ is the transmission through P2 for parallel orientation of P1 and P2 in the case of $\phi = 0$.

The transmission ratio, $T_{\perp}/T_{\parallel}$, is

$$\frac{T_{\perp}(\phi)}{T_{\parallel}(\phi)} = \frac{\sin^2(\phi)}{\cos^2(\phi)} = \tan^2(\phi). \qquad (2)$$

Solving Equation (2) for the rotation angle ϕ of optical polarization gives

$$\phi = \arctan\left[\pm\left(\frac{T_{\perp}}{T_{\parallel}}\right)^{1/2}\right]. \qquad (3)$$

The positive sign applies to *dextro-rotatory* (= d-rotatory = right-rotatory) samples. The negative sign applies to *laevo-rotatory* (= l-rotatory = left-rotatory) samples. The rotation sign is determined by rotating the analyzer P2 slightly out of the crossed orientation by an angle δ ($|\phi| > \delta > 0$) and measuring the transmission ratio (see illustration in lower part of **Figure 1(a)**). For d-rotatory behavior one gets

$$T_{\text{d-rot}}(\angle(\text{P2},\text{P1}) = 90^\circ + \delta, \phi) < T_{\text{d-rot}}(\angle(\text{P2},\text{P1}) = 90^\circ, \phi), \quad (4a)$$

and for l-rotatory behavior one gets

$$T_{\text{l-rot}}(\angle(\text{P2},\text{P1}) = 90^\circ + \delta, \phi) > T_{\text{d-rot}}(\angle(\text{P2},\text{P1}) = 90^\circ, \phi). \quad (4b)$$

The analysis above is unambiguous for optical rotation $|\phi| < 90^\circ$ which is accomplishable by reduction of sample length and/or concentration of optical active molecules.

The experimental arrangement for the determination of the optical rotation of polarization as a function of wavelength (optical rotatory dispersion) is shown in **Figure 1(b)**. The P1-S-P2 polarizer-sample-analyzer assembly is mounted in the sample chamber of the applied UV-Vis spectrophotometer Cary 50 from Varian. The wavelength range of this spectrophotometer is 190 nm to 1100 nm.

In our experiments calcite Glan polarizers were used for P1 and P2. The polarizer P1 is oriented for horizontal polarized light transmission (after Cary 50 spectrometer horizontal polarized light component is more intense than vertical polarized light component). The light transmissions in the Cary 50 spectrophotometer through the horizontal oriented polarizer P1 alone (dashed curve), through the horizontal oriented polarizers P1 and P2 (solid curve), and through the horizontal oriented polarizer P1 and vertical oriented polarizer P2 (dotted curve) are shown in **Figure 2**. Below 390 nm the transmission through calcite begins to decrease remarkably and below 218 nm the transmission through calcite becomes diminishingly small (transmission behavior of calcite is found in [7,8]). For optical rotatory dispersion measurement in the wavelength range from 200 nm to 390 nm the calcite Glan polarizers should be replaced by α-BBO Glan polarizers with better transmission performance (BBO = barium borate = BaB_2O_4) [9].

The Cary 50 spectrophotometer with polarimeter accessory of **Figure 1(b)** was applied to determine the molar optical rotation [1-5] of grape sugar (D-glucose) in Millipore water in the wavelength range from 220 nm to 820 nm. Grape sugar from Frießinger Mühle, Bad Wimpfen (www.friessinger-muehle.de) was purchased in a supermarket and used without further purification. Its composition is D-glucose with crystal water, the sum formula is $C_6H_{12}O_6 \cdot H_2O$ (molar mass M_m = 198.18 g·mol^{-1}). It was dissolved in Millipore water. A mass concentration of C_m = 0.3 g·cm^{-3} was prepared (near saturation concentration) and used in the experiments. The corresponding grape sugar molar density was $C_{Mol} = C_m / M_m = 1.51\ \mathrm{mol \cdot dm^{-3}}$ and the corresponding molecule number density was $N_m = C_{Mol} N_A = 9.12 \times 10^{20}\ \mathrm{cm^{-3}}$ (N_A is the Avogadro constant). The experiments were carried out at room temperature (ϑ = 21.5 ± 0.5°C). The solution was measured in fused silica cells. In the wavelength range from 300 nm to 820 nm a cell length of ℓ = 5 cm was used. In the wavelength range from 220 nm to 300 nm the measurements were carried with a cell length of ℓ = 1 cm.

Light transmissions, $T(\lambda)$, through the applied grape sugar solution in cell lengths of ℓ = 1 cm (solid curve) and ℓ = 5 cm (dashed curve) in the wavelength range from 200 nm to 820 nm are displayed in **Figure 3**. The inset shows the absorption cross-section spectrum $\sigma(\lambda)$ of the applied grape sugar in the UV spectral region. It was calculated by the relation $\sigma = -\ln(T)/(N_m \ell)$. Absorption shoulders at 325 nm and 215 nm, and an absorption peak at 268 nm with σ (268 nm) = 5.64 × 10^{-22} cm^2 are observed.

The angle (in radian) of which the plane of polarization rotates in an optical active sample of length ℓ, vacuum wavelength λ, and refractive indices n_r and n_l is given by [1-5]

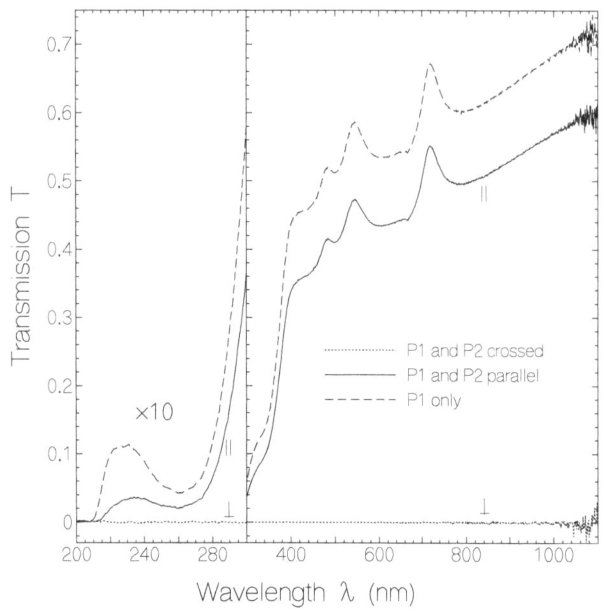

Figure 2. Transmission behavior of applied calcite Glan-polarizers in Carry 50 UV-Vis spectrophotometer. Transmission plane of polarizer P1 is oriented horizontal. For left part curves are expanded a factor of 10.

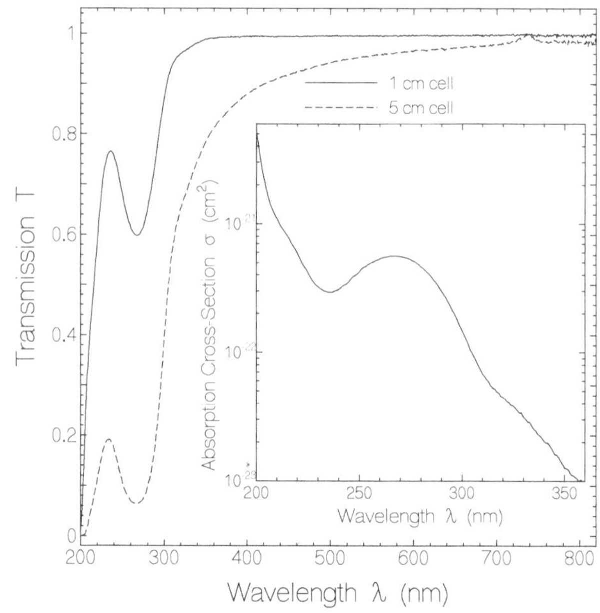

Figure 3. Transmission $T(\lambda)$ of grape sugar (D-glucose) in Millipore water. Mass concentration: C_m = 0.3 g·cm^{-3} of $C_6H_{12}O_6 \cdot H_2O$. Molar mass of $C_6H_{12}O_6 \cdot H_2O$: M_m = 198.18 g mol^{-1}. Sample lengths ℓ = 1 cm (solid curve) and ℓ = 5 cm (dashed curve). Inset shows absorption cross-section spectrum $\sigma(\lambda)$ of grape sugar in Millipore water ($\sigma = -\ln(T) M_m/(\ell C_m N_A)$ where N_A is the Avogadro constant).

$$\phi = \frac{\phi_r - \phi_l}{2} = -\frac{k_r - k_l}{2}\ell = \frac{n_l - n_r}{2}k_0\ell = \frac{\pi\ell}{\lambda}(n_l - n_r), \quad (5)$$

where $k_i = n_i k_0 = n_i 2\pi/\lambda$ is the wavevector of right ($i = r$) or left ($i = l$) circular polarized light $(k_0 = 2\pi/\lambda)$.

The specific rotation $[\phi]$ of an optical active species in a sample is expressed in degrees (1 rad = 180°/π = 180/π deg) and normalized to mass density or concentration ρ of the optical active species in $g \cdot cm^{-3}$ and sample length ℓ in dm. The corresponding numerical value equation is

$$[\phi] = \frac{180}{\pi}\frac{\phi}{\rho\ell}. \quad (6)$$

($[\phi]$ is polarization rotation in deg for sample of density or concentration of $\rho = 1\ g \cdot cm^{-3}$ and sample length of $\ell = 1$ dm; ϕ is in rad).

The molar optical rotation $[m]$ of optical active molecules in a sample is expressed in degrees and normalized to the molar density or concentration $\rho_{Mol} = 1000\rho/M_{Mol}$ in mol dm^{-3} and sample length ℓ in m (ρ in $g \cdot cm^{-3}$ and M_{Mol} in $g \cdot mol^{-1}$). The corresponding numerical value equation is [3-5]

$$[m] = \frac{180}{\pi}\frac{\phi}{\rho_{Mol}\ell} = \frac{180}{\pi}\frac{\phi M_{Mol}}{1000\rho\ell} = [\phi]\frac{M_{Mol}}{100}. \quad (7)$$

([m] is polarization rotation in deg for sample of density or concentration of $\rho_{Mol} = 1\ mol \cdot dm^{-3}$ and sample length of $\ell = 1$ m). The specific optical rotation as well as the molar optical rotation depends on the wavelength λ, the temperature ϑ, the solvent, and in the case of sample dissociation or aggregation on the used concentration. The complete specification is therefore expressed in $[\phi]_\lambda^\vartheta$ (solvent, concentration in g per 100 cm^3) or $[m]_\lambda^\vartheta$ (solvent, concentration in mol per 1 dm^3).

The wavelength dependence of the optical rotation in the transparency region is related to the absorption spectrum of the optical active sample by the Drude expression [1-5]

$$\phi(\lambda) = \sum_j \frac{A_j}{\lambda^2 - \lambda_j^2}, \quad (8)$$

where λ_j is the wavelength of transition from the ground state to the excited state j, and A_j is a constant depending on the absorption strength of this transition.

In the absorption region the absorption band linewidths $\Delta\lambda_j$ (FWHM) have to be considered and the Drude expression changes to the Cotton effect equation [2,4]

$$\phi(\lambda) = \sum_j \frac{A_j}{\lambda^2 - \lambda_j^2 + \frac{\lambda_j^4}{\lambda^2(\lambda^2 - \lambda_j^2)}\left(\frac{\Delta\lambda_j}{2}\right)^2}. \quad (9)$$

(Equation 2.407 in [2] rewritten to λ dependence using $d\nu/d\lambda = d(c/\lambda)/d\lambda = -c/\lambda^2$ leading to $\gamma_j = -(c/\lambda^2)(\Delta\lambda_j/2)$).

3. Results and Discussion

The transmission measurement results through the polarizer—grape sugar solution—analyzer arrangement in the Cary 50 spectrophotometer are depicted in **Figure 4**. The concentration was $C_m = 0.3\ g \cdot cm^{-3}$. The right part shows the transmissions through the cell of 5 cm length and the left part shows the transmission through the cell of 1 cm length. The solid curves belong to $T_\perp(\lambda)$ and the dashed curves belong to $T_\parallel(\lambda)$. The dotted curves shows the ratio $T_\perp(\lambda)/T_\parallel(\lambda)$. The shape of $T_\parallel(\lambda)$ is mainly determined by the transmission behavior of the calcite polarizers. The finite transmission $T_\perp(\lambda)$ is caused by the rotation of the polarization direction by the optical active D-glucose solution. The transmission ratio $T_\perp(\lambda)/T_\parallel(\lambda)$ is relevant for the calculation of optical rotation of polarization (Equations (2) and (3)).

The determined optical rotatory dispersion $\phi \times 180/\pi$ (in degree) of the investigated grape sugar solution is depicted in **Figure 5**. It was obtained from **Figure 4** by application of Equation (3). The left part belongs to a cell length of 1 cm and the right part belongs to a cell length of 5 cm. The dash-dotted curve on the right part is the single-term regression fit of the Drude expression Equation (8) $(\phi(\lambda) = A_0/(\lambda^2 - \lambda_0^2))$ to the experimental

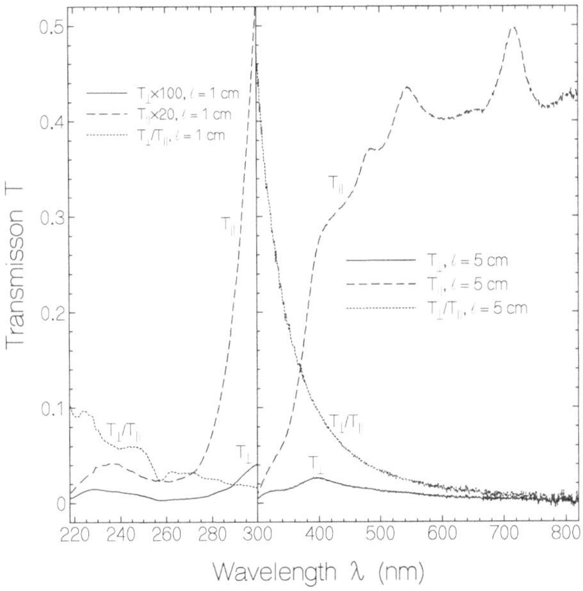

Figure 4. Transmission behavior of polarizer arrangement with inserted grape sugar sample. Mass concentration: C_m = 0.3 $g \cdot cm^{-3}$ of $C_6H_{12}O_6 \cdot H_2O$. Temperature ϑ = 21.5°C ± 0.5°C. Left part: Sample length ℓ = 1 cm; $T_\perp$ (solid curve) is expanded a factor of 100; $T_\parallel$ (dashed curve) is expanded a factor of 20. Right part: sample length ℓ = 5 cm.

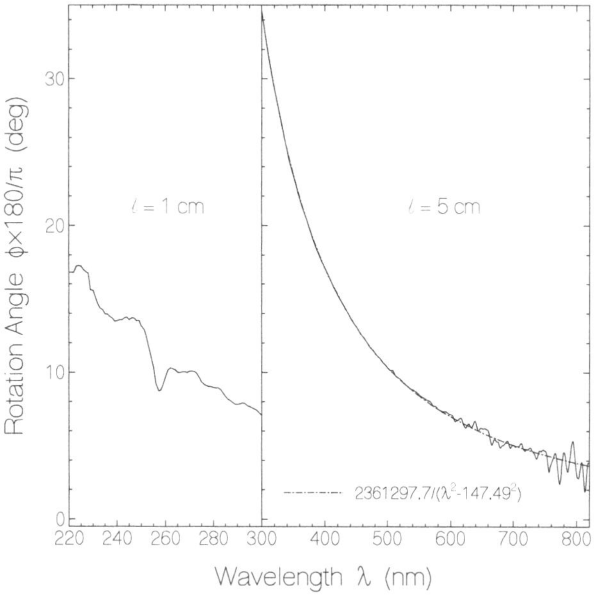

Figure 5. Optical polarization rotation ϕ of grape sugar samples in Millipore water. Mass concentration: C_m = 0.3 g·cm^{-3} of $C_6H_{12}O_6 \cdot H_2O$. Left part: Sample length ℓ = 1 cm. Right part: Sample length ℓ = 5 cm. Curves calculated from $T_\perp/T_\parallel$ of Figure 4 (Equation (3)).

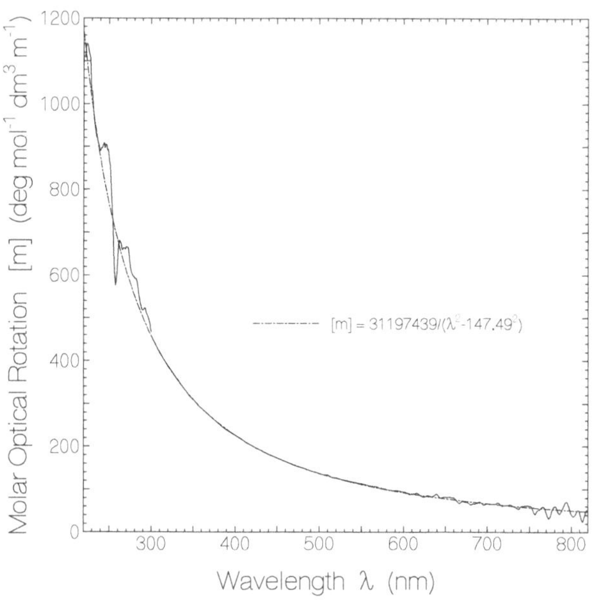

Figure 6. Molar optical rotation [m] of grape sugar in Millipore water. Curve calculated from curves of Figure 5 by use of Equations (6) and (7).

data. In the right region ($\lambda \geq 300$ nm) D-glucose is transparent and the Drude relation fits well to the experimental curve. The apparent single oscillator transition wavelength is λ_0 = 147.49 nm indicating a dominant apparent absorption band there. In the left region ($\lambda < 300$ nm) the transmission through the calcite polarizers is small reducing the accuracy of rotation angle dispersion determination. There D-glucose is slightly absorbing (see **Figure 3**) causing a slight structuring of the optical rotatory dispersion curve due to the Cotton effect dispersion [1-5].

The obtained molecule specific molar optical rotatory dispersion $[m(\lambda)]$ (Equation (7)) of D-glucose is displayed in **Figure 6**. It shows the optical rotation in degree (°) of 1M (1 mol·dm^{-3}) glucose in a cell of 1 m length. Over the whole displayed wavelength range from 220 nm to 820 nm the dispersion fits well to the single-term Drude formula. Some Cotton effect dispersion is seen in the D-glucose absorption region. The D-glucose absorption in the 220 to 300 nm region (see inset of **Figure 3**) is too weak to influence markedly the polarization rotatory dispersion which is determined by strong absorption in the vacuum UV spectral range.

The obtained optical rotatory dispersion of D-glucose in Millipore water is in good agreement with published results [10]. The here determined molar optical rotation at λ_D = 589.3 nm is $[m]_D^{21.5^\circ C}$ (Millipore water, 1.514 mol·dm^{-3}) = 95.8 ± 1 deg mol^{-1}·dm^3·m^{-1}. The corresponding specific rotation is $[\phi]_D^{21.5^\circ C}$ (Millipore water, 27.27 g/100 cm^3) = 53.2 ± 0.6° (M_m = 180.16 g·mol^{-1} for D-glucose) is in good agreement with results found in the literature ($[\phi]_D^{20^\circ C}$ (water, 10 g/100 cm^3) = 52.7° [11]).

4. Conclusion

A simple fixed-polarizer-analyzer polarimeter and spectropolarimeter were designed and applied to the measurement of the optical rotatory dispersion of D-glucose. In the arrangements the polarizer and analyzer are aligned perpendicular ($T_\perp$) and parallel ($T_\parallel$), and the rotation of the polarization plane is calculated from the transmission ratio $T_\perp/T_\parallel$. Besides two polarizers, no extra polarimeter and spectropolarimeter equipment are required for the determination of the specific optical rotation at a fixed wavelength or for the determination of the optical rotatory dispersion over a wide wavelength range.

5. Acknowledgements

The author thanks Prof. F. J. Gießibl for his kind hospitality.

REFERENCES

[1] L. D. Barron, "Molecular Light Scattering and Optical Activity," 2nd Edition, Cambridge University Press, Cambridge, 2009.

[2] E. Charney, "The Molecular Basis of Optical Activity. Optical Rotatory Dispersion and Circular Dichroism," John Wiley & Sons, New York, 1979.

[3] P. Grabbé, "Optical Rotatory Dispersion and Circular Dichroism in Organic Chemistry," Holden-Day, San Francisco, 1965.

[4] A. Moscowitz, "Theoretical Aspects of Optical Activity. Part One: Small Molecules," In: I. Prigogine, Ed., *Advances in Chemical Physics*, Vol. IV, Interscience Publishers, New York, 1972, pp. 67-112.

[5] S. F. Mason, "Molecular Optical Activity and the Chiral Discriminations," Cambridge University Press, Cambridge, 1982.

[6] E. M. Slayter, "Optical Methods in Biology," Wiley-Interscience, New York, 1970, Chapter 27.

[7] D. E. McCarthy, "Transmittance of Optical Materials from 0.17 μ to 3.0 μ," *Applied Optics*, Vol. 6, No. 8, 1967, pp. 1896-1898.

[8] F. Wu, G. Li, J. Huang and D. Yu, "Calcite/Barium Fluoride Ultraviolet Polarizing Prism," *Applied Optics*, Vol. 34, No. 19, 1995, pp. 3668-3670.

[9] V. P. Solntsev, E. G. Tsvetkov, V. A. Gets and V. D. Antsygin, "Growth of α-BaB_2O_4 Single Crystals from Melts at Various Compositions: Comparison of Optical properties," *Journal of Crystal Growth*, Vol. 236, 2002, pp. 290-296.

[10] I. Listowsky, G. Avigad and S. Englard, "Optical Rotatory Dispersion of Sugars. I. Relationship to Configuration and Conformation of Aldopyranoses," *Journal of the American Chemical. Society*, Vol. 87, No. 8, 1965, pp. 1765-1771.

[11] S. Budavari, "The Merck Index. An Encyclopedia of Chemicals, Drugs, and Biologicals," 13th Edition, Entry# 4353 Glucose, Merck & Co., Inc., Rahway, 1989, p. 699.

A New Metal Tag for Highly Selective and Sensitive Analyses of Amino Acids and Dipeptides by HPLC/ICP-MS

Daigo Iwahata, Kazuki Nakamura, Rie Yamada, Hiroshi Miyano, Naoyuki Yamada

Institute for Innovation, Ajinomoto Co., Inc., Kawasaki, Japan.

ABSTRACT

We have developed a novel metal tag, bis(ethylenediamine)-4'-methyl-4-carboxybipyridine-ruthenium N-succinimidyl ester (ECRS) for sensitive analysis of amino acids using high performance liquid chromatography/inductively coupled plasma mass spectrometry (HPLC/ICP-MS). ECRS is a functional reagent, containing an ester group at one end that can be activated to bind to amino group and a chelated ruthenium at the other. The activated ester was reacted briefly with amino groups under weakly alkaline conditions. The ruthenium was detected sensitively by ICP-MS. ECRS was reacted with 17 proteinogenic amino acids in borate buffer. The derivatives were separated by reversed phase HPLC and identified by quadrupole-based ICP-MS. ECRS was suitable for speciation; low molecular weight compounds containing amino groups. We have thus established a quantitative analytical method for amino acids and dipeptides. The detection limits of branched amino acids (signal-to-noise ratio of 3) were 1.5 nmol·L^{-1} in the standard solution (100 amol per injection).

Keywords: Metal Tag; HPLC/ICP-MS; Amino Acid; Dipeptide; Derivatization; Speciation

1. Introduction

Amino acids are the fundamental components of proteins and are essential for life. They are obtained from the diet and utilized throughout the body [1-3]. In addition, amino acids work as hub compounds in metabolic pathways [4], cell signaling, and other cellular phenomena. Therefore, the concentrations of amino acids in biofluids can reflect various diseases (hepatocellular carcinoma [1], malignant melanoma [5], liver disease [6,7], coronary heart disease [8], rheumatoid arthritis [9], chronic renal failure [10], diabetes mellitus [11], Alzheimer's disease [12,13]) and metabolic conditions.

However, the volumes of biological samples are usually very small and contain minimal amounts of free amino acids. Therefore, several highly sensitive analytical methods for amino acids have been developed [14-18]. In 1958, Stein, Moore and Spackmann developed an amino acid analyzer system for the colorimetric determination of amino acids. Their method involved the production of a purple color with ninhydrin reagent [19], which was first reported by Ruhemann in 1911 [20], after stepwise separation of the amino acids on a cation exchange resin by raising the pH of the citric acid buffer solution. The analyzer is now capable of assaying not only protein hydrolysates but also amino acids in biological fluids, through progressive improvements to HPLC pumping performance and the development of more effective column resins. However, the sensitivity is not very high and the detection limits are a few picomoles.

In addition to the ninhydrin method, a variety of other techniques for converting amino acids to sensitively analyzable fluorescent derivatives have been developed since the late 1970s [21]. The reagents used to produce the derivatives include ortho-phthalaldehyde (OPA) [22], 9-amino quinolyl-N-hydroxysuccinimidyl carbamate (AQC) [23], and 4-fluoro-7-nitrobenzofrazane (NBD-F). [24] Their detection sensitivities are within the sub-picomole to femtomole range. Recently, mass spectrometry has also been used for this analysis, by either detecting the amino acids directly [25] or after pre-column derivatization [26,27].

The first report of the combination of immunoreaction

and ICP-MS was described in 2001 by Zhang *et al.*, for the determination of thyroid hormones in human serum. [28] The authors used biotinylated antibodies and Eu^{3+} labeled streptavidin. In 2002, the same group described the determination of thyroxin using a competitive immunoassay [29]. In the same year, Baranov *et al.* implemented a gel-filtration-based ICP-MS immunoassay to separate and quantify various labeled antibodies [30,31], while Zhang *et al.* used antibodies conjugated with colloidal gold nanoparticles for a sandwich-type immunoreaction [32]. Zhang and Baranov analyzed the metals by off-line ICP-MS after digestion.

In 2004, Whetstone *et al.* introduced a cysteine-specific labeling strategy, referred to as an element-coded affinity tag for peptides and proteins [33]. The reagent consists of a derivative of the bi-functional chelating agent 1,4,7,10-tetraazacyclododecane-N,N',N",N"'-tetraacetic acid (DOTA), and contains a bromoacetamido reactive group to specifically target cysteine residues. In recent approaches, two strong metal chelators (DOTA and diethylenetriamine pentaacetic acid (DTPA)) were used as metal tags with HPLC/ICP-MS [34-36]. These metal tags were suitable for size exclusion column chromatography.

In 2008, we established a highly sensitive amino acid analytical method, using reversed phase HPLC/ICP-MS. We derivatized amino acids with the commercial reagent, bis (2,2'-bipyridine)-4'-methyl-4-carboxybipyridine-ruthenium N-succinimidyl ester (BCRS). The derivatized amino acids were detected by ICP-MS. BCRS was reacted with 10 primary amino acids and the derivatives were identified [37]. However, the other proteinogenic amino acids were not separated. The BCRS derivatives, which included two bipyridines in the compounds, were too bulky to separate by reversed phase HPLC. This reagent was not enough for speciation. In this study, we have designed and synthesized a new metal tag, bis (ethylenediamine)-4'-methyl-4-carboxybipyridine-ruthenium N-succinimidyl ester (ECRS). ECRS was reacted with 17 proteinogenic amino acids (except for unstable amino acids in the stock and the buffer solutions; glutamine, asparagine and tryptophan) and 3 common dipeptides, in the presence of borate buffer. It was suitable for speciation.

2. Experimental

2.1. Chemicals and Reagents

Reagent grade acetonitrile, acetic acid, ammonium hydroxide, ammonium acetate and sodium borate were purchased from Wako Pure Chemical Industries, Ltd. (Osaka, Japan). 4'-Methyl-2,2'-bipyridine-4-carboxylic acid was obtained from Rubypy Scientific Inc. (Ottawa, Canada). Bis(2,2-bipyridine)-4-methyl-4-carboxybipyridine-ruthenium N-succinimidyl ester-bis(hexafluorophosphate) (=BCRS), all amino acids, diamines, and other reagents for synthesis were obtained from Fluka Chemie AG/Sigma-Aldrich (St Louis, MO, USA). Distilled deionized water (18.3 MO) was obtained from a Millipore water purifying system (Bedford, MA, USA).

2.2. Design of a New Metal Tag

The image of the metal tag for low molecular compounds is shown in **Figure 1**. This metal tag contains the reactive group that can be activated to react with amino acids and the chelated metal. The nature of the chelating group is hydrophobic. The hydrophobicity of the chelation moiety is very important. The suitable hydrophobicity is needed to retain and separate the low molecular compounds in the reverse phase column.

2.3. Synthesis of the New Metal Tag

The synthesis scheme for the new metal tag reagent is shown in **Figure 2**, and has been previously described [38]. Ruthenium chloride (1.0 g = 3.82 mmol) was dissolved in water/ethanol (v/v = 16 mL/4 mL), and the solution was heated to 50°C. 4'-Methyl-2,2'-bipyridine-4-carboxylic acid (778 mg = 3.6 mmol) was dissolved in 20 mL DMF and dripped into the ruthenium solution slowly, under an argon atmosphere. The mixture was heated to 80°C and stirred for 4 h. It was cooled to room temperature and stored at 4°C overnight. On the following day, it was warmed to room temperature, and ethylene diamine (2.55 mL) was added. The reaction solution was heated to 80°C and stirred for 6 h, then concentrated under reduced pressure, and boiled with 10 mL isopropanol. The residue was combined with 50 mL pure water and 8 mL 60% hexafluorophosphoric acid, followed by 20 mL saturated hexafluorophosphoric acid. The separated crystals were washed with 20 mL pure water and dried under reduced pressure (Compound 1, 2.89 mmol, yield: 76%). Compound 1 (2.0 g = 2.75 mmol) was dissolved in 40

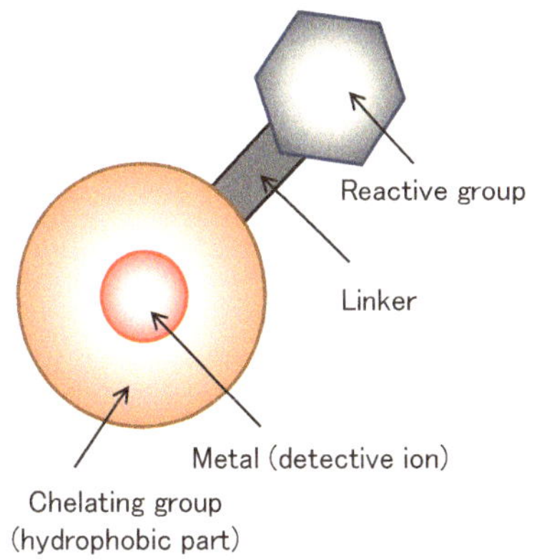

Figure 1. Constitution of the metal-coded affinity tag for low molecular compounds (schematic view).

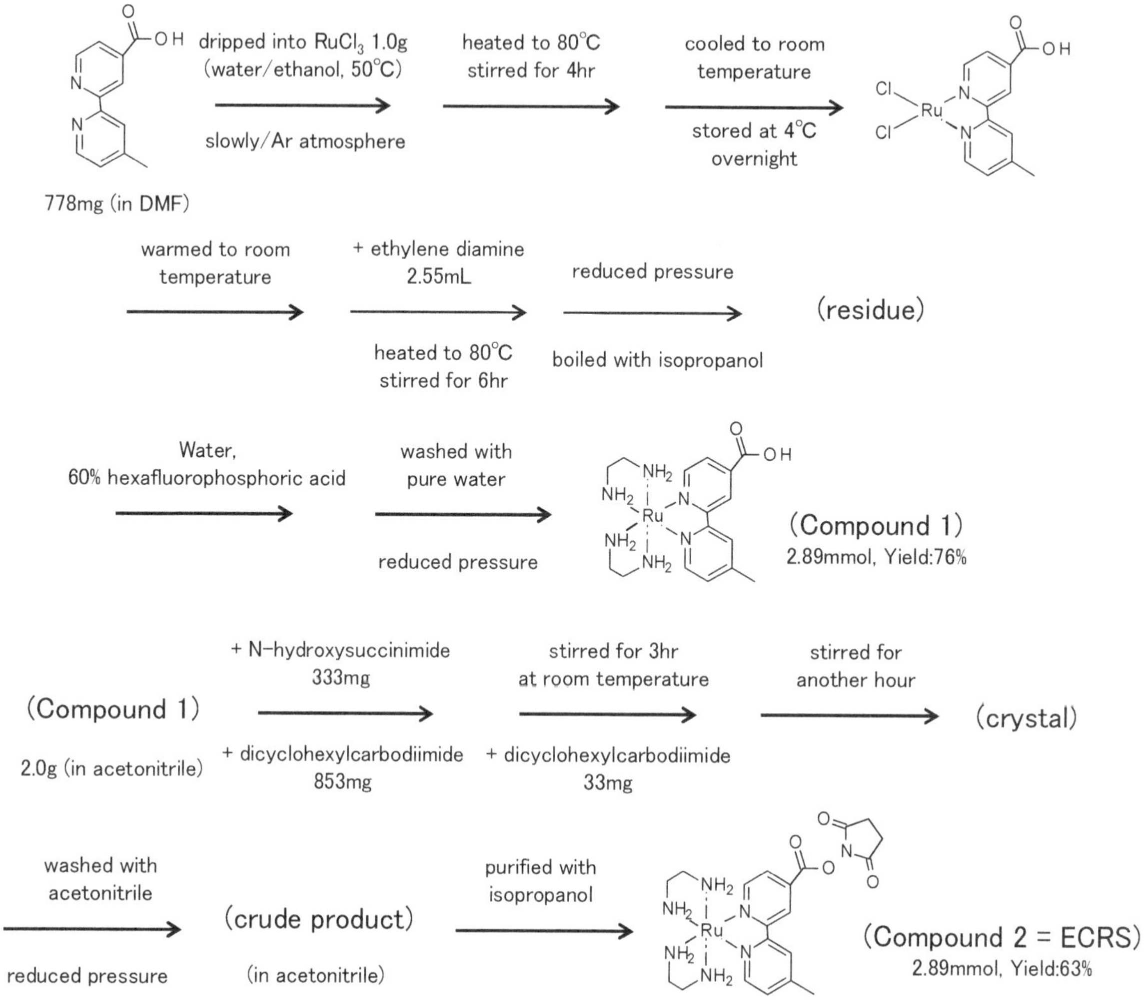

Figure 2. Synthesis scheme of the metal-coded affinity tag.

mL acetonitrile, then 333 mg N-hydroxysuccinimide and 853 mg dicyclohexylcarbodiimide were added. The solution was stirred at room temperature for 3 h, and after dicyclohexylcarbodiimide (33 mg) was added, it was stirred for another hour. The remaining crystals were washed with acetonitrile and concentrated under reduced pressure. The crude product was dissolved in 5 mL acetonitrile and purified with isopropanol (Compound 2 = ECRS, 2.89 mmol, yield 63%).

2.4. Instrumentation

HPLC experiments were performed with a Shimadzu modular system equipped with a control unit (CBM-20A), autosampler (SIL-20AC HT UFLC), UV detector (SPD-20A UFLC), column oven (CTO-20AC), two pumps (LC-20ADXR UFLC) and degasser (DGU-20A3). The system was coupled to a quadrupole-based ICP-MS (7700, Agilent Co., Inc., USA). An LC/MS system was used for Q-TOF MS (Q-TOF Premier, Waters Corporation, MA, USA).

2.5. ICP-MS Parameters

The sampling and skimmer cones were both composed of nickel. The rates of plasma gas flow, auxiliary gas flow and carrier gas flow were 15.0, 1.0 and 1.0 - 1.2 $L{\cdot}min^{-1}$, respectively. The collision gas (He) flow rate was 4.3 $mL{\cdot}min^{-1}$. High temperature He mode was effective in decreasing the back ground value. The radio frequency (RF) power was maintained at 1590 W. The sampling depth was set between 6 and 7 mm. The ICP-MS parameters were optimized for each investigation to achieve the best signal to noise ratio, using a standard solution containing 1 ppb each of lithium, yttrium and thallium. The sensitivity of the yttrium ion was about 150,000 cps/ppb. Detection was performed using ICP-MS, with ruthenium ion-selective detection at $m/z = 101$ or 102.

2.6. HPLC Conditions

The derivatized amino acids were injected into an Inertsil ODS-3 column, 3 μm particle size, 1.0 mm i.d. × 150 mm (GL Science Instruments, Inc. Tokyo, Japan). Mobile phase A was 10 mmol·L^{-1} acetate buffer (pH 4. 8). The acetate buffer was prepared as follows: 120 μL of acetic acid and 3.7 g of ammonium acetate were mixed and dissolved in 1 L of distilled, deionized water. Mobile phase B was acetonitrile-water (v/v = 60:40). Each mobile phase was filtered through a 0.45 μm membrane (Nalge Nunc International, Rochester, NY, USA).

The flow rate was 50 μL·min^{-1}, and the injection volume was 1 μL. The following gradient program was used: 0 - 5 min 5%, 5 - 40 min to 35%, 40 - 55 min 100% and 55 - 70 min 5% mobile phase B. The column was run at room temperature (about 25°C). The sample tray temperature was 4°C.

2.7. Q-TOFMS Parameters

Mass spectrometry was performed in the ESI positive mode. The desolvation gas was set to 700 L·h^{-1} at a desolvation temperature of 120°C, the cone gas was set to 50 L·h^{-1}, and the source temperature was set to 80°C. The capillary and cone voltages were set to 3000 and 20 V, respectively. The MCP detector voltage was set to 1, 600 V. Argon was employed as the collision gas. The collision gas flow was 0.6 mL·min^{-1} and the collision energy was 4 V. The scan range was from 100 to 1000 m/z.

2.8. Amino Acid and Dipeptide Derivatization with the New Metal Tag (ECRS)

The derivatization of glutamic acid with ECRS is shown in **Figure 3**. ECRS was prepared by dissolution to a concentration of 10 μmol·L^{-1} in acetonitrile, just before use. It was stored in a case at room temperature. Stock solutions of 0.1 μmol·L^{-1} of each amino acid and dipeptide, dissolved in 0.1 mol·L^{-1} HCl, were prepared. They were properly diluted and mixed before use. A 10 μL aliquot of the sample solution was mixed with 60 μL of 0.2 mol·L^{-1} sodium tetraborate (pH 8.0) and 20 μL of acetonitrile, and then a 10 μL portion of the tag solution was added. The metal tag was present in excess, relative to the amino acids and dipeptides in the reaction mixture. After mixing for 1 min using a vortex mixer, the mixture was heated at 55°C for 5 min in an aluminum bath. A 50 μL aliquot of 1 mol·L^{-1} ammonium acetate buffer (pH 4.8) was added to the reaction mixture, which was then thoroughly mixed with a vortex mixer. The mixture was stored in a tightly closed container at 4°C until it was subjected to HPLC/ICP-MS.

2.8. Analyses of Amino Acid and Dipeptide Derivatives

The amino acid and dipeptide derivatives were detected by HPLC, HPLC/ICP-MS, and LC-Q-TOFMS. Each amino acid and dipeptide derivatized with ECRS was detected at 254 nm, using the UV detector. Mixed amino acids and dipeptides derivatized together were discovered at m/z = 101, using HPLC/ICP-MS. To combine HPLC with ICP-MS, the HPLC outlet (0.05 mL·min^{-1}) was connected with PEEK tubing via a ceramic valve to a micro concentric nebulizer within a spray chamber. The mass spectra of the derivatives were obtained using LC-Q-TOFMS.

3. Results

We designed and synthesized a new metal tag (ECRS). ECRS is a functional reagent with reactivity to amine group and it contains the chelating group, too. The chelate of ECRS with ruthenium had suitable hydrophobicity and size for analysis by HPLC, using a C-18 reversed phase column. It was smaller than the strong hydrophobic tag (BCRS) used in our previous study.

In total, 1.4 g (yield 63%) of ECRS was obtained using our synthesis scheme. The ECRS was kept at room temperature until use. Its activity remained constant over a year.

We derivatized branched amino acids (BCAAs) with ECRS at various incubation times (5, 30, 60, 90, 120 min). The amino acids and dipeptides rapidly derivatized within 5 min with ECRS. In contrast, the time needed to react amino acids with BCRS was 90 min.

The chromatograms of 17 primary and secondary amino acids and 3 dipeptides derivatized with ECRS are shown in **Figure 4**. Each amino acid and dipeptide sample was prepared independently. We have recorded the UV chromatogram at 254 nm for each sample. The excess unreacted metal tag eluted before many amino compound derivatives, except for the acidic amino acids (glutamic acid and aspartic acid), formed. Lysine, which has two

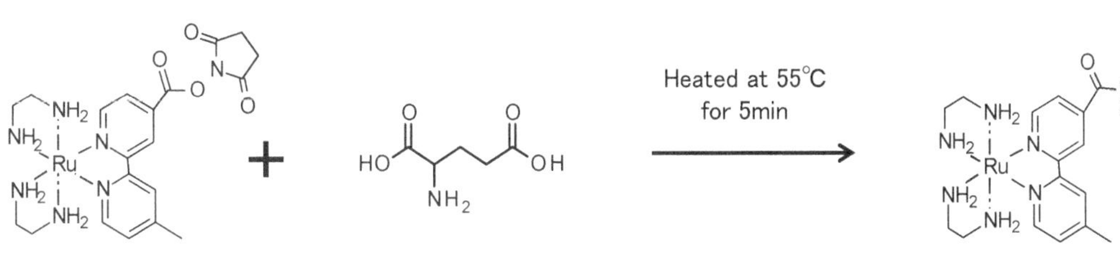

Figure 3. Derivatization of glutamic acid with the ECRS.

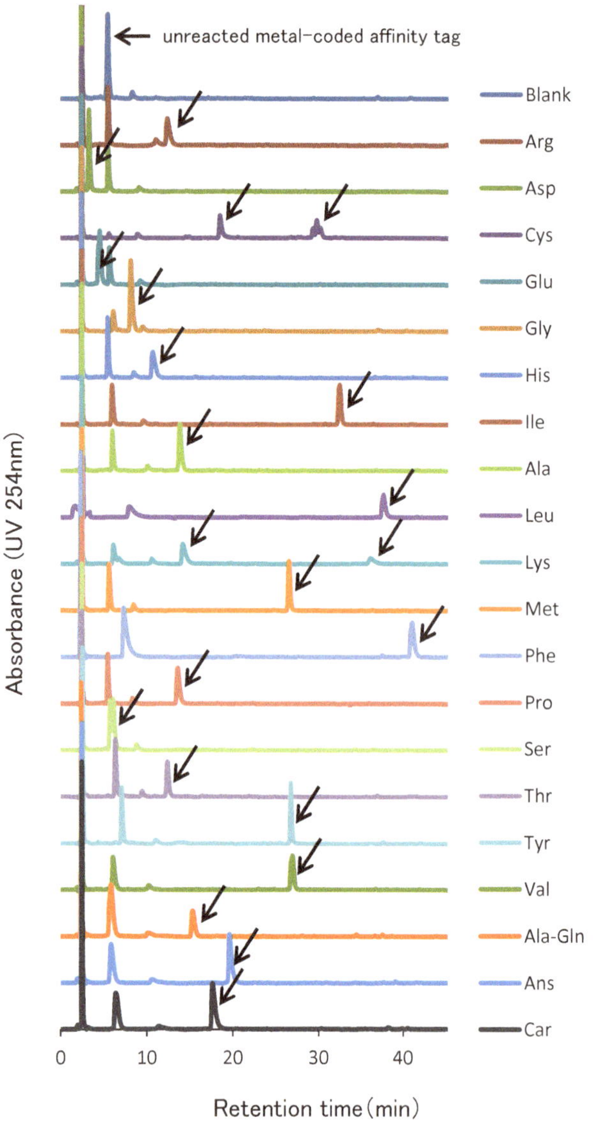

Figure 4. The Chromatograms of amino acids and dipeptides.

amino groups in the compound, formed two types of derivatives. We acquired the mass spectra of the derivatives by LC-Q-TOFMS, as shown in **Figure 5**.

The mixture of amino acids and dipeptides was derivatized with ECRS. The chromatogram for this solution, detected by HPLC/ICP-MS (m/z = 101), is shown in **Figure 6**. We used to detect m/z = 101 or 102. The signal intensity at m/z = 102 increased than m/z = 101, but background value was higher, too. We confirmed that the amino acids and the dipeptides were simultaneously derivatized.

The derivatization with ECRS was suitable for the quantitative analysis of amino acids. The calibration curves of BCAAs, in the concentration range of 0.75 - 750 μmol·L^{-1} in the standard solution (0.05 - 50 pmol per injection), are shown in **Figure 7**.

The detection limit of BCAAs was 1.5 nmol·L^{-1} in the standard solution (100 amol per injection) (signal to noise ratio = 3). The peak shapes of the dipeptides and threonine were broader than those of the BCAAs. The detection limit of the dipeptides and threonine was 15 nmol·L^{-1} under the standard solution conditions (1 fmol per injection). The correlation coefficients of the derivatized BCAAs were r^2 = 0.999. The relative standard deviation of 1.5 μmol·L^{-1} of the BCAA adducts was 1.5% (n = 5).

4. Discussion

To develop a new metal tag for HPLC/ICP-MS, we had to maintain the bond between the amino acid and the metal moiety in an aqueous solution, such as the HPLC mobile phase, and to separate the amino acid derivatives by reversed phase HPLC.

Most metal tags are a bi-functional reagent with a group that can be activated towards amino acids and a metal-binding group. Metal binding can, in general, be achieved with a chelator. Our novel metal tag was also the chelate compound, because we chose ethylenediamine as the metal chelator. The ruthenium ion remained associated with the complex under our HPLC conditions, because the ion was not detected in the void volume. We tried to synthesize the metal tag with the different metals (ex. indium and rare earth elements), but they were not stable.

The separation ability by HPLC was also important. The first major metal tag employed was the DOTA cys-specific labeling technique. It was recently reported that DTPA reacted with the amino terminus and the internal Lys. Both of these methods derivatized proteins and peptides. They are not suitable to analyze low molecular weight compounds, such as amino acids and dipeptides. It is too hydrophilic to separate with reverse phase column. On the other hand, reversed phase HPLC methods for the quantification and separation of amino acids were developed and we subsequently designed our new metal tag. Our tag was more hydrophobic than those previously used for proteins. Furthermore, it was superior because the metal chelator was not too large. The ethylenediamine moiety was more favorable for separation, compared with the bipyridine moiety.

We confirmed the structures of the derivatives by LC-Q-TOFMS (scan range 100 to 1000 m/z). We detected the precursor ion of the amino acid and dipeptide derivatives (**Figure 5**). The double charged ion was stronger than the single charged ion (**Figures 5(a)**, **(c)** and **(e)**). The chemical formula and formula weight of delivertized amino acids and dipeptide are shown in **Table 1**. All single charged precursor ions of the derivatives were in good agreement with the exact masses (**Figures 5(b)**,

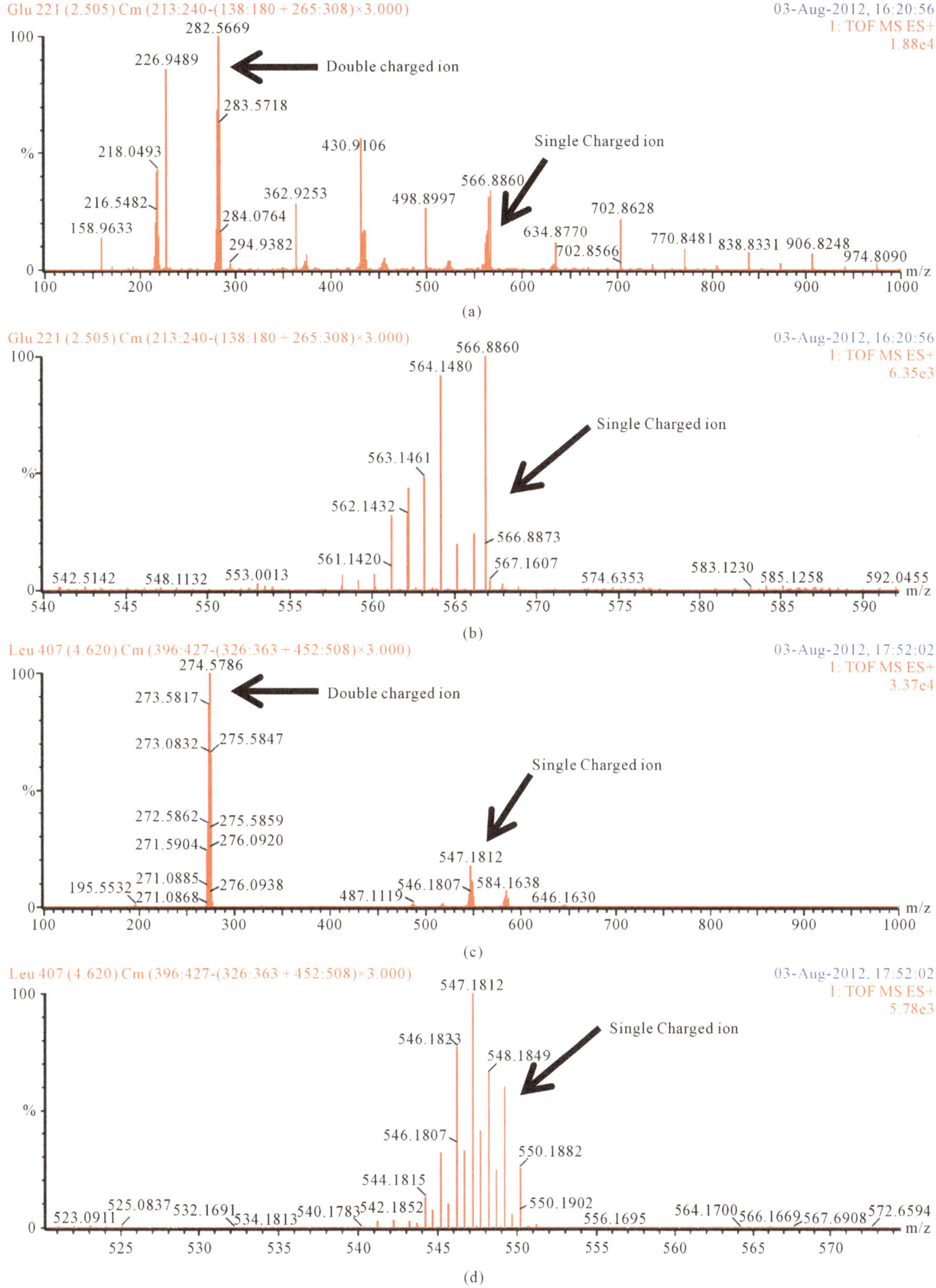
Glu 221 (2.505) Cm (213:240-(138:180 + 265:308)×3.000)
03-Aug-2012, 16:20:56
1: TOF MS ES+
1.88e4
Double charged ion
Single Charged ion
282.5669
226.9489
283.5718
218.0493
216.5482
158.9633
284.0764
294.9382
362.9253
430.9106
498.8997
566.8860
634.8770
702.8628
702.8566
770.8481
838.8331
906.8248
974.8090
%
m/z
(a)
Glu 221 (2.505) Cm (213:240-(138:180 + 265:308)×3.000)
03-Aug-2012, 16:20:56
1: TOF MS ES+
6.35e3
Single Charged ion
566.8860
564.1480
563.1461
562.1432
566.8873
561.1420
567.1607
542.5142
548.1132
553.0013
574.6353
583.1230
585.1258
592.0455
m/z
(b)
Leu 407 (4.620) Cm (396:427-(326:363 + 452:508)×3.000)
03-Aug-2012, 17:52:02
1: TOF MS ES+
3.37e4
Double charged ion
Single Charged ion
274.5786
273.5817
273.0832
275.5847
272.5862
275.5859
271.5904
276.0920
271.0885
276.0938
271.0868
195.5532
487.1119
546.1807
547.1812
584.1638
646.1630
m/z
(c)
Leu 407 (4.620) Cm (396:427-(326:363 + 452:508)×3.000)
03-Aug-2012, 17:52:02
1: TOF MS ES+
5.78e3
Single Charged ion
547.1812
546.1823
548.1849
546.1807
550.1882
544.1815
550.1902
523.0911
525.0837
532.1691
534.1813
540.1783
542.1852
556.1695
564.1700
566.1669
567.6908
572.6594
m/z
(d)

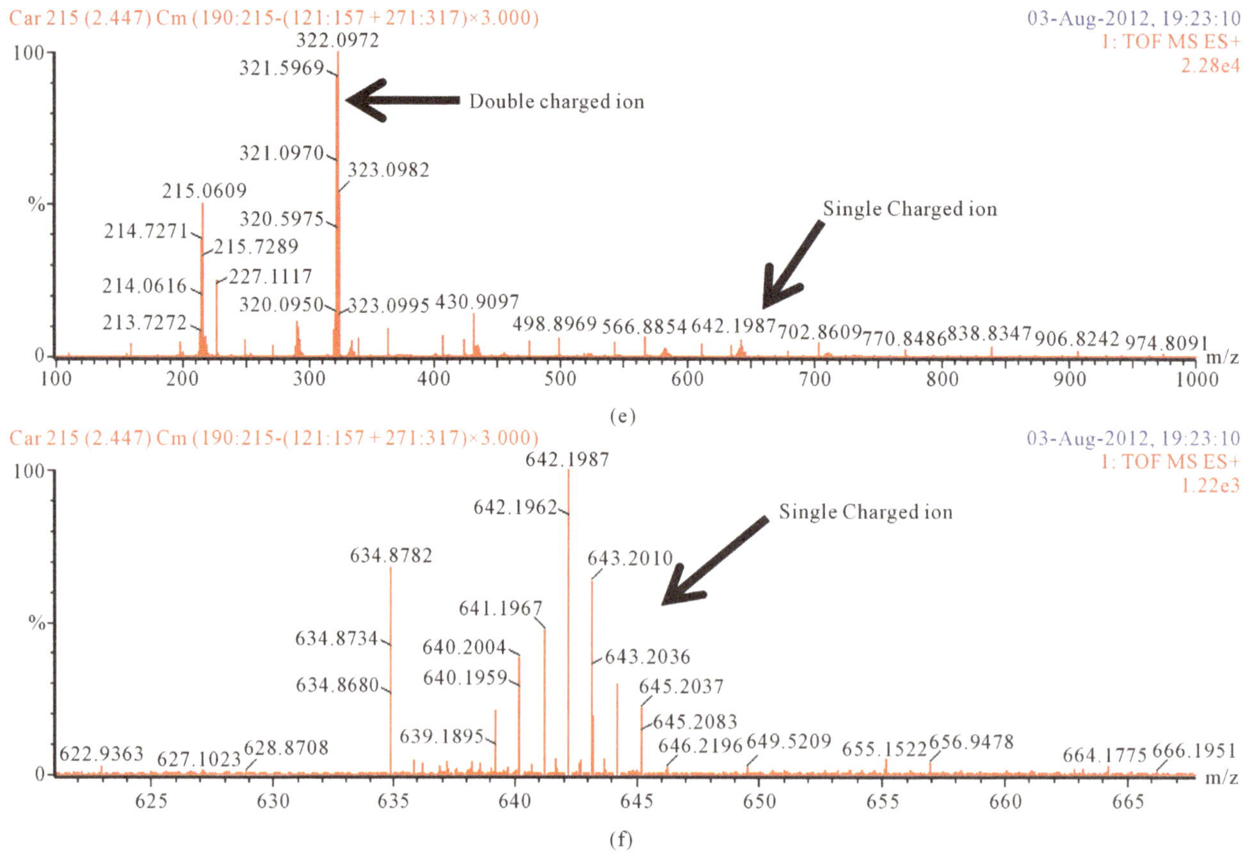

Figure 5. The mass spectra of delivertized amino acids and dipeptide. (a) Glutamic acid; scan range from 100 to 1000 m/z; (b) Glutamic acid; scan range from 540 to 595 m/z; (c) Leucine; scan range from 100 to 1000 m/z; (d) Leucine; scan range from 520 to 575 m/z; (e) Carnosine; scan range from 100 to 1000 m/z; (f) Carnosine; scan range from 620 to 570 m/z,

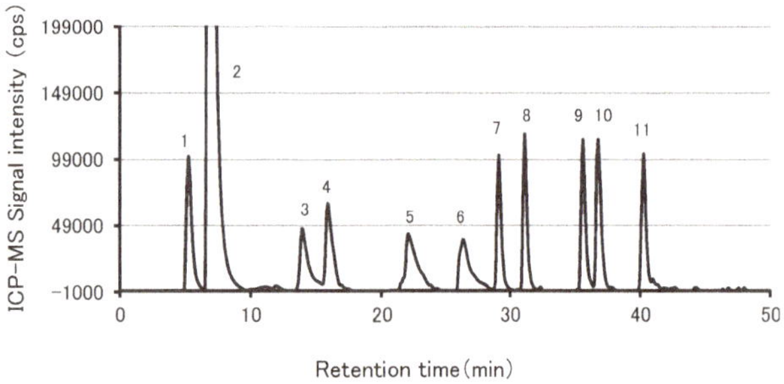

Figure 6. The chromatogram of 8 amino acids and 2 dipeptides derivatized with ECRS (m/z = 101). The concentration of all amino compounds in the stock solution is 75 μmol·L^{-1}. 1. glutamic acid, 2. unreacted ECRS, 3. threonine, 4. L-alanine, 5. carnosine, 6. anserine, 7. methionine 8. valine, 9, isoleucine, 10. leucine, 11. phenylalanine.

Table 1. The chemical formula and formula weight of delivertized amino acid and dipeptide.

	Chemical Formula	Formula Weight
Glu	$C_{22}H_{34}N_6O_5Ru$	563.621
Leu	$C_{23}H_{38}N_6O_3Ru$	547.666
Car	$C_{27}H_{40}N_8O_4Ru$	641.739

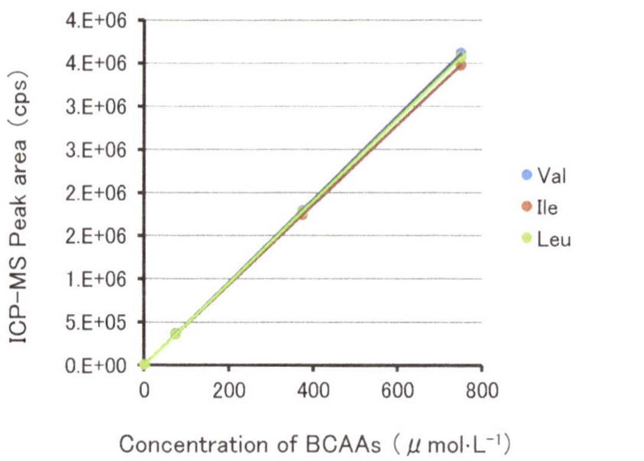

Figure 7. The calibration curves of the BCAAs. The vertical axis was the peak area with ICP-MS. The horizontal axis is concentration of BCAAs in the analytical solution. The correlation coefficients are $r_2 = 0.999$ (from 0.75 - 750 μmol·L^{-1}). The relative standard deviations through all analytical process are 1.5 % (at 1.5 μmol·L^{-1}, n = 5).

(d) and **(f)**). It was identified that the amino acids and dipeptides were deivatized as expected.

We separately derivatized each amino acid and dipeptide, and monitored them with a HPLC/UV detector (**Figure 4**). All derivatives were separated from the unreacted metal tag. We could detect 17 proteinogenic amino acids. The secondary amino acid, proline, was also derivatized. Proline was not derivatized by previous metal tag, BCRS. BCRS only derivatized primary amino acids. The lysine and cysteine derivatives generated two peaks, because they had two reaction sites. Although we derivatized lysine and cysteine at some incubation times, two kinds of derivatives, bound with one or two ECRS moieties, were observed. In this study, one mole of lysine did not react completely with two moles of the metal tag at the same condition of the other amino acids. To achieve the optimal derivatization condition for lysine and cysteine, the concentration of the metal tag, the pH of the borate buffer and the temperature of reaction should be individually explored.

The reaction time for the derivatization with ECRS was very short. The process was completed within 5 min. In a previous report, the derivatization reaction with BCRS took 1.5 h. The small chelator probably had a positive influence on reducing the reaction rate. In addition, we only considered the reaction time, but whereas ECRS can be left at room temperature, BCRS, in contrast, must be kept at −20°C.

In our analytical method, the amino acids and dipeptides were combined with the ruthenium atom by ECRS, and were detected with same sensitivity as the ruthenium ion. The actual detection limits of the amino acids through all processes were about 100 amol, which is better than the fluorescent analytical method [14,15], equivalent to the LC/MS/MS method (after pre-column derivatization) [17,18] and our previous method [37,39]. The calibration curves for the BCAAs are shown in **Figure 7**. The three curves exhibit equal slopes and intensities. In **Figure 4**, the peak heights of the derivatives differed but the peak areas were the same. The detection limits of dipeptides and threonine were higher than those of the other amino acids, because these peaks were broadened under our HPLC conditions for the comprehensive analysis. The sensitivity might be improved by optimizing the gradient elution method for each compound.

5. Conclusion

The new analytical method using the metal tag with HPLC/ICP-MS is not significantly affected by the sample matrix. ICP is one of the most powerful ion sources and it has high temperature (between 6000 and 10,000 K) and high electron number density (between 6.4×10^{14} and 1.2×10^{15} cm^{-1}). All analytes and the sample matrix are ionized at once. On the other hand, LC-MS is powerful tool that is very sensitive and selective, but it is often affected by impurities. When the analyte and the impurity elute at the same time, the intensity of the analyte is decreased, obviously. In some cases, we cannot obtain analytical values. In our method, we detected the ruthenium ion in the metal tag. This will be useful, because all analytes are detected with high accuracy and the same sensitivity regardless of sample matrix. In this study, we have designed and synthesized a new ruthenium complex, and have shown that our method could be developed for speciation further. For example, by changing the reactive functional group of the metal tag, it could be applied to another low molecular weight compounds, such as carboxylic acids.

6. Acknowledgements

The authors thank the synthesis team at the Nard Institute, Ltd., for support in the synthesis of the new metal tag compound.

We are grateful to Ms. Sachise Karakawa for providing the MS spectra from LC/MS measurements.

REFERENCES

[1] A. Watanabe, T. Higashi, T. Sakata and H. Nagashima, "Serum Amino Acid Levels in Patients with Hepatocellular Carcinoma," *Cancer*, Vol. 54, No. 9, 1984, pp. 1875-1882.

[2] K. J. Carpenter, A. E. Harper and R. E. Olson, "Experiments that Changed Nutritional Thinking," *The Journal of Nutrition*, Vol. 127, No. 5, 1997, pp. 1017S-1053S.

[3] R. M. Jennie, "Protein and Amino Acid Metabolism during and after Exercise and the Effects of Nutrition," *Annual Review of Nutrition*, Vol. 20, 2000, pp. 457-483.

[4] A. Wagner and D. Fell, "The Small World inside Large Metabolic Networks," *Proceedings of Royal Society*, Vol. 268, No. 1478, 2001, pp. 1803-1810.

[5] G. Weinlich, C. Murr, L. Richardsen, C. Winkler and D. Fuchs, "Decreased Serum Tryptophan Concentration Predicts Poor Prognosis in Malignant Melanoma Patients," *Dermatology*, Vol. 214, 2007, 214, pp. 8-14.

[6] E. Holm, O. Sedlaczek and E. Grips, "Amino Acid Metabolism in Liver Disease," *Current Opinion in Clinical Nutrition & Metabolic Care*, Vol. 2, No. 1, 1999, pp. 47-53.

[7] M. Holecek, "Three Targets of Branched-Chain Amino Acid Supplementation in the Treatment of Liver Disease," *Nutrition*, Vol. 26, No. 5, 2010, pp. 482-490.

[8] O. A. Obeid, "Plasma Amino Acid Concentrations in Patients with Coronary Heart Disease: A Comparison between U.K. Indian Asian and Caucasian Men," *International Journal for Vitamin and Nutrition Research*, Vol. 75, No. 4, 2005, pp. 267-273.

[9] D. Gerber, "Decreased Concentration of Free Histidine in Serum in Rheumatoid Arthritis, an Isolated Amino Acid Abnormality Not Associated with Generalized Hypoaminoacidemia," *The Journal of Rheumatology*, Vol. 2, 1975, pp. 384-392.

[10] Q. Zhang, M. Takahashi, Y. Noguchi, T. Sugimoto, *et al.*, "Plasma Amino Acid Profiles Applied for Diagnosis of Advanced Liver Fibrosis in Patients with Chronic Hepatitis C Infection," *Hepatology Research*, Vol. 34, 2006, pp. 170-177.

[11] Y. Ohtsuka and Y. Agishi, "Abnormal Amino Acid Metabolism in Diabetes Mellitus," *Nippon Rinsho*, Vol. 50, 1992, pp. 1631-1636.

[12] A. Mochizuki, A. Tamaoka, A. Shimohata, Y. Komatsuzaki and S. I. Shoji, "Ab42-Positive Non-Pyramidal Neurons around Amyloid Plaques in Alzheimer's Disease," *Lancet*, Vol. 355, No. 9197, 2000, pp. 42-44.

[13] A. R. Hipkis, "Carnosine, Diabetes and Alzheimer's Desease," *Expert Review of Neurotherapeutics*, Vol. 9, No. 5, 2009, pp. 583-585.

[14] T. Santa, A. Takeda, S. Uchiyama, T. Fukushima, *et al.*, "N-(4-Nitro-2,1,3-benzoxadiazoyl-7-yl)-N-methyl-2-amin oacetohydrazide (NBD-CO-Hz) as a Precolumn Fluorescent Derivatization Reagent for Carboxylic Acids in High-Performance Liquid Chromatography," *Journal of Pharmaceutical and Biomedical Analysis*, Vol. 17, No. 6-7, 1998, pp. 1065-1070.

[15] S. Uchiyama, T. Santa, N. Okiyama, T. Fukushima and K. Imai, "Fluorogenic and Fluorescent Labeling Reagents with a Benzofurazan Skeleton," *Biomedical Chromatography*, Vol. 15, No. 5, 2001, pp. 295-318.

[16] S. Iwatani, S. V. Dien, K. Shimbo, K. Kubota, *et al.*, "Determination of Metabolic Flux Changes during Fed-Batch Cultivation from Measurements of Intracellular Amino Acids by LC-MS/MS," *Journal of Biotechnology*, Vol. 128, No. 1, 2007, pp. 93-111.

[17] K. Shimbo, A. Yahashi, K. Hirayama, M. Nakazawa and H. Miyano, "Multifunctional and Highly Sensitive Precolumn Reagents for Amino Acids in Liquid Chromatography/Tandem Mass Spectrometry," *Analytical Chemistry*, Vol. 81, No. 13, 2009, pp. 5172-5179.

[18] K. Shimbo, T. Oonuki, A. Yahashi, K. Hirayama and H. Miyano, "Precolumn Derivatization Reagents for High-Speed Analysis of Amines and Amino Acids in Biological Fluid Using Liquid Chromatography/Electrospray Ionization Tandem Mass Spectrometry," *Rapid Communications in Mass Spectrometry*, Vol. 23, No. 10, 2009, pp. 1483-1492.

[19] D. H. Spackman, W. H. Stein and S. Moore, "Automatic Recording Apparatus for Use in the Chromatography of Amino Acids," *Analytical Chemistry*, Vol. 30, No. 7, 1958, pp. 1190-1206.

[20] S. Ruhemann, "Triketohydrindene Hydrate. Part IV. Hydrindantin and Its Analogues," *Journal of the Chemical Society*, Vol. 99, 1911, pp. 1306-1310.

[21] S. A. Cohen and D. J. Strydom, "Amino Acid Analysis Utilizing Phenylisothiocyanate Derivatives," *Analytical Biochemistry*, Vol. 174, No. 1, 1988, pp. 1-16.

[22] B. S. Glaeser and T. A. Hare, "Measurement of GABA in Human Cerebrospinal Fluid," *Biochemical Medicine*, Vol. 12, 1975, pp. 274-282.

[23] S. A. Cohen and K. M. D. Antonis, "Applications of Amino Acid Derivatization with 6-Aminoquinolyl-N-hydroxysuccinimidyl Carbamate Analysis of Feed Grains, Intravenous Solutions and Glycoproteins," *Journal of Chromatography A*, Vol. 661, No. 1-2, 1994, pp. 25-34.

[24] Y. Watanabe and K. Imai, "Serum Amino Acid Levels in Patients with Hepatocellular Carcinoma," *Analytical Biochemistry*, Vol. 116, No. 2, 1981, pp. 471-472.

[25] M. Piraud, C. Vianey-Saban, K. Petritis, C. Elfakir, *et al.*, "ESI-MS/MS Analysis of Underivatised Amino Acids: A New Tool for the Diagnosis of Inherited Disorders of Amino Acid Metabolism. Fragmentation sTudy of 79 Molecules of Biological Interest in Positive and Negative Ionisation Mode," *Rapid Communications in Mass Spectrometry*, Vol. 17, No. 12, 2003, pp. 1297-1311.

[26] W. C. Yang, H. Mirzaei, X. Liu and F. Regnier, "Enhancement of Amino Acid Detection and Quantification by Electrospray Ionization Mass Spectrometry," *Analytical Chemistry*, Vol. 78, No. 13, 2006, pp. 4702-4708.

[27] P. Ross, Y. Huang, J. Marchese, B. Williamson, *et al.*, "Multiplexed Protein Quantitation in Saccharomyces Cerevisiae Using Amine-Reactive Isobaric Tagging Reagents," *Molecular & Cellular Proteomics*, Vol. 3, 2004, pp. 1154-1169.

[28] C. Zhang, F. Wu, Y. Zhang, X. Wang and X. Zhang, "A Novel Combination of Immunoreaction and ICP-MS as a Hyphenated Technique for the Determination of Thyroid-Stimulating Hormone (TSH) in Human Serum," *Journal of Analytical Atomic Spectrometry*, Vol. 16, No. 12, 2001, pp. 1393-1396.

[29] C. Zhang, F. Wu and X. Zhang, "ICP-MS-Based Competitive Immunoassay for the Determination of Total Thyroxin in Human Serum," *Journal of Analytical Atomic Spectrometry*, Vol. 17, No. 10, 2002, pp. 1304-1307.

[30] Z. A. Quinn, V. I. Baranov, S. D. Tanner and J. L. Wrana, "Simultaneous Determination of Proteins Using an Element-Tagged Immunoassay Coupled with ICP-MS Detection," *Journal of Analytical Atomic Spectrometry*, Vol. 17, No. 8, 2002, pp. 892-896.

[31] V. I. Baranov, Z. Quinn, D. R. Bandura and S. D. Tanner, "A Sensitive and Quantitative Element-Tagged Immunoassay with ICPMS Detection," *Analytical Chemistry*, Vol. 74, No. 7, 2002, pp. 1629-1636.

[32] S. Zhang, C. Zhang, Z. Xing and X. Zhang, "Simultaneous Determination of Alpha-Fetoprotein and Free Beta-Human Chorionic Gonadotropin by Element-Tagged Immunoassay with Detection by Inductively Coupled Plasma Mass Spectrometry," *Clinical chemistry*, Vol. 50, No. 7, 2004, pp. 1214-1221.

[33] P. A. Whetstone, N. G. Butlin, T. M. Corneillie and C. F. Meares, "Element-Coded Affinity Tags for Peptides and Proteins," *Bioconjugate Chemistry*, Vol. 15, No. 1, 2004, pp. 3-6.

[34] A. Tholey and D. Schaumlöffel, "Metal Labeling for Quantitative Protein and Proteome Analysis Using Inductively-Coupled Plasma Mass Spectrometry," *TrAC Trends in Analytical Chemistry*, Vol. 29, No. 5, 2010, pp. 399-408.

[35] S. Crotti, C. Granzotto, W. R. Cairns, P. Cescon and C. Barbante, "Elemental Labeling for the Identification of Proteinaceous-Binding Media in Art Works by ICP-MS," *Journal of Mass Spectrometry*, Vol. 46, No. 12, 2011, pp. 1297-1303.

[36] D. Esteban-Fernandez, C. Scheler and M. W. Linscheid, "Absolute Protein Quantification by LC-ICP-MS Using MeCAT Peptide Labeling," *Analytical and Bioanalytical Chemistry*, Vol. 401, No. 12, 2011, pp. 657-666.

[37] D. Iwahata, K. Hirayama and H. Miyano, "A Highly Sensitive Analytical Method for Metal-Labelled Amino Acids by HPLC/ICP-MS," *Journal of Analytical Atomic Spectrometry*, Vol. 23, No. 8, 2008, pp. 1063-1067.

[38] B. M. Peek, G. T. Ross, S. W. Edwards, G. J. Meyer, *et al.*, "Synthesis of Redox Derivatives of Lysine and Related Peptides Containing Phenothiazine or Tris(2,2'-bipyridine)ruthenium(II)," *International Journal of Peptide and Protein Research*, Vol. 38, No. 2, 1991, pp. 114-123.

[39] D. Iwahata, M. Tsuda, T. Aigaki and H. Miyano, "Quantitative Analysis of Branched Amino Acids in a Single Fruit Fly by LC-ICP-MS after Pre-Column Derivatization with a Metal Tag Reagent," *Journal of Analytical Atomic Spectrometry*, Vol. 26, No. 12, 2011, pp. 2461-2466.

17

Estimation of Vitamins B-Complex (B_2, B_3, B_5 and B_6) of Some Leafy Vegetables Indigenous to Bangladesh by HPLC Method

Md. Nazmul Hasan[1], M. Akhtaruzzaman[1], Md. Zakir Sultan[2*]

[1]Institute of Nutrition and Food Science, University of Dhaka, Dhaka, Bangladesh; [2]Centre for Advanced Research in Sciences, University of Dhaka, Dhaka, Bangladesh.

ABSTRACT

The current experiment was conducted for the simultaneous determination of several water-soluble vitamins like riboflavin (vitamin B_2), niacin (vitamin B_3), pantothenic acid (vitamin B_5) and pyridoxine (vitamin B_6) in five highly consumed local leafy vegetables named as bottle gourd leaves (*Lagenaria vulgaris*) (local name Lau shak), green amaranth leaves (*Amaranthus viridis*) (local name Data shak), red amaranth leaves (*Amaranthus gangeticuss*) (local name Lal shak), Indian spinach (*Basella alba*) (local name Pui shak) and bitter gourd leaves (*Momordica charantia*) (local name Korola shak). The analyses were performed by HPLC using an analytical reversed phase C-18 (ODS column, 250 × 4.6 mm, 5 μm, Phenomenex, Inc.) column with the mobile phase consisting of a mixture of buffer (hexane sulphonic acid sodium, potassium dihydrogen phosphate and triethylamine, pH 3.0) and methanol in the ratio of 96:4 (v/v) at a flow rate of 1 mL/min with UV detection at 210 nm. The retention times for the vitamins were obtained as 3.61 min, 6.37 min, 9.51 min and 11.51 min for Vitamins B_2, B_3, B_5 and B_6, respectively. These obtained values of the vitamins were compared with the values available in published literatures of Deshio Khaddar Pustiman (DKPM), Indian food value (IFV) and United States Department of Agriculture (USDA).

Keywords: HPLC Analysis; Vitamin B_2; Vitamin B_3; Vitamin B_5; Vitamin B_6; Leafy Vegetables

1. Introduction

Leafy vegetables are those plants whose leaves or aerial parts are used as food [1]. Vegetables constitute essential components of the diet such as protein, vitamins, iron, calcium and other nutrients which are usually in very small quantity [2]. Increased awareness of the usefulness of vegetable inclusions in human food has enhanced their consumption as part of the daily diet. The therapeutic potential of the vegetables is usually attributed to the high content of vital vitamins as well as micronutrients [3]. Vegetables are highly recommended because of the relatively high nutritional value and their consumption gives diversity to daily food intake, adding flavor and taste to the diet [4]. Leafy vegetables are good source of vitamins and minerals. Even though vitamin is required a small amount per day in health, it plays a vital role in our health [5]. Vitamins are reported to reduce the damage by free radicals and check degenerative disease [6]. The vitamins B are a group of water soluble vitamins that play important roles in cell metabolism. The term B-group vitamins usually refer to thiamine, riboflavin, pyridoxine, niacin, pantothenic acid, biotin, cyanocobalamine and folic acid. The deficiency syndromes of some B vitamins are beri-beri (cardiac and dry), peripheral neuropathies, pellagra, and oral and genital lesions (related to riboflavin deficiency)—which were once major public health problems in parts of the world [7]. Several vitamins of the B-group act mainly as coenzymes in the metabolism of foodstuffs to produce energy [8].

As their chemicals structures are not related, a considerable number of papers have been published describing different physical, chemical and biological methods to analyze the vitamins [9]. The simultaneous determination of several water-soluble vitamins is difficult and often many different analyses have to be performed. Different instrumental methods have been used for the determina-

*Corresponding author.

tion of B-group vitamins, including electrochemical methods, *spectrophotometry*, derivative UV spectrophotometry, spectrofluorimetry, normal phase and reversed phase TLC and HPLC procedures as well as capillary electrophoresis. The determination of B-complex mainly in tablets using HPLC methods has been extensively described. The most widely used methods for the determination of B-group vitamins are reversed-phase HPLC, using a C18 column and aqueous–organic mobile phases, in acidic media [9]. However, this study investigated the B-vitamin profile namely riboflavin (vitamin B_2), niacin (vitamin B_3), pantothenic acid (vitamin B_5) and pyridoxine (vitamin B_6) in five highly consumed local leafy vegetables named as bottle gourd leaves (*Lagenaria vulgaris*) (local name Lau shak), green amaranth leaves (*Amaranthus viridis*) (local name Data shak), red amaranth leaves (*Amaranthus gangeticuss*) (local name Lal shak), Indian spinach (*Basella alba*) (local name Pui shak) and bitter gourd leaves (*Momordica charantia*) (local name Korola shak). These B vitamins were ana- lyzed with a reversed phase HPLC method equipped with UV detector and auto-injector using the modified meth- ods described by Aslam *et al.* [10]. To the best of our knowledge, there is no available data published in the literature regarding the vitamins content in these leafy vegetables grown in Bangladesh.

2. Materials and Methods

2.1. Collection of Samples

The current experiment was conducted on five leafy vegetable samples for the determination of their B vitamins content. The vegetables were bottle gourd leaves (*Lagenaria vulgaris*) (local name Lau shak), green amaranth leaves (*Amaranthus viridis*) (local name Data shak), red amaranth leaves (*Amaranthus gangeticuss*) (local name Lal shak), Indian spinach (*Basella alba*) (local name Pui shak) and bitter gourd leaves (*Momordica charantia*) (local name Korola shak).

Each of leafy vegetables was collected from nine regions of Bangladesh during the months of December 2011 to January 2012. The specimen samples were deposited in the laboratory of Institute of Nutrition and Food Science, University of Dhaka, Dhaka-1000, Bangladesh.

Samples were collected from different central markets of Dhaka city where the vegetables come from the different district of Bangladesh. Procedure for collection of food sample is important to get reliable representative nutrient values. Care was taken to avoid risk of inadvertent damage and deterioration of nutrients during transport from collection point to the laboratory. The vegetable samples were water sprayed to keep moisten, well packed in clean dark plastic poly bags to prevent water loss and damage by light, and then transported to lab within the shortest time span.

In the laboratory samples were prepared for analysis. The samples were washed well with double distilled water and the water was removed by blotting paper. The edible portion of the samples were then separated and weighted. These were chopped out into small pieces and homogenized. About 10 g sample from the chopped weight were taken for the estimation of B-vitamins. Some freshly chopped samples were preserved in refrigerator.

2.2. Reagents and Solvents

Standards of vitamins were purchased from Sigma-Aldrich, Taufkirchen, Germany. HPLC grade acetonitrile and methanol were purchased from Active Fine Chemicals Ltd., Dhaka, Bangladesh. Glacial acetic acid and orthophosphoric acid were purchased from BDH, Bengaluru-560025, India; triethylamine and hexane sulphonic acid sodium salt were purchased from Merck, Germany.

2.3. Instrumentation

High Performance Liquid Chromatographic system (Shimadzu-UFLC Prominence), equipped with an auto sampler (Model-SIL 20AC HT) and UV-Visible detector (Model-SPD 20A) was used for the analysis. The data were recorded using LC-solutions software.

2.4. Preparation of Extraction Solutions

Extraction solution was made by mixing 50 mL of acetonitrile with 10 mL of glacial acetic acid and the volume was finally made up to 1000 mL with double distilled water.

2.5. Preparation of Buffer and Mobile Phase

To prepare buffer, 1.08 g of hexane sulphonic acid sodium salt and 1.36 g of potasium dihydrogen phosphate were dissolved in 940 mL HPLC water and 5 mL of triethylamine was added to it and the pH was adjusted to 3.0 with orthophosphoric acid. For the preparation of mobile phase, buffer and methanol were mixed in a ratio of 96:4 and filtered through 0.22 μm membrane filter and sonicated for degassing in an ultrasonic bath.

2.6. Standard Preparation

Standard stock solution for vitamin B_2 (riboflavin) was prepared by dissolving 6.9 mg of riboflavin in 100 mL of extraction solution. Standard stock solution for vitamin B_3 (niacin) was prepared by dissolving 41.5 mg of niacin in 25 mL of double distilled water. Standard stock solution for vitamin B_5 (calcium salt of pantothenic acid) was prepared by dissolving 21.4 mg of calcium d-pantothenate in 25 mL of double distilled water. Standard stock solution for vitamin B_6 (pyridoxine) was prepared by dissolving 20.8 mg of pyridoxine hydrochloride in 25 mL of double distilled water.

2.7. Preparation of Samples

10 g of each sample was weighed and made into homogenized in mortar with pestle and transferred into conical flask and 25 mL of extraction solution was added, kept on shaking water bath at 70°C for 40 min. Thereafter, the sample was cooled down, filtered and finally the volume was made up to 50 mL with extraction solution. Then sample filtered through 0.45 μm filter tips and aliquots of 20 μL from this solution was injected into the HPLC by using auto-sampler.

2.8. Chromatographic Conditions

Analytical reversed phase C-18 column (ODS column, 250 × 4.6 mm, 5 μm, Phenomenex, Inc.) was used for the separation. Mobile phase consisting of a mixture of buffer and methanol in the ratio of 96:4 (v/v) was delivered at a flow rate of 1 mL/min with UV detection at 210 nm. The mobile phase was filtered through 0.22 μm membrane filter, sonicated and degassed before use. Analysis was performed at room temperature (~26°C) temperature. All the prepared sample solutions were first chromatographed to ensure that interfering peaks were not present. 20 μL aliquots of the standard solutions and sample solutions were injected.

3. Results and Discussion

HPLC is one of the most convenient and accurate analytical techniques. To analyze the vitamin B complex from leafy vegetables, a RP-HPLC method was used which was described in the literature [10], by using the mobile phase comprising of buffer and methanol in the ratio of 96:4 (v/v) over C-18 column (250 × 4.6 mm, 5 μm, Phenomenex, Inc.) at ambient temperature. The flow rate was at 1 mL/min and the eluent was monitored by UV detector at 210 nm. The retention times of the vitamins were obtained as described in **Table 1**.

Table 1. Retention times of B-vitamins at 210 nm.

Name	Rt (minutes)
Vitamin B_2	3.61
Vitamin B_3	6.37
Vitamin B_5	9.51
Vitamin B_6	11.51

The method was found to be linear for the concentration range described in **Table 2**. When average peak areas were plotted against concentration levels a good correlation coefficients (r^2) were obtained as 0.99978, 0.99987, 0.99863 and 0.99996 for riboflavin, niacin, pantothenic acid and pyridoxine, respectively (**Table 2**). The slopes (m) of the calibration curve were found to be as 513339.785, 71752.58, 6583.039 and 67668.38774 for riboflavin, niacin, pantothenic acid and pyridoxine, respectively. The intercepts (c) of the calibration curve also were obtained to be as 119358.225, 96524.8, 24465.6 and 51841.6 for riboflavin, niacin, pantothenic acid and pyridoxine, respectively (**Table 2**, **Figure 1**).

The HPLC method was successfully performed for the estimation of riboflavin (vitamin B_2), niacin (vitamin B_3), pantothenic acid (vitamin B_5), and pyridoxine (vitamin B_6) in leafy vegetables.

The amount of vitamin B_2 (riboflavin) in five leafy vegetables samples ranged from 0 to 0.523 mg/100 g (**Table 3**). The content of vitamin B_2 was found highest in green amaranth leaves (Data shak) (0.523 mg/100 g). Red amaranth leaves (Lal shak) contained second highest amount of vitamin B_2 (0.442 mg/100 g). Indian spinach (Pui shak) and bottle gourd leaves (Lau shak) have 0.397 mg/100 g and 0.321 mg/100 g vitamin B_2, respectively. Among the five leafy vegetables vitamin B_2 was lowest in bitter gourd leaves (Korola shak) (0.137 mg/100 g).

The highest amount of vitamin B_3 0.512 mg/100 g was estimated in bitter gourd leaves (Korola shak). Red amaranth leaves (Lal shak) contained 0.016 mg/100g; it was second the highest position of vitamin B_3. Indian spinach (Pui shak) and green amaranth leaves (Data shak) contained slight amount of niacin. Vitamin B_3 was not found in bottle gourd leaves (Lau shak).

The content of vitamin B_5 was not identified in green amaranth leaves (Data shak), red amaranth leaves (Lal shak), Indian spinach (Pui shak), bitter gourd leaves (Korola shak) and bottle gourd leaves (Lau shak), respectably (**Table 3**).

Availability of vitamin B_6 is comparatively less/absent in the leafy vegetables. The highest content of vitamin B_6 was present in bottle gourd leaves (Lau shak) which were 0.755 mg/100 g. Red amaranth leaves (Lal shak) contained 0.152 mg/100 g of vitamin B_6. Green amaranth leaves (Data shak) contained slight amount of pyridoxine (vitamin B_6), where it could not identify in bitter gourd leaves (Korola shak) and Indian spinach (Pui shak).

Comparison of B vitamin values between the present studies and the literature data is presented in **Table 4**. Literature data for vitamin B_5 and B_6 in selected leafy vegetables are unavailable. In the present study vitamin B_2 was detected in bottle gourd leaves 0.321 mg/100 g where Deshio Khaddar Pustiman (DKPM) [11], Indian

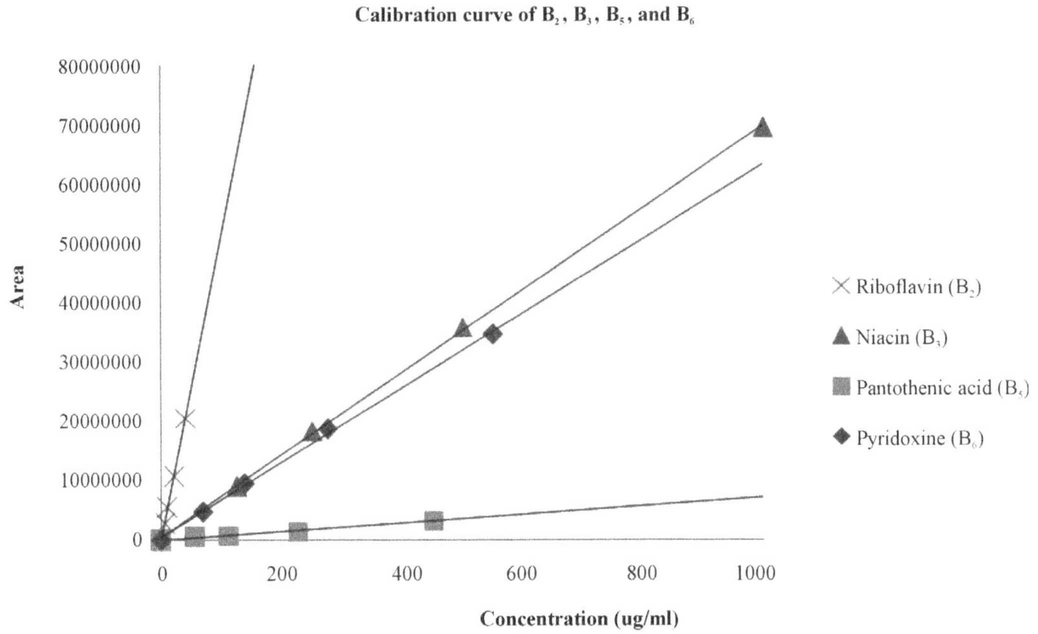

Figure 1. Calibration curve of vitamin B_2 (riboflavin), vitamin B_3 (niacin), vitamin B_5 (pantothenic acid) and vitamin B_6 (pyridoxine).

Table 2. The important parameters of the linearity curves for the standards.

Vitamin		$y = mx + c$	r^2	Conc. range (μg/mL)
B_2	Riboflavin	$y = 513339.785x + 119358.225$	0.99978	5 - 40
B_3	Niacin	$y = 71752.58x + 96524.8$	0.99987	125 - 500
B_5	Pantothenic acid	$y = 6583.039x + 24465.6$	0.99863	56.25 - 225
B_6	Pyridoxine	$y = 67668.38774x + 51841.6$	0.99996	68.75 - 275

m = slope; c = intercept; r = correlation coefficient.

Table 3. Vitamin B content of leafy vegetables.

English name	Scientific name	Bengali or local name	Average content of vitamin B (mg/100g of edible portion)			
			$B_2 \pm$ SD	$B_3 \pm$ SD	$B_5 \pm$ SD	$B_6 \pm$ SD
Bottle gourd leaves	*Lagenaria vulgaris*	Lau shak	0.321 ± 0.008	0.000	0.000	0.755 ± 0.116
Green amaranth leaves	*Amaranthus viridis*	Data shak	0.523 ± 0.003	0.051 ± 0.037	0.000	0.07 ± 0.002
Red amaranth leaves	*Amaranthus gangeticuss*	Lal shak	0.442 ± 0.009	0.016 ± 0.002	0.000	0.152 ± 0.006
Indian spinach	*Basella alba*	Pui shak	0.397 ± 0.007	0.023 ± 0.002	0.000	0.000
Bitter gourd leaves	*Momordica charantia*	Korola shak	0.137 ± 0.019	0.512 ± 0.103	0.000	0.000

"SD = Standard deviation".

food value (IFV) [12] and United States Department of Agriculture (USDA) [13] did not show the result. Bitter gourd leaves contain 0.137 mg/100 g B_2, which was not done in DKPM, IFV and USDA, respectively. Indian spinach contains 0.397 mg/100g of vitamin B_2, while the Indian study reported that Indian spinach contains 0.26 mg/100 g, the value was not found in DKPM and USDA. In the present study the content of B_2 was almost double in green amaranth leaves (0.523 mg/100 g) compare to the IFV and DKPM (0.18 mg/100 g). On the other hand, the value obtained from red amaranth Leaves 0.442 mg/100g was nearer to values of IFV but there was little difference with DKPM (**Table 4**).

The content of vitamin B_3 was investigated for all of selected leafy vegetable samples. Vitamin B_3 content of bitter gourd leaves was found 0.512 mg/100g whereas the value was not found IFV, DKPM and USDA, respectively. In this study vitamin B_3 was not found in bottle

Table 4. Comparison with published data.

Sample	Reference	Vitamin B_2 (riboflavin)	Vitamin B_3 (niacin)	Vitamin B_5 (pantothenic acid)	Vitamin B_6 (pyridoxine)
Bottle gourd leaves	A.V	0.321	0	0	0.755
	DKPM	-	-	-	-
	IFV	-	-	-	-
	USDA	-	-	-	-
Green amaranth leaves	A. V	0.523	0.051	0	0.07
	DKPM	0.18	0.18	-	-
	IFV	0.18	0.18	-	-
	USDA	0.158	0.658	-	0.192
Red amaranth leaves	A. V	0.442	0.0162	0	0.152
	DKPM	0.13	0.13	-	-
	IFV	0.30	0.30	-	-
	USDA	-	-	-	-
Indian spinach	A. V	0.397	0.023	0	0
	DKPM	-	-	-	-
	IFV	0.26	0.26	-	-
	USDA	-	-	-	-
Bitter gourd leaves	A. V	0.137	0.512	0	0
	DKPM	-	-	-	-
	IFV	-	-	-	-
	USDA	-	-	-	-

DKPM = Deshio Khaddar Pustiman [11], IFV = Indian Food Value [12], USDA = United States Department of Agriculture [13], AV = Average Value, and "-" = Not Done.

gourd leaves, where IFV, DKPM and USDA were not done the estimation before. Beside these the content of vitamin B_3 of red amaranth leaves was found 0.0162 mg/100 g, where DKPM and IFV got 0.13 and 0.30 mg/100 g, respectively. Green amaranth leaves were found 0.051 mg/100 g of vitamin B_3 where DKPM, IFV and USDA were found 0.18, 0.18 and 0.658 mg/100 g, respectively. In Indian spinach, the value was 0.023 mg/100 g where IFV was got 0.26 mg/100 g but DKPM and USDA were not done the estimation (**Table 4**).

The five leafy vegetable samples was not contained any vitamin B_5. Literature data for vitamin B_5 in selected leafy vegetables are unavailable (**Table 4**).

Literature data for vitamin B_6 in selected leafy vegetables are also unavailable. We found that bottle gourd leaves and red amaranth leaves contained vitamin B_6 as 0.755 mg/100g and 0.152 mg/100g, respectively, where green amaranth leaves contained slight amount and it was absent in Indian spinach and bitter gourd leaves (**Table 4**). Some values obtained from the present study were found to be different for different samples than the previous reports. This variation may occur due to various factors such as the environmental aspects like season, rainfall, climate, geographical and geological. Maturity of vegetables may also be partly responsible for this variation.

4. Conclusions

Health is very important to survival. A healthy and balanced diet is required to maintain a good health. To have a healthy diet everybody should know the nutrient composition of the foods they eat as well as the important nutrient content and amount of vitamin on their foods. In Bangladesh no recent data is available about the composition of foods that are eaten by the local and ethnic people. Moreover, Bangladeshi and Indian frequently suffer from lack of B vitamins in their daily diets.

Vitamin B deficiency is one of the major dietary problems in developing countries like Bangladesh. Poor people cannot intake vitamin B from rich foods like liver, egg, meat, fish etc. They depend on vegetables and fruits for vitamins and minerals. The study will provide the vitamins B composition of some leafy vegetables. Here we reported of vitamins B in a limited number of samples. We hope that, these new values will serve as a useful means to calculate dietary intake of B vitamins in the general population. These data will also be helpful in the preparation of a complete food composition table which will then be used everywhere, e.g. for the preparation of diet therapy, for food based dietary guidelines, for nutrition education, for food security, safety and regulation, for the labeling of food in food industry, for nutritional survey and also for other research purposes. Encouraging regular intake of those leafy vegetables containing B vitamins would remove vitamin B deficiency in Bangladesh. There is worldwide call for food composition database. With the increasing concern of the relationships between diet, food habits and degenerative diseases, development of FCD/FCT is a time demand issue. FCD provides support to variety of research activities in the food security mapping [14]. The third world countries are far behind to address this issue. South Asian countries, Particularly SAARC countries appear to be still in the initial stage of development of food composition database (FCD) as well as of food composition tables.

REFERENCES

[1] T. A. Yekeen, O. A. Adetiba, M. A. Azeez, M. A. Falodun, S. I. Akintaro and T. A. Yekeen, "Studies on the Proximate Analysis of *Solanum aethiopicum*, *Lactuca taraxacifolia* and *Talinum triangulare* and Potential Cytotoxic Effects of Their Aqueous Extract Using *Allium cepa* Assay," *Annals of Biological Research*, Vol. 2, No. 5, 2011, pp. 696-706. http://scholarsresearchlibrary.com/ABR-vol2-iss5/ABR-2011-2-5-696-706.pdf

[2] H. C. Thomson and W. C. Kelly, "Vegetable Crops," 5th Edition, McGraw Hill Publishing Co. Ltd., New Delhi, 1990.

[3] A. Adetutu, A. A. Adesokan and A. A. Ezekiel, "Determination of Vitamins in Five Selected West African Green Leafy Vegetables," *Journal of American Science*, Vol. 9, No. 2, 2013, pp. 40-43.

[4] M. Modi, A. T. Modi and S. Hendriks, "Potential Role for Wild Vegetables in Household Food Security: A Preliminary Case Study in Kwazulu-Natal, South Africa," *African Journal of Food Agriculture Nutrition and Development*, Vol. 6, No. 1, 2006, pp. 1-13. http://www.ajfand.net/Volume6/No1/Modi1850.pdf

[5] A. A. Olaniyi, "Essential Medical Chemistry," 2nd Edition, Shaneson, 2000.

[6] R. A. Jacab and G. Sotoudeh, "Vitamin C Function and Status in Chronic Disease," *Nutrition in Clinical Care*, Vol. 5, No. 2, 2002, pp. 66-74.

[7] FAO/WHO, "Thiamin, Riboflavin, Niacin, Vitamin B6, Pantothenic Acid and Biotin," Chapter 3, 2003.

[8] G. M. F. Ball, "Vitamins in Foods. Analysis, Bioavailability, and Stability," CRC Press, Boca Raton, 2006.

[9] R. Amidzic, J. Brboric, O. Cudina and S. Vladimirov, "RP-HPLC Determination of Vitamins B1, B3, B6, Folic Acid and B12 in Multivitamin Tablets," *Journal of the Serbian Chemical Society*, Vol. 70, No. 10, 2005, pp. 1229-1235.

[10] J. Aslam, M. S. Mohajir, S. A. Khan and A. Q. Khan, "HPLC Analysis of Water-Soluble Vitamins (B1, B2, B3, B5, B6) in *in Vitro* and *ex Vitro* Germinated Chickpea (*Cicer arietinum* L.)," *African Journal of Biotechnology*, Vol. 7, No. 14, 2008, pp. 2310-2314.

[11] INFS, Deshio Khaddyer Poostiman (Food Value of Local Foods) or DKPM," University of Dhaka, Dhaka, 1992.

[12] G. Gopalan, B. V. Ramasastri and S. C. Balasubramaniam, "Nutritive Value of Indian Foods [IFV]," National Institute of Nutrition, Indian Council of Medical Research, Hyderabad, 2007.

[13] United States Department of Agriculture (USDA), Agriculture Research Service, Food Composition Database. http://ndb.nal.usda.gov/ndb/search/list

[14] H. Greenfield and D. A. T. Southgate, "Selection of Foods," In: FAO, Ed., *Food Composition Data: Production, Management and Use*, Chapter 1, Rome, 2003, pp. 33-46.

Evaluation of Major Factors Affecting Spatial Resolution of Gamma-Rays Camera

Hongwei Xie*, Jianhua Zhang, Jinchuan Chen, Faqiang Zhang, Linbo Li, Jianming Qi, Yanyun Chu

Institute of Nuclear Physics and Chemistry, China Academy of Engineering Physic, Mianyang, China.

ABSTRACT

The spatial resolution of the gamma-rays camera was measured on a ^{60}Co gamma-rays source with edge method. The gamma-rays camera is consisting with rays-fluorescence convertor, optical imaging system, MCP image intensifier, CCD camera, electronic control system and other devices, and is mainly used in the image diagnostics of the intense pulse radiation sources [1]. Due to the relatively big quantum detective efficiency (DQE) and quantum gain of the gamma-rays, etc., the experimental data were processed by averaging multiple images and fitting curves. According to the experimental results, the spatial resolution MTF (modulation transfer function) at the 10% intensity was about 2 lp/mm. Meanwhile, because of the relatively big dispersion effects of the fluorescence transmissions in the scintillator and the optical imaging system, the maximal single-noise ratio (SNR) of the camera was found to be about 5:1. In addition, the spatial resolution of the camera was measured with pulse X-rays with 0.3 MeV in average energy and exclusion of the effects of secondary electrons from consideration. Accordingly, the spatial resolution MTF at the 10% intensity was about 5 lp/mm. This could be an additional evidence to verify the effects of secondary electrons induced by the 1.25 MeV gamma-rays in the scintillator upon the spatial resolution. Based on our analysis, the dispersion sizes of the secondary electrons in the scintillator are about 0.4 mm - 0.6 mm. Comparatively, as indicated by the detailed analysis of the spatial resolutions of the MCP image intensifier and CCD devices, both of them have little effect on the spatial resolution of the gamma-rays camera that could be well neglected.

Keywords: Gamma-Rays Camera; Point Spread Function (PSF); Gamma-Rays Image Detector; Scintillator; MCP Image Intensifier; CCD

1. Introduction

A γ-rays camera was developed for the image diagnostics of the intense pulsed γ-rays radiation sources, which was consisting with rays-fluorescence convertor, optical imaging system, MCP + CCD, electronic control system and other devices. Due to its good performances in various tunable (adjustable) parameters including exposure time spot, exposure time duration and gain, etc., the camera has been widely used in the framing image diagnostics of the pulse radiation fields [1,2].

The point spread function (PSF) of the image diagnostic system is not only the major technical specification of image system performances, but also a major factor to cause the image degradation. For a given or a certain PSF, it could be considered as a major criteria for the uncertainty evaluation of a given image and for the image super-resolution reconstruction. Other parameters to describe the system spatial resolution are including modulation transfer function (MTF) and the linear spread function (LSF), etc. The integration of PSF along the 1-D direction would then provide the LSF, which would then provide MTF after the Furrier Transform. All of three parameters (PSF, LSF and MTF) could reflect the spatial resolution of the rays system [3].

The edge method is a common method for the measurement of spatial resolution of γ-rays cameras [3]. In its application, the edge would be placed in a 2° - 3° with the pixel array to minimize the effects of the sub-pixels upon the spatial resolution, and the data would be fitted to the curves to proved a very smooth measurement data at the same time. However, in dealing with the spatial resolution measurement of the γ-rays cameras with thick

*Corresponding author.

scintillator and/or high energy, such a data processing might be technically difficult in various aspects as follows [4]: 1) the big DQE fluctuations induced by the interactions between the high energy γ-rays and scintillator; 2) the relatively big quantum gain of the γ-rays-fluorescence convertor; 3) the relatively more measurement steps of the system. Another important factor to affect the spatial resolution is the measurement condition. Presumably, an ideal parallel radiation source with step distribution would be most desirable for the measurement. However, the measurement of the spatial resolution would inevitably affected by various factors, including the size of radiation source, the system structure and the working conditions of system, etc. Moreover, the data process from edge spread function (ESF) to LSF and then to LSF would also amplify the noise. All of the factors add up to lead a relatively big uncertainty in the spatial resolution of the high energy γ-rays cameras. In order to contain the DQE effects upon the measurement results, serial procedures are applied as follows. Firstly, multiple ESFs would be averaged. Then, the averaged ESF would then be subject to linear fitting to provide a relatively uniform and smooth ESF. And the LSF would be obtained after the integration of the ESF. Finally, the MTF and PSF could be well available after a certain process.

The geometrical spatial resolution of the γ-rays camera is about 0.1 mm/pixel. And in order to provide a higher DQE, some special methods were introduced, e.g. to use the optical structure with short distance and big field of view and to increase the scintillator, etc. The spatial resolution of the γ-rays camera was measured on a Co radiation source with edge method. At the same time, the spatial resolutions of the camera components were studied respectively for the scintillator, MCP image intensifier and CCD camera, etc., to provide some valuable reference for the further improvement of the spatial resolution of the γ-rays camera.

2. Experimental Set-Ups

2.1. Gamma-Rays Camera

The γ-rays camera was developed by Institute of Nuclear Physics and Chemistry, China Academy of Engineering Physic (INPC, CAEP) for scientific experiments. Characterized in high sensitivity, high resolution and big dynamic range, the camera is mainly used in the image diagnostics of the high energy γ-rays sources. The limit field of view of the camera is Φ 100 mm. The process could roughly be interpreted as follows. Firstly, the radiation source is imaged onto the imaging plane by the γ-rays. Then the fluorescence image is imaged onto the incident plane of the MCP image intensifier through the optical imaging system and conversion system. Finally, the intensified image is recorded by CCD device, and the MCP and CCD is coupled with fiber plate. Thus, the major tow parts of the system are γ-rays imagines system and the image recording system. During the process, a 10 mm YAG crystal is used to convert the γ-rays into the fluorescence image. Besides, a copper reflector with an efficiency of over 95% is placed in a 45° angle with γ-rays direction to reflect the fluorescence image. Another purpose of the reflector is to avoid the direct γ-rays irradiation onto the MCP image intensifier and CCD camera. The specially developed imaging system could provide an amplification factor of 5:1 and a light collection efficiency of over 95%. The MCP image intensifier was manufactured by Proxitronic Co. Ltd with a spatial resolution of over 37 lp/mm. Besides, the intensifier is also characterized in tunable shutter time and electron gain, which could facilitate the imaging of the γ-rays at different time. Finally, the amplification factor of the fiber plate used to couple the MCP and CCD is 1.5:1, providing a stable and reliable recording system. And the CCD camera used in the experiment was provided by Andor Co. Ltd specifically for scientific purpose, with a pixel array of 1024 × 1024, pixel size of 13.3 μm × 13.3 μm, and a digital data output in 16 bit. Moreover, the CCD could be used in a very low temperature even below −65°C, making it a favorite option for the imaging diagnostic with low signal intensity. The total system is placed into an iron container to provide a sealed and shielded environment to avoid the effects of the electromagnetic effects upon the measurement result.

2.2. Spatial Resolution Measurement with Edge Method

The experiment was carried out on the intense Co source in CAEP. The schematic experimental set-up is given in **Figure 1**. The 60 Co γ-rays source used in the experiment is a line source with a geometrical parameter of Φ

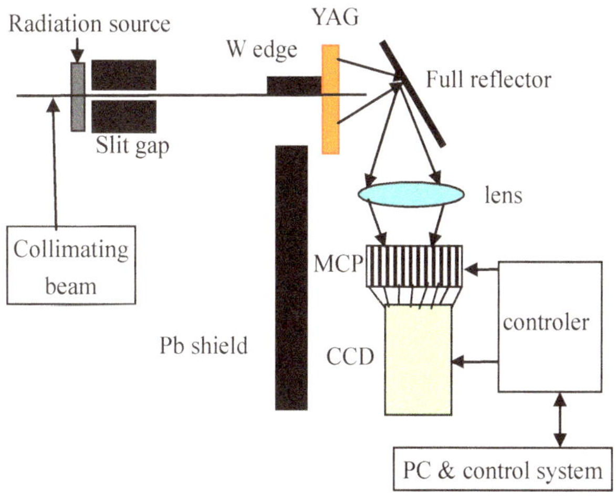

Figure 1. The schematic diagram for the PSF measurement of the γ-rays camera.

11 mm × 450 mm. And the source intensity for every line is about 10^4 Ci (1 Ci = 3.7×10^{14} Bq). A 5 mm × 50 mm Pb silt gap (200 mm in thickness) was used for the beam collimation and constraint. A 100 mm tungsten edge collimator was placed in front of but closely contacted with scintillator at the same time. The radiation flux of the γ-rays at the scintillator was about 9.0×10^6 $\gamma/cm^2 \cdot s$. As for the system collimation and adjustment, the procedures might be described as follows. Firstly, a laser beam 0.5 mm in diameter was transmitted to transverse the front center and back center of the slit gap collimator. Then the beam would be irradiated onto the scintillator surface perpendicularly. Finally, the 100 mm-thick tungsten edge would be placed parallel with the laser beam, and the edge image would be adjusted to be at the center of the field of view of the system. In addition, due to high sensitivity of the MCP image intensifier and CCD devices to the γ-rays, another 10 cm-thick Pb shield was used to protect the system. After installation and adjustment, the radiographic system could be operated with PC remote control.

3. Experimental Data and Data Processing

3.1. Performance Calibration of Radiographic System

Firstly, the flat field response is a major standard to evaluate the response uniformity of the radiographic system. A Pb collimating aperture (100 mm in diameter) was used to provide beam constraint and shield for the radiation source, and a YAG scintillator was placed 400 cm away from the radiation source. With such a set up, the γ-rays could presumably have a homogenous spatial distribution. The image signals corresponding to various radiation doses were experimentally calibrated. According to the results, the flat field responses under various radiation doses were same, the intensities in the central region were relatively high, and the non-homogeneities in the central region and edge region were about 30%. The flat field images are mainly used for the homogenous correction of the radiographic images. After the correction, the images could convey information about the intensity distribution of the incident γ-rays.

Secondly, the geometrical distortion of the radiographic system was also calibrated with black-and-white squares (5 mm × 5 mm in size). The image signals were obtained and analyzed with the image processing programs. As indicated by the results, the geometrical distortion of the radiographic system was in agreement with the experimental requirements.

Finally, the dynamic sensitivity of the radiographic system was calibrated. The homogenous γ-rays were used to irradiate the YAG scintillator, and the radiation images corresponding to various radiation doses were recorded. After data processing, the relationship curves between the radiation doses and image signal intensities could be well obtained. As indicated by the curves, with a radiation dose rage of $1.0 \times 10^6 - 2.0 \times 10^8$ $\gamma \cdot cm^{-2}$, the radiographic system could demonstrate a fairly good linearity.

3.2. Experimental Data

The edge image experimentally obtained is given in **Figure 2**. Despite of the fact that a 100 mm-thick W collimator was placed in region with relatively lower intensity to provide an attenuation (a decay) factor of more than 10^{-6}, the SNR of the system was found to be only 5:1. The major reason for this might be attributed to the fluorescence dispersion effects through out the fluorescence transmittance, including: 1) the dispersion in the scintillator; 2) the dispersion in the optical system; 3) the dispersion in the MCP image intensifier; 4) the dispersions in the coupling fiber plate between the MCP and CCD and the ambient materials. Despite of the Pb shield of more than 10 cm in thickness, the scattered γ-rays along the optical path still acted as a non-negligible source to induce the noise.

Due to the relatively big non-homogeneity of the system, only an effective region 100 × 100 in size was adopted for data processing. As shown in **Figure 2**, the image signals fluctuated in relatively big amplitude with a standard deviation of about 3753. The reasons for this might be attributed to the single-particle detecting effects of the scintillator and the relatively high quantum gains of the scintillator and MCP image intensifier, etc. Thus, during the data processing, the average was made for 100-line data. After that, the averaged data were subject to Gaussian fitting. Finally, the ESF curves could be obtained to be very uniform and smooth (as shown in **Figure 3**).

3.3. Spatial Distribution of System

Both of the MTF and PSF are important parameters to evaluate the spatial resolution of the radiographic system, but very difficult to be measured directly. However, they are functions of the LSF. LSF could be determined by

Figure 2. The experimentally obtained edge fluorescence image.

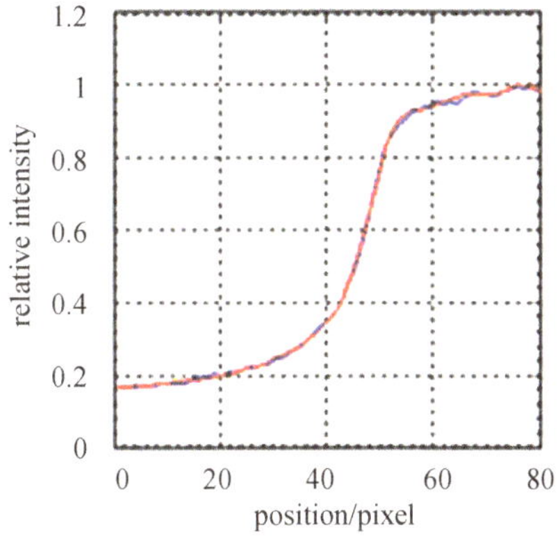

Figure 3. The experimentally obtained ESF distribution.

differentiating ESF. And the relationships among LSF, MTF and PSF could be described in the following equations. For a given ESF, the LSF could be determined in Equation (1) [4]:

$$LSF(x) = \frac{d\left[ESF(x)\right]}{dx} \quad (1)$$

Since MTF is the result of the Furrier Transform of LSF, we have:

$$MTF(f) = \left| \frac{\int_{-\infty}^{\infty} LSF(x)\exp(-j2\pi fx)\,dx}{\int_{-\infty}^{\infty} LSF(x)\,dx} \right| \quad (2)$$

And the relationship between PSF and LSF could be given as follows: [4]

$$PSF(r) = -\frac{1}{\pi}\int_{r}^{\infty} \frac{\left\{d\left[LSF(x)\right]/dx\right\}}{\left(x^2 - r^2\right)^{\frac{1}{2}}}\,dx \quad (3)$$

Based on the equations mentioned above, after processing of the edge image, the LSF, MTF and PSF of the γ-rays camera could be obtained as given in **Figure 4**. As shown in the figure, with 10% intensity, the MTF is only 2 lp/mm, and the point spread size is about 2 mm.

4. Major Factors Affecting Spatial Resolution

The spatial resolution of the γ-rays camera was measured with edge method. In the following parts, the effects of YAG scintillator, MCP image intensifier and CCD camera upon the system spatial resolution will be analyzed, respectively.

4.1. PSF of CCD Device

The spatial resolution of CCD device is mainly dependent on the pixel size and the electron diffusion in the silica substrate. In the silica substrate, the electron-hole pairs would be induced by photons. And the electrons would spread in the substrate and move driven by the superposed voltage before stored in the potential wells. The spread size or dimension of the electrons in the silica substrate is dependent on the superposed voltage [5,6]. And the superposed voltage is then dependent on the substrate thickness and manufacturing process. With a superposed voltage ranging in 2.84 V - 115.4 V, the electron spread sizes are within 48.8 μm to 3.7 μm, and the electron spread function is basically in accordance with the Gaussian function distribution. The LSF of the electrons could be described in the following equation [6]:

$$LSF(x) = a + b \cdot \exp\left(-\frac{(x - x_0)^2}{2\sigma^2}\right) \quad (4)$$

where: a, b, x_0 and σ are fitting constants. The spread of the electrons in the silica substrate could be obtained by using a point light source with a negligible size, sub-pixel sampling method and Gaussian function fitting [7]. In the high precision image measurement, the electron

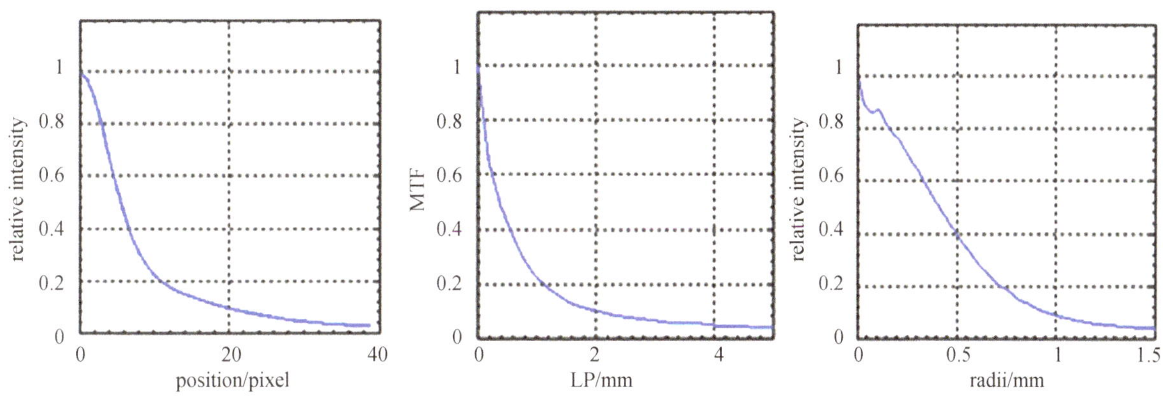

LSF Spread Function MTF Spread Function PSF Spread Function

Figure 4. The experimentally obtained LSF, MTF and PSF of the γ-rays camera.

spread in the silica substrate is a non-negligible factor to cause the degradation. Thus, in the common design of the scientific-level high sensitivity CCD, the back illumination method is used, and the thickness of the silica substrate is deliberately determined to be 10 μm - 15 μm, which is roughly equal to the electron spread sizes (5 - 10 μm) [7]. The typical electron PSF is given in **Figure 5**. As shown in the figure, the electron spread size is about 20 μm corresponding to 10% intensity. Specifically, for a CCD with a pixel size of 13.3 μm × 13.3 μm, the electron spread might be about 1 to 2 pixels.

4.2. PSF of MCP Image Intensifier

The MCP image intensifier is mainly consisting with optical cathode, micro-channel plate (MCP) and fluorescence screen. The MCP location is between the optical cathode and fluorescence screen. The MCP input plane is 0.1 mm - 0.3 mm away from the optical cathode and the output plane is about 0.5 mm - 1.3 mm away from the screen. Firstly, the photoelectrons are excited from the cathode by the incident beams. Then, driven by forward accelerating electric field between the cathode and MCP input plane, the electrons would be injected into the MCP input plane almost along a straight direction before being multiplied in the micro-channels. Then the electrons would be transmitted out of the MCP output plane and accelerated again by the electric field between the output plane and the screen. Finally, the electrons would impact onto the screen to generate the corresponding fluorescence image.

The spatial resolution of the MCP image intensifier is dependent on multiple factors, including various superposed voltages, the exposure quantity of the optoelectronic cathode and the manufacturing process, etc. Besides, the dispersion effect of the electrons in the cathode and fluorescence screen would be another important factor to damage the spatial resolution. The MCP image intensifier used in the experiment could provide a spatial resolution of more than 37 lp/mm. And the PSF could be given as follows [8]:

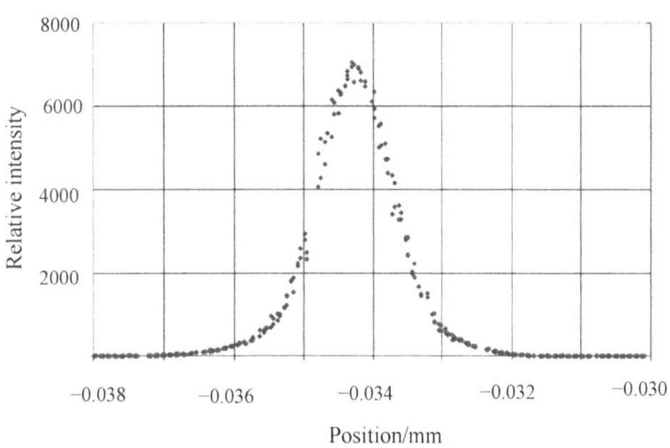

Figure 5. The spread size of electrons in the silica substrate.

$$PSF(r) = \exp\left[-(r/0.02)^2\right] \tag{5}$$

where: r stands for the radius (mm). The electron spread function of a typical MCP image intensifier is given in **Figure 6**. As shown in the figure, the diameter is about 0.06 mm corresponding to 10% of PSF. Take account the minification factor of 1.5, and the spatial resolution of the MCP image intensifier is basically equal to that of the CCD camera.

4.3. Fluorescence Dispersion Effects

The fluorescence dispersion could be divided into two parts. One is the fluorescence dispersion in the scintillator; another one is due to the defocusing in the optical imaging system. The secondary electrons would be generated by the interactions between the γ-rays and the scintillator. And the electrons would move within the scintillator and induce energy deposition. Part of this energy would be converted into fluorescent lights. Each of the electron trajectories would be acting as a point light source. Due to the homogenous spatial distribution of the lights, only some of the lights could be collected by the light collecting system, while most of them would be dispersed within the scintillator. During the dispersion, in the interface, half penetration half refection and full reflection occur, while in the scintillator, only scattering takes place. The deliberately designed optical imaging system used in the experiment was made up of 12 lenses to provide a light collecting efficiency 30°. At the same time, the defocusing effects induced by the fluorescence dispersion would also affect the spatial resolution.

Similarly, the edge method was again used to measure the fluorescence spatial resolution in the radiographic system. The spatial resolution of the fluorescence was calibrated on a pulsed X-rays source with energy of about 0.3 MeV. The distance between the YAG scintillator with the radiation source was about 250 cm, and the spatial resolution of X-rays was presumably homoge-

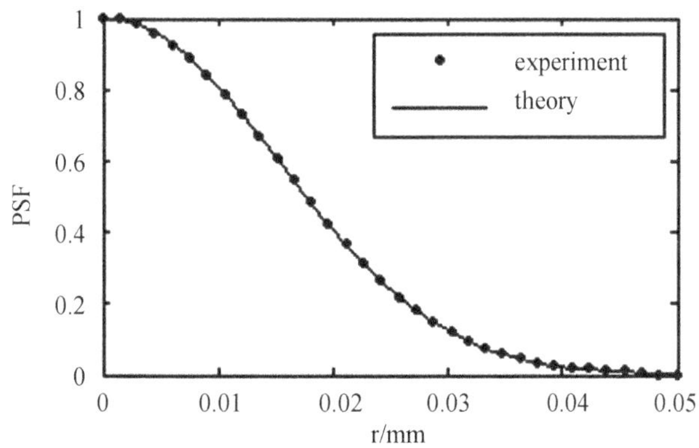

Figure 6. The experimentally obtained electron spread function of the MCP image intensifier.

nous. In addition, the effects of transmittance range the secondary electrons was neglected, which had been rationalized experimentally and theoretically. During the experiment, a 0.1 mm-thick copper film was tightly contacted with the scintillator surface, and the film was reasonably considered as the edge in the experiment. The SNR of the optical imaging system was finally found to be 10:1. After data processing, the MTF of the spatial resolution corresponding to 10% intensity was about 5 lp/mm (as shown in **Figure 7**), which was obtained with the additional contributions from the effects if the spatial resolutions of the MCP image intensifier and CCD device upon the fluorescence transmission.

4.4. Transmittance Range of Secondary Electrons in Scintillator

The spatial resolution of the γ-rays in the YAG crystal is mainly dependent on the transmittance range of the secondary electrons and the fluorescence dispersion in the scintillator, etc. [9,10]. The interactions between the γ-rays and the scintillator would generate secondary electrons. Each of the electrons would be a small light source, whose size would be decisive for the resolution limit. In order to get the fluorescence size of a single secondary electron, an optical imaging system with high light collecting efficiency was adopted in addition to a high sensitivity camera consisting with a MCP image intensifier and cooler-equipped CCD camera. And the movement trajectories of the secondary electrons were experimentally obtained.

The experiment was carried out on a Co radiation source, whose principles are same with those given in **Figure 1**. The distance between the scintillator and the radiation source was 400 cm, and the radiation dose on the scintillator was 3.7×10^7 $cm^{-2}\cdot s^{-1}$. At the same time, a shading plate 30 mm in diameter was place behind the scintillator to contain the effective image size within this range. The exposure time was 0.1 ms. The fluorescence image is given in **Figure 8**. As shown in the figure, there're miscellaneous speckle effects.

As for the miscellaneous speckle effects in the image, the speckle images are in array distribution, and there's nearly no signs for the linear propagation of the electrons. The reason for this might be attributed to the limit of the spatial resolution of the optical system and the fluorescence dispersion effects in the scintillator and transmission system. Due to their effects, only the degraded electron trajectories could be observed in the image.

With energy higher than 0.1 MeV, the ratios between the energy depositions in the scintillator with the luminescent efficiencies could be considered to be equal. In this extent, for any of the speckle arrays, approximately, the signal intensities over 50% higher than the peak would be classified as the electron trajectories, while those below those level would be the results of the scattered fluorescence. According to the estimation of the miscellaneous speckle sizes, the electron dispersion size is predictably 2 to 3 pixles, which, in dealing with the PSF of about 4 to 6 pixels, corresponds to a spread size of 0.4 mm to 0.6 mm size.

For a detailed description of the PSF of the γ-rays in the scintillator, MCNP program was introduced to track and simulate the transportation process of the γ-rays in the scintillator. And for the theoretical computation, the electronic fluorescence efficiency was supposed to be proportional to the electronic energy deposition efficiency. Other conditions for the computation were including the scintillator thickness (10 mm) and the interval of the radiuses of the concentric cylinders (0.010 mm). The energy disposition distributions in the scitillator were calculated for the γ-rays respectively 0.3 MeV and 1.25 MeV in energy, whose results are given in **Figure 9**. As shown in the figure, the maximal transmittance range of the 0.3 MeVγ-rays on the projecting plane is about 0.2 mm, while the range increases to be 0.8 mm for 1.25 MeV rays. And the PSF sizes corresponding to the 1% intensity are 0.44 mm.

5. Conclusion

The spatial resolution of the γ-rays camera was measured

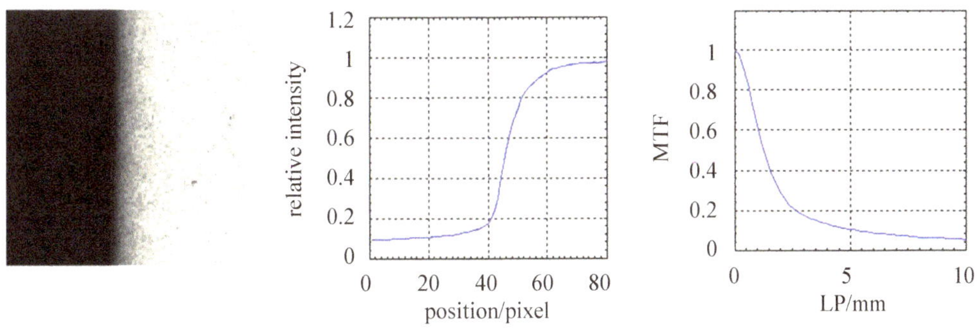

The knife image | The edge spread function | MTF

Figure 7. The spatial resolution of the fluorescence transmission in the system.

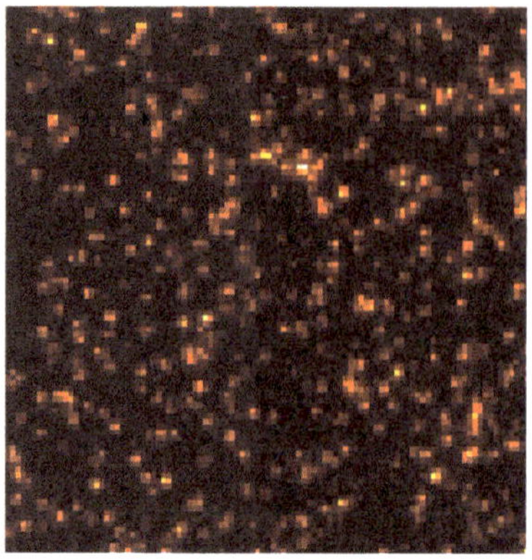

Figure 8. The speckle effects of the secondary electrons in the scintillator.

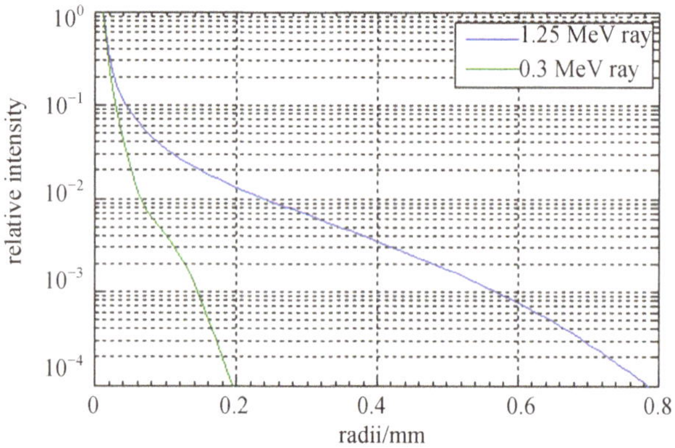

Figure 9. The PSF of the γ-rays in the scintillator.

on a 1.25 MeV source with edge method. Due to the relatively big quantum detective efficiency (DQE) and quantum gain of the gamma-rays, etc., the experimental data were processed by averaging multiple images and fitting curves. According to the experimental results, the MTF (modulation transfer function) of spatial resolution at the 10% intensity was about 2 lp/mm. Based on this, further analysis was made for the spatial resolutions of the MCP image intensifier and CCD device. Besides, the spatial resolution of the scintillating fluorescence in the γ-rays camera was measured as well as the dispersion effects of the secondary electrons in the scintillator. As indicated by the results, both the fluorescence dispersion and the dispersion of the secondary electrons are the major factors that affect the spatial resolution.

6. Acknowledgements

The study presented in this paper is sponsored jointly by CAEP Sci. & Tech. Development Foundation (Contract No.: 2010B0103006 and 2011B0103017) and National Natural Science Foundation (Contract No.11005095 and No.11305155). And the experimental study was made on the pulsed X-rays source in Northwest Institute of Nuclear Technology and with great help from Prof. Guo Jianming. The authors owe great gratitude to all of them.

REFERENCES

[1] S. Watson, T. Kauppila, *et al.*, "The High-Energy Radiographic Machine Emitting X-Rays Flash Radiographic Camera," *SPIE Proceedings*, Vol. 2869, 1997, pp. 920-928.

[2] G. J. Yates and N. S. P. King, "High-Frame Intensified Fast Optically Shuttered TV Cameras with Selected Imaging Applications," *SPIE Proceedings*, Vol. 2273, 1997, pp. 126-149.

[3] P. B. Gree and T. van Doorn, "Evaluation of an Algorithm for the Assessment of the MTF Using an Edge Method," *Medical Physics*, Vol. 27, No. 9, 2000, pp. 2048-2059.

[4] H. Q. Zhu, "Study on the Image Restoration of Y-Ray Pinhole Imaging System," Ph.D. Thesis, Tsinghua University, Beijing, 2008.

[5] P. Z. Takacs and I. Kotov, "PSF and MTF Measurement Methods for Thick CCD Sensor Characterization," *SPIE Proceedings*, Vol. 7742, 20120, Article ID: 774207.

[6] A. Karcher and C. J. Bebek, "Measurement of Lateral Charge Diffusion in Thick Fully Depleted, Back-Illuminated CCDs," *IEEE Transactions on Nuclear Science*, Vol. 51, No. 5, 2004, pp. 2231-2237.

[7] H. Schwarzer and A. Boerner, "Dynamic PSF and MTF Measurement on a 9k TDI CCD," *SPIE Proceedings*, Vol. 7106, 2008.

[8] H. Q. Zhu and K. L. Wang, "MTF Measurement and Analysis of Microchannel Plate Image Intensifiers," *Acta Photonica Sinica*, Vol. 36, No. 11, 2007, pp. 1983-1986.

[9] J. M. Ma and K. L. Wang, "Research on PSF of Scintillator as γ-Ray Fluorescenc Image Converter," *Nuclear Electronics & Detection Technology*, Vol. 31, No. 4, 2011, pp. 473-478.

[10] J. H. Zhu and S. L. Niu, "Energy Deposition of Gamma Rays in LSO Crystal," *High Power Laser and Particle Beams*, Vol. 22, No. 6, 2010, pp. 1351-1354.

Exhaled Breath Condensates as a Source for Biomarkers for Characterization of Inflammatory Lung Diseases

Puneet Bajaj[1], **Faoud T. Ishmael**[1,2]

[1]Department of Medicine, Division of Pulmonary, Allergy, and Critical Care Medicine, Section of Allergy and Immunology, Milton S. Hershey Medical Center, The Pennsylvania State University, Hershey, USA; [2]Department of Biochemistry and Molecular Biology, College of Medicine, The Pennsylvania State University, Hershey, USA.

ABSTRACT

Inflammatory lung diseases such as asthma and chronic obstructive pulmonary disease are common and difficult to diagnose and characterize. This is due in large part to difficulty in obtaining samples directly from the inflamed lung. The collection of lung secretions by traditional methods including bronchoalveolar lavage and induced sputum collection are limited by their invasive nature. Exhaled breath condensate (EBC) is a simple and non-invasive technique of collecting fluid samples, which are representative of airway lining fluid. Advances in collection methods and evolving molecular techniques have led to development of more sensitive assays for existing biomarkers and identification of new biomarkers, which can be potentially useful in monitoring lung inflammation. In this review, we present the current understanding of various biomarkers including small molecules (H_2O_2, pH and nitric oxide related biomarkers), lipid mediators (8-isprostane, leukotrienes and prostaglandins), small proteins (cytokines and chemokines) and nucleic acids (DNA and microRNAs). We also discuss the differential profile of biomarkers in recognizing different patterns of lung inflammation. As the sensitivity of methods of EBC improves, this biofluid will play an increasing role in diagnosis and monitoring of lung diseases.

Keywords: Exhaled Breath Condensate, Inflammation; Biomarkers; Hydrogen Peroxide; Ph; Micrornas; Nitric Oxide; Leukotrienes; Prostaglandins; Metabolomics

1. Introduction

Diseases of the lung such as asthma and chronic obstructive lung disease (COPD) are characterized by airway inflammation. These are common diseases with a US prevalence of 8.4% for asthma [1] and between 3% - 7% for COPD [2]. However, these diseases are difficult to diagnose, distinguish from each other, and characterize phenotypically. Inflammation in the airway involves an interplay between environmental stimuli, airway epithetlial cells, and leukocytes (**Figure 1**). Stimuli such as infections, allergens, pollutants, oxidants, or irritants directly damage the airway cells or induce them to produce inflammatory mediators such as cytokines, reactive oxygen species (ROS), or acidic products. These mediators subsequently recruit various leukocytes to the lung, which can produce additional products, including cytokines, ROS, nitrates, and lipid mediators that contribute to the chronic inflammatory state. These mediators are found in airway lining fluid (ALF) and their measurement may serve as biomarkers to identify different inflammatory patterns in lung diseases. Their role in the diagnoses and management of variety of lung diseases is now being recognized.

However, collection of ALF has always posed methodological challenges. The most reliable method is collection of fluid by bronchoalveolar lavage (BAL), an invasive technique that requires bronchoscopy to collect samples directly from the lower lung. Other methods, such as induced sputum, may be less invasive. However, it is difficult to reproducibly isolate high quality samples, salivary contamination is common, and it is a difficult and uncomfortable procedure for the subjects.

An emerging non-invasive technique to isolate ALF is exhaled breath condensates (EBC). Physiologically, the exhaled breath is constituted predominately by water vapor and aerosolized particles, generated by ALF. By cooling breath vapor, EBC can be collected and its biochemical composition has been found to be very similar to ALF [3]. One of the main advantages of EBC is the non-invasive collection technique, which can be conveniently performed by the patients in most age groups. The ease of collection of samples, combined with increasing

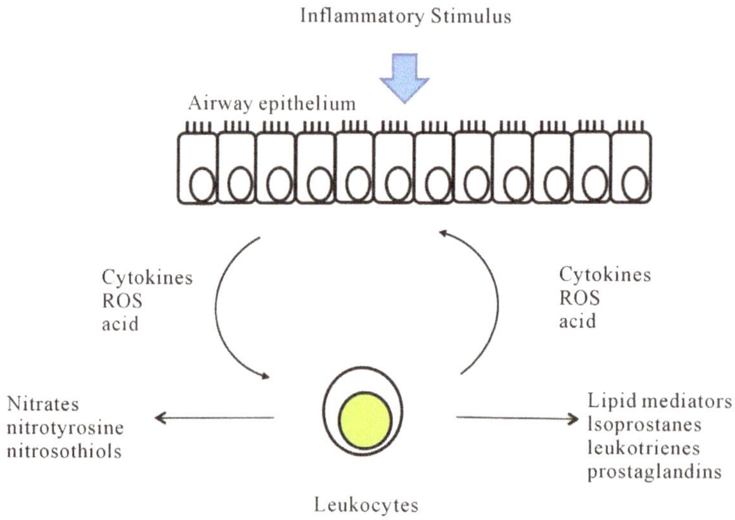

Figure 1. Inflammatory mediators produced in airways. Stimulation ofairway cells by inflammatory stimuli leads to production of inflammatory mediators such as cytokines, reactive oxygen species (ROS), or acidic products. These mediators subsequently recruit various leukocytes to the lung, which can produce additional products, including cytokines, ROS, nitrates, and lipid mediators that contribute to the chronic inflammatory state.

sensitivity of various biomarker detection makes EBC a novel and potentially important diagnostic tool.

2. Collection of EBC

The principle of collecting EBC involves cooling the exhaled air below the dew point by transferring the heat to a chilled condenser surface (**Figure 2**). This leads to condensation of water vapor content in the breadth over aerosolized ALF particles, leading to formation of enlarged EBC droplets on condenser wall [3]. The volume of EBC collected depends on multiple factors including amount of expired air, material and temperature of condenser etc. There are two commercial devices available for collection of EBC at this point-ECoScreen (VIASYS Healthcare, Hoechberg, Germany) and RTube (Respiratory Research Inc., VA) [3]. The ECoScreen uses an electric cooler, while the Rtube utilizes a metal tube that has been pre-cooled and then placed around the collection vessel. Studies have shown that level of biomarkers obtained from different collection devices varies and hence they cannot be compared directly [4]. Each has advantages, such as greater potential sensitivity of the EcoScreen versus enhanced portability and simplicity of the Rtube [5]. According to recent American Thoracic Society (ATS)/European Respiratory Society (ERS) guidelines, the subject should breath orally for around 10 minutes through the collection device, in a tidal pattern and this should allow collection of 1 - 3 ml of condensate [6]. The apparatus should have a salivary trap to minimize the salivary contamination of the condensate and contamination can further be evaluated by checking the amylase level in the collected sample [7].

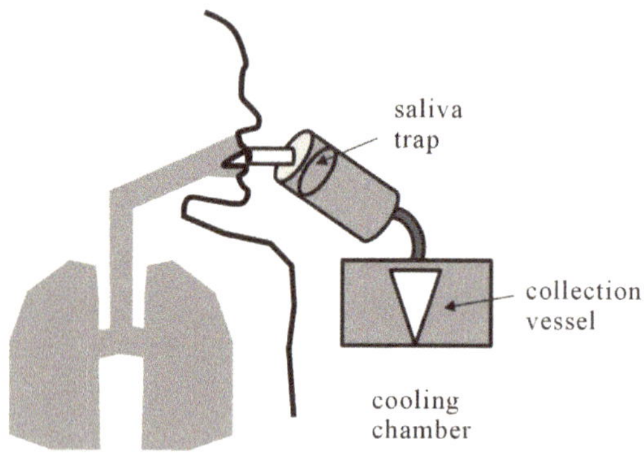

Figure 2. Schematic of EBC collection apparatus. The collection system consists of a salivary trap, a cooling mechanism, and collection vessel.

3. Components of EBC and Their Role as Bio-Markers

Expired air, saturated with water vapor deposits in cooling chamber as distilled water and forms the main constituent of EBC. In addition, water-soluble volatile organic compounds (VOCs) present in gaseous phase in exhaled breath dissolve into the condensate. The third component is derived from aerosolized micro-particles which originate directly from ALF in the lower airways and contain both non-volatile constituents and dissolved VOCs [8]. Though the exact source of aerosolized particle in exhaled breath is not very clear studies have indicated a predominantly lower airway origin [9,10]. Detection of VOCs depends on multiple factors including water solubility, gas-liquid partition coefficient and temperature of condenser. The non-volatile compounds form a broad range of molecules ranging from inorganic ions like sodium, to macromolecules such as small protein molecules like cytokines or nucleic acids.

EBC has been recognized as a source of multitude of organic and inorganic compounds including small inorganic molecules (H_2O_2, pH and nitric oxide related biomarkers), lipid mediators (8-isprostane, leukotrienes and prostaglandins), small proteins (cytokines and chemokines) and nucleic acid derivatives [11]. These biomarkers can reflect the underlying state of pulmonary inflammation and can be altered in various pulmonary diseases like asthma, COPD, bronchiectasis, infections, and lung cancer (**Table 1**).

4. Small Inorganic Molecules in EBC

Hydrogen peroxide (H_2O_2) is a reactive oxygen species, generated in inflammatory cells from metabolism of superoxide anion (O_2^-) by superoxide dismutase. Increased oxidative stress during inflammation induces respiratory bursts resulting in marked production of O_2^- which leads to increased production of H_2O_2 [12]. Its levels in EBC have been measured by spectrofluorimetry [13], spectrophotometry [14], and chemiluminescence [15]. As compared to healthy nonsmokers, H_2O_2 have been reported to be elevated in smokers [13], asthmatics [16] [17], COPD patients [14], bronchiectasis [18] and Acute respiratory distress syndrome (ARDS) [19]. Levels of H_2O_2 have been shown significantly elevated in acute worsening of asthma [11], COPD exacerbations [20], pulmonary infection related exacerbations of cystic fibrosis [21] and may have potential role in guiding therapy in these situations.

Despite recognition of H_2O_2, as an indicator of oxidative stress, it has not been established as a reliable EBC biomarker due to several limitations. The mean levels of H_2O_2 have shown a wide variability in EBC of healthy nonsmoking adults (0.01 - 0.45 micromoles) [13] and healthy children between 8 - 13 years (median H_2O_2 levels was reported as 0.13 μM, with reference range of <0.01 - 0.48 μM) [22]. Additionally, variability of levels were found on different days of measurement on same subjects and at different times of the day [13]. The concentrations of H_2O_2 in EBC was also found to be related to expiratory flow rate, suggesting a predominantly bronchial origin of H_2O_2, as compared to alveolarorigin, further limiting its usefulness as a pulmonary inflammatory marker [23].

EBC pH reflects the acid/base balance of the airways. The major source of acidification of EBC includes intrinsic pulmonary inflammatory burden but there could be additional contribution from gastroesophageal reflux disease [24]. EBC pH from samples collected from oral breathing were found to be similar to endotracheal samples but different from salivary pH, supporting a predominantly lower airway origin of EBC pH [25]. Standard pH analysis is performed after de-aeration of carbon dioxide (CO_2) with an inert gas like argon to minimize artifacts [24]. One of the main advantage of EBC pH is the fact that it has been found to be an extremely reproducible and robust parameter and does not show deviation with type of ventilation, gender, time of day collected, volume of EBC collected, duration of EBC collection, degree of hyperventilation [24] or temperature or duration of sample storage [24,25]. Normal values of gas-standardized (CO_2-free) EBC pH have been reported between 7.5 and 8.1 [24,25]. Decrease in EBC pH has been reported in both stable and unstable asthma patients as compared to healthy controls, with a greater decrease seen in unstable asthmatics treated with inhaled corticosteroids [26]. Similarly, EBC pH levels have been reported to be lower in patients with COPD compared to age and gender matched controls and correlate with disease severity, as expressed by GOLD stages, especially in ex-smokers [27]. EBC pH may have potential role in diagnosis of upper respiratory symptoms like chronic cough due to underlying acid reflux [28]. A system for continuous monitoring of EBC pH in intubated patients on mechanical ventilation has also been developed to monitor lung health in critically ill patients [24,29].

5. Nitric Oxide Related Biomarkers

Nitric Oxide (NO) is an important bioactive mediator that plays important roles in multiple cellular functions. The levels of NO increase during inflammatory conditions due to activation of inducible NO synthase in response to proinflammatory cytokines [30]. NO can be measured in exhaled breath in form of Fractional excretion of NO (FeNO) and has been shown to correlate with active lower airway inflammation in asthmatics [31]. Nitrates and nitrites are products of NO metabolism that can be detected in EBC and their levels have been reported to be elevated in smokers with COPD as compared to healthy smokers [32]. NO can combine with reactive oxygen molecules to form highly reactive nitrogen species (RNS) which further covalently interact with various biomolecules to form nitroso- and nitro-molecules [33]. S-Nitrosothiols are formed by interaction of RNS with low molecular thiols and have reported to be elevated in severe asthma, COPD and cystic fibrosis [34]. 3-Nitrosotyrosine is another covalent adduct molecule generated by interaction of RNS with tyrosine with in the proteins [33] and has been shown to be elevated in patients with asthma [35] and cystic fibrosis [36]. NO and related molecules are non-specific markers of oxidative stress and may be useful in monitoring various pro-inflammatory pathologies of lower airways.

Table 1. Biomarkers in EBC.

Type	*Biomarker*	*Utility*	*Method*	*Disease associations*
Small inorganic molecules	H_2O_2	Non-specific marker of Oxidative stress	Spectrophotometry [13], SpectroFluorimetry [14], Chemiluminescence [15]	Asthma [16,17] COPD [14] Smokers [13] Bronchiestasis [18] ARDS [19]
	pH	Non-specific marker of Oxidative stress	pH meter after gas standardization [24]	Asthma [26], COPD [27], cough related to GERD [28] Acute lung injury (ALI) and ARDS [29]
Nitric oxide related biomarkers	Nitrates	Non-specific marker of Oxidative stress	Fluorimetry [32]	Elevated in smokers with COPD [32]
	S-Nitroso-thiols	Non-specific marker of Oxidative stress	Spectophotometry [34]	Asthma Cystic Fibrosis Asthma, COPD and cystic fibrosis [34]
	3-Nitroso-tyrosine		liquid chromatography–tandem mass spectrometry [35], EIA[36]	Asthma [35], cystic fibrosis [36]
Lipid mediators	8- IP	Non-specific marker of Oxidative stress	EIA [41], Gas Chromatography/mass spectrometery [81]0]	Asthma [39], Smoking and COPD [40], ALI and ARDS [42] Cystic Fibrosis [41] Exercise induced Bronchospasm (EIB) [43]
	LTB4	Neutrophil predominant inflammation	EIA [45]	COPD [45], cystic fibrosis [46], asthma [47], NSCLC [48]
	CysLTs; LTC4, LTD4, LTE4	Eosinophils and mast cell pre-dominant inflammation	EIA [49]	Allergic asthma [49], EIB [50]
	PGD2	Mast cell predominant inflammation	EIA [49]	Allergic asthma [49]
	PGE2	Unclear role- inhibitory vs. proinflammatory	EIA [45]	COPD [45] and CF [55]
	TxA2	Broncho-constriction	RIA [56]	Asthma [56]
Cytokines and chemokines	IL-2	Acute Inflammatory cytokine	multiplex immunoassay [68]	Asthma [68]
	IL-4	Th2 Inflammatory pathway	multiplex immunoassay [68]	Asthma [68]
	IL-5	Th2 Inflammatory pathway	multiplex immunoassay [68]	Asthma [68]
	IL-6	Th17 Inflammatory pathway	multiplex immunoassay [68]	Asthma [68], COPD exacerbation [69]
	IFN-γ	Th1 Inflammatory pathway	multiplex immunoassay [68]	Asthma [68]
	IL-18	Th1 Inflammatory pathway	multiplex immunoassay [68]	Asthma [68]
	IL1β	Acute Inflammatory cytokines	multiplex immunoassay [69]	COPD exacerbation [69]
	IL-8	Neutrophilic Inflammation	multiplex immunoassay [68]	Asthma [68], COPD exacerbation [69]
	IL-10	Treg Inflammatory pathway	multiplex immunoassay [68]	Asthma [68] , COPD exacerbation [69]8]
	IL-12	Th1 Inflammatory pathway	multiplex immunoassay [69]	COPD exacerbation [69]
	RANTES	Allergic inflammation	EIA [70]	Asthma [70]

Continued

	MIP1α	Migration of T cells, NK cells, monocytes	multiplex immunoassay [68]	Asthma [68]
	eotaxin	Eosinophilic inflammation	EIA [71]	Asthma [71]
Nucleic acids	DNA	Infection	PCR [73,75]	COPD exacerbation [73], Detection of HPV in Lung Cancer [75]
	miRNAs	Allergic vs. non allergic inflammation	qPCR	Asthma and COPD

6. Lipid Mediators

Lipid mediators are produced in leukocytes from the enzymatic metabolism of phospholipids (**Figure 3**) and have numerous physiologic and pathologic functions. Levels may be altered in a variety or lung diseases underscoring their potential as biomarkers. Isoprostanes are prostaglandin like substances produced *in vivo* by free radical-induced peroxidation of arachidonic acid, independent of cyclooxygenase (COX) enzymes [37]. They are chemically stable molecules and their levels in body fluids have been reported as a reliable indicator of oxidative stress [38]. Elevated levels of 8-Isoprostane (8-IP) have been reported in asthma [39], healthy smokers and smokers with COPD [40], cystic fibrosis [41], acute lung injury and ARDS [42], and exercise induced bronchospasm [43].

Leukotrienes are derivatives of arachidonic acid through the action of 5-Lipoxigenase (5-LO) [44]. Leukotriene (LT) B4, formed from hydrolysis of LTA4 has been shown to be a major neutrophil chemoattractant. Elevated LTB4 levels in EBC have been reported in neutrophilic predominant inflammatory lung pathologies including COPD [45] and cystic fibrosis [46]. Its levels have been reported to be significantly elevated in moderate to severely uncontrolled asthma patients whereas in mild episodic asthmatics, its role is not clear [47]. A recent study has also reported elevated LTB4 levels in patients with Non-Small Cell Lung Cancer (NSCLC), indicating presence of neutrophilic inflammation in airways of these patients [48].

Cysteinyl-leukotrienes (CysLTs; LTC4, LTD4, and LTE4), are produced in mast cells and eosinophils and their levels have been shown to be elevated in EBC from patients with asthma induced by allergic inflammation [49]. Another study reported elevation of CysLTs in asthma patients with exercise induced bronchospasm (EIB), with levels correlating with decrease in FEV1 [50], whereas no difference was seen in levels of LTB4 in EIB patients. In a recent study, multiple eicosanoids were measured in EBC samples of aspirin sensitive and aspirin tolerant asthma patients using gas chromatography/mass spectrometry and high-performance liquid chromatography [51]. The authors did not find significant difference in concentration of leukotrienes, prostaglandins, and thromboxanes at baseline or after aspirin challenge in aspirin sensitive asthma patients as compared to aspirin tolerant asthma patients. The eicosanoids that were found to be elevated baseline in aspirin sensitive patients were 5 and 15-hydroxyeicosatetraenoic acid (HETE). Another study reported significant increase in the level ofprostaglandin D2 and E2 metabolites and 5-, 15-HETE in aspirin intolerant asthma patients by usingcomplementary high performance liquid chromatography (HPLC) and gas chromatography-massspectrometry (GC/MS) [52]. The authors proposed that highly sensitive eicosanoid profiling in EBC can be used to identify aspirin intolerant asthma phenotypes.

Prostaglandins are lipid mediators derived from metabolism of arachidonic acid by action of cyclooxygenase (COX) enzyme. Prostaglandin D2 (PGD2) is predominantly produced by mast cells and its levels were reported to increase in EBC after allergen-induced bronchoconstriction in asthmatic patients [49]. The levels were quantified using enzyme immunoassays (EIA) and increase in levels correlated with degree of bronchoconstriction. The role of Prostaglandin E2 (PGE2) is not very clear. Earlier studies reported a predominantly inhibitory role in lung inflammation [53] but more recent studies have postulated that it may also have pro-inflammatory action [54]. A significant increase of PGE2 was reported in COPD patients, both steroid naive and steroid treated, as compared to healthy controls [45]. PGE2 levels were also reported to be elevated in patients with stable and unstable cystic fibrosis [55]. The levels of thromboxane A2 (TxA2), were reported to be increased in asthmatics as compared to healthy controls [56]. These authors also reported an improved TxA2 detection rate with radioimmunoassay (RIA) as compared to standard EIA.

7. Cytokines and Chemokines

Cytokines and chemokines are small protein molecules involved in intercellular signaling and play an important in inflammatory pathways in lungs and other organs. Elevated levels of cytokines have been reported in BAL specimens of patients with acute lung injury and ARDS

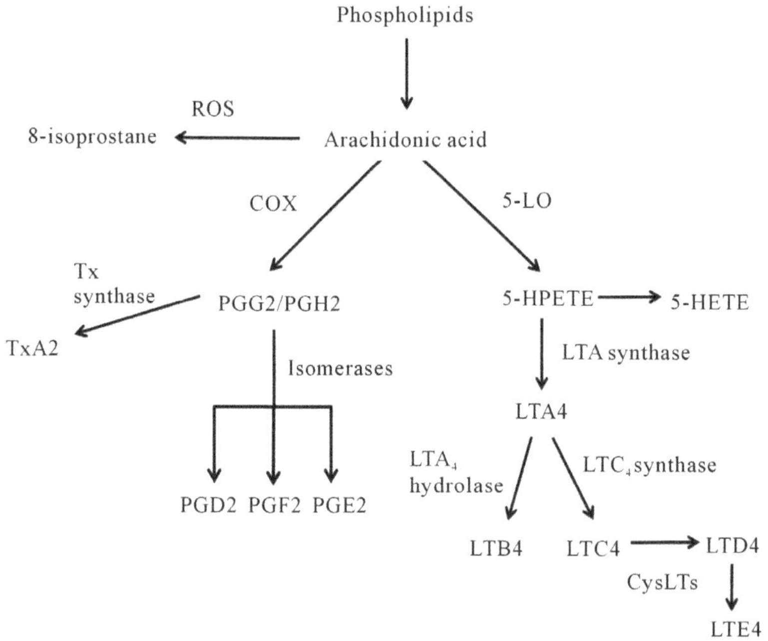

Figure 3. Lipid Mediators. Prostaglandins (PGD2, PGE2 and PGF2) and Thromboxane A2 (TxA2) are derived from arachidonic acid by COX mediated pathway. 8-Isprostane (8-IP) is formed by free radical peroxidation of arachidonic acid, independent of cyclooxygenase (COX) enzymes. Leukotrienes (LTB4 and Cysteinyl-leukotrienes-LTC4, LTD4 and LTE4) are produced by action of 5-lipoxigenase (5-LO).

[57,58], asthma [59], COPD [60] and cystic fibrosis [61]. Elevated cytokines levels have also been reported in induced sputum samples in patients with COPD [62] and asthma [63]. EBC provides a non-invasive alternative for obtaining ALF samples and has been increasingly studied as a source for monitoring cytokines. Unlike other non-specific markers of oxidative stress, cytokines signatures can be useful in defining type of inflammation and etiology. Interaction of naïve T-cells (Th0) with other inflammatory cells and molecules leads to their differential into either T helper (Th)1 T-cells which are involved in cell mediated immunity or Th2 T-cells, which are involved with antibody mediated immunity and allergic inflammation. Increased levels of IL-4 (Th2 cytokine) and decreased interferon-γ (Th1 cytokine) have been reported in EBC samples of asthmatic children indicating a Th2 predominant inflammation [64]. Other authors reported extremely low concentration of cytokines when collected by standard collecting apparatus and using EIA based analysis [65]. Improved detection of cytokines was reported with use of flow cytometry but this group did not find any difference in cytokine profile between asthma and COPD patients [66]. More recently, the use of glass condenser system has been reported to improve the yield of cytokines and other biomarkers in EBC samples [67]. The use of this system along with use of a more sensitive multiplex immunoassay has been reported to significantly improve the detection of cytokines in EBC [68]. This group reported increased levels of IL-2, IL-4, IL-5, IL-6, IFN-γ, IL-18, MIP1α, RANTES, IL10, and IL-8 in EBC samples from asthmatic patients as compared to controls. Elevated levels of cytokines IL1β, IL-6, IL-8, IL-10, IL-12p70 have also been in reported in patients with acute COPD exacerbation as compared to stable COPD, healthy smokers and healthy nonsmokers [69]. More recent studies have also reported increased levels of eosinophil specific chemokines-RANTES [70] and eotaxin [71] in EBC of asthma patients. Another potential use for cytokine monitoring can be in diagnosing airway inflammation in preschool children with wheezing [72].

8. Nucleic Acids

One of the main limitations of EBC is the low concentration of biomolecules, which pushes the detection limits of many assays. An emerging use of EBC is the measurement of nucleic acids, which can be amplified and detected in low numbers by PCR techniques. A few reports have demonstrated the presence of microbacterial DNA in EBC, and the impact on this a methodology for diagnosis of infection is emerging [73-76].

Another useful biomarker may be microRNAs (miRNAs), which are small 20 - 25 nucleotide, non-coding RNAs. MiRNAs are involved in post-transcriptional regulation of gene expression and are emerging as regulatory molecules of inflammation [77]. Differential expression of specific miRNAs in sputum of non-small cell lung cancer patients [78], especially in adenocarcinoma [79] can potentially serve as biomarkers for early detection of lung cancers and highlights their potential role in disease diagnosis.

We recently demonstrated that miRNAs are expressed in EBCs, and they can be reproducibly isolated and quantified by real time PCR (**Figure 4(a)**). In this methodology, an adapter is added to the 3' end of miRNA, and specific miRNA are amplified by using a forward primer specific to miRNAs of interest and a universal reverse primer to the adapter. After real time PCR analysis, the miRNAs were subjected to gel electrophoresis which confirmed the presence of one species at the expected size of the miRNA plus adapter sequence.

MiRNAs in can be secreted into bodily fluids from cells in exosomes, vesicles that contain miRNA, mRNA, and proteins. Exosomes can be isolated in the pellets of serum or saliva subjected to ultracentrifugation. We found that in contrast to saliva, exosomes are not present in EBC, and miRNAs exist free in solution (**Figures 4(b)** and **(c)**). We profiled miRNAs from EBC samples of asthmatics (n = 10), COPD patients (n = 10) and healthy controls (n = 14), and found that miRNAs were differenttially expressed between the groups (**Figure 4(d)**). For instance, in asthmatics, miR-1248 and miR1291 expression were lower compared to COPD patients and healthy controls. These miRNAs are predicted to regulate allergic mediators, such as IL-13, IL-5, GATA3, and the high affinity IgE receptor. Thus, miRNAs may not only be used as biomarkers to differentiate between types of lung disease, but may also offer insight into the pathogenesis of disease.

We also found that small nucleolar RNAs such as SNORD44 is expressed in EBC. These RNAs have served as "housekeeping" genes to normalize expression of miRNAs between samples. Methods of normalization between subjects are essential components of EBC biomarker measurement. Additionally, since as little as 50 - 100 molecules/ml of miRNA can be reproducibly quantified using qPCR, this may be an especially promising technique.

9. Limitations and Future Directions

EBC is a promising technology that allows insight into pulmonary pathophysiology in a non-invasive way. The concentration of biomolecules in EBC is highly diluted (up to 2000 to 10,000 fold) by condensed water vapor [6],

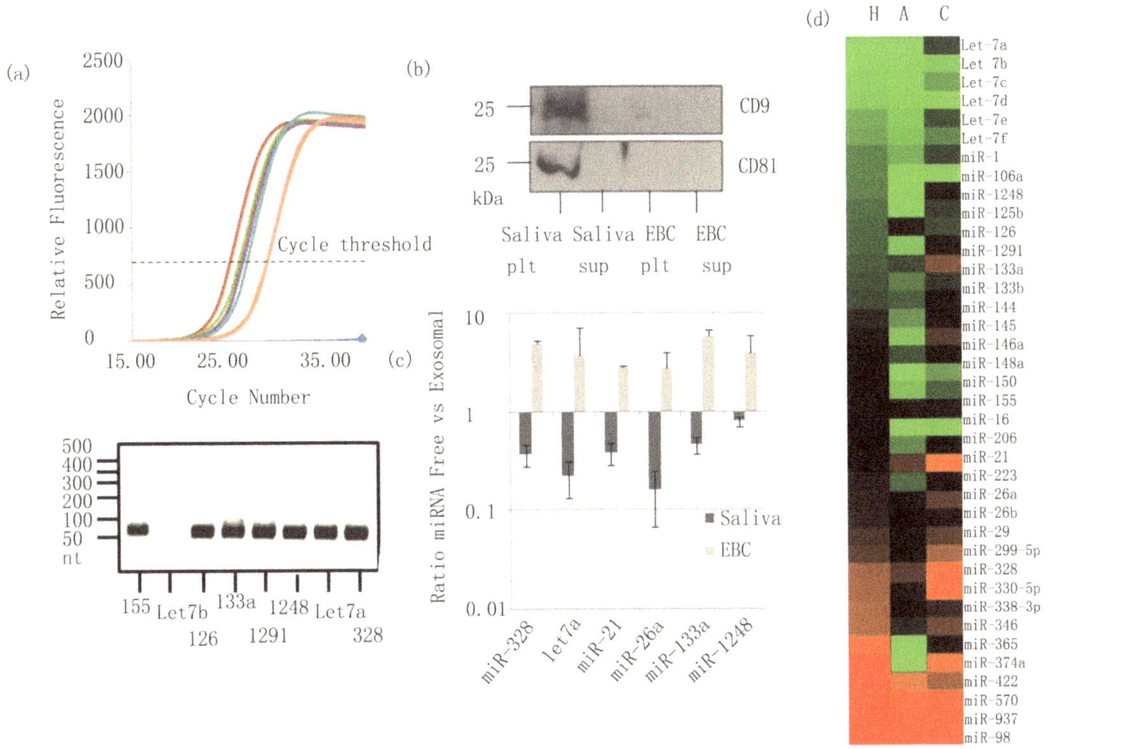

Figure 4. MiRNA expression in EBC: (a) real time PCR analysis of miRNAs and analysis of the products by gel electrophoresis; (b) Saliva or EBC were subjected to ultracentrifugation to pellet exosomes. Western blot for the exosome-specific proteins CD81 and CD9 showed their presence in the ultracentrifugation pellet (plt) of the saliva but not pellet of EBC or the supernatant (sup) or EBC or saliva; (c) Real time PCR analysis of miRNA expression demonstrates that they are present in the exosomal fraction of saliva but in the supernatant of EBC; (d) Heat map showing average expression of miRNA in healthy subjects (H), asthmatics (A), or COPD patients (C).

barely at lower detection limit by standard immunoassay techniques. The role of dilution indicators including conductivity of sample [80], urea and electrolytes [10] have been proposed for standardization of EBC samples but have not yet been established. In the absence of standardization, the wide variability of biomolecules concentration has become one of the main limitations of this technology. Using ratios of related biomarkers instead of absolute values, or normalization of data to housekeeping genes in the case of miRNA, may be useful as standardization measures.

With the development of newer analytical methods like liquid chromatography/mass spectrometry (LC/MS) [81], GC/MS, HPCL, radioimmunoassay (RIA) [56] and multiplex immunoassays [68] and use of glass condenser system for EBC collection [68], there have been significant improvement in the yield and detection threshold of various biomarkers. Another important development is application of metabolomics, which is study of measuring biomolecules in metabolic pathway in body fluids. In a recent study [82], nuclear magnetic resonance technique (NMR) spectroscopy was applied to analyze metabolomics in EBC samples from healthy subjects, COPD patients and patients with laryngectomy and significant difference in their metabolic fingerprints were reported in-between study groups.

In summary, EBC is an exciting technology, allowing an insight into biomolecule milieu of airway lining fluid. It is an important source of biomarkers, both non-specific markers of oxidative stress and specific biomarkers indicating different aspects of pulmonary inflammation. As the ability to comprehensively measure many markers increase via high-throughput methods, EBC will likely play a prominent role in disease diagnosis and characterization. At present, it is primarily used in research, but is emerging as a novel tool in diagnosing and managing inflammatory lung conditions and developing personalized pharmacological strategies.

REFERENCES

[1] Centers for Disease Control and Prevention, "Vital Signs: Asthma Prevalence, Disease Characteristics, and Self-Management Education—United States, 2001-2009," 2011. http://www.cdc.gov/mmwr/preview/mmwrhtml/mm6017a4.htm

[2] Centers for Disease Control and Prevention, "Chronic Obstructive Pulmonary Disease among Adults—United States, 2011," 2012. http://www.cdc.gov/mmwr/preview/mmwrhtml/mm6146a2.htm

[3] J. Hunt, "Exhaled Breath Condensate: An Evolving Tool for Noninvasive Evaluation of Lung Disease," *The Journal of Allergy and Clinical Immunology*, Vol. 110, No. 1, 2002, pp. 28-34.

[4] J. Liu, D. H. Conrad, S. Chow, V. H. Tran, D. H. Yates and P. S. Thomas, "Collection Devices Influence the Constituents of Exhaled Breath Condensate," *European Respiratory Journal*, Vol. 30, No. 4, 2007, pp. 807-808.

[5] O. U. Soyer, E. A. Dizdar, O. Keskin, C. Lilly and O. Kalayci, "Comparison of Two Methods for Exhaled Breath Condensate Collection," *Allergy*, Vol. 61, No. 8, 2006, pp. 1016-1018.

[6] I. Horváth, J. Hunt, P. J. Barnes, K. Alving, A. Antczak, E. Baraldi, G. Becher, W. J. C. van Beurden, M. Corradi, R. Dekhuijzen, R. A. Dweik, T. Dwyer, R. Effros, S. Erzurum, B. Gaston, C. Gessner, A. Greening, L. P. Ho, J. Hohlfeld, Q. Jöbsis, D. Laskowski, S. Loukides, D. Marlin, P. Montuschi, A. C. Olin, A. E. Redington, P. Reinhold, E. L. J. van Rensen, I. Rubinstein, P. Silkoff, K. Toren, G. Vass, C. Vogelberg and H. Wirtz, "Exhaled Breath Condensate: Methodological Recommendations and Unresolved Questions," *European Respiratory Journal*, Vol. 26, No. 3, 2005, pp. 523-548.

[7] R. M. Effros, M. B. Dunning 3rd, J. Biller and R. Shaker, "The Promise and Perils of Exhaled Breath Condensates," *American Journal of Physiology Lung Cellular and Molecular Physiology*, Vol. 287, No. 6, 2004, pp. L1073-L1080.

[8] M. D. Davis, A. Montpetit and J. Hunt, "Exhaled Breath Condensate: An Overview," *Immunology and Allergy Clinics of North America*, Vol. 32, No. 3, 2012, pp. 363-375.

[9] G. I. Sidorenko, E. I. Zborovskiĭ and D. I. Levina, "Surface-Active Properties of the Exhaled Air Condensate (a New Method of Studying Lung Function)," *Terapevticheskiĭ Arkhiv*, Vol. 52, No. 3, 1980, pp. 65-68.

[10] R. M. Effros, B. Peterson, R. Casaburi, J. Su, M. Dunning, J. Torday, J. Biller and R. Shaker, "Epithelial Lining Fluid Solute Concentrations in Chronic Obstructive Lung Disease Patients and Normal Subjects," *Journal of Applied Physiology*, Vol. 99, No. 4, 2005, pp. 1286-1292.

[11] C. Caffarelli, E. Calcinai, L. Rinaldi, C. Povesi Dascola, L. Terracciano and M. Corradi, "Hydrogen peroxide in exhaled breath condensate in asthmatic children during acute exacerbation and after treatment," *Respiration*, Vol. 84, No. 4, 2012, pp. 291-298.

[12] A. B. Tonnel and B. Wallaert, "Oxidants and Bronchial Inflammatory Processes," *The European Respiratory Journal*, Vol. 3, No. 9, 1990, pp. 987-988.

[13] D. Nowak, S. Kalucka, P. Białasiewicz and M. Król, "Exhalation of H_2O_2 and Thiobarbituric Acid Reactive Substances (TBARs) by Healthy Subjects," *Free Radical Biology and Medicine*, Vol. 30, No. 2, 2001, pp. 178-186.

[14] F. De Benedetto, A. Aceto, B. Dragani, A. Spacone, S. Formisano, R. Cocco and C. M. Sanguinetti, "Validation of a New Technique to Assess Exhaled Hydrogen Peroxide: Results from Normals and COPD Patients," *Monaldi Archives for Chest Disease*, Vol. 55, No. 3, 2000, pp. 185-188.

[15] Y. Hu, Z. Zhang and C. Yang, "A Sensitive Chemiluminescence Method for the Determination of H_2O_2 in Exhaled Breath Condensate," *Analytical Science*, Vol. 24, No. 2, 2008, pp. 201-205.

[16] A. H. Al-Obaidy and A.-G. M. Al-Samarai, "Exhaled Breath Condensate pH and Hydrogen Peroxide as Non-Invasive Markers for Asthma," *Saudi Medical Journal*, Vol. 28, No. 12, 2007, pp. 1860-1863.

[17] A. Emelyanov, G. Fedoseev, A. Abulimity, K. Rudinski, A. Fedoulov, A. Karabanov and P. J. Barnes, "Elevated Concentrations of Exhaled Hydrogen Peroxide in Asthmatic Patients," *Chest*, Vol. 120, No. 4, 2001, pp. 1136-1139.

[18] S. Loukides, I. Horvath, T. Wodehouse, P. J. Cole and P. J. Barnes, "Elevated Levels of Expired Breath Hydrogen Peroxide in Bronchiectasis," *American Journal of Respiratory and Critical Care Medicine*, Vol. 158, No. 3, 1998, pp. 991-994.

[19] D. Kietzmann, R. Kahl, M. Müller, H. Burchardi and D. Kettler, "Hydrogen Peroxide in Expired Breath Condensate of Patients with Acute Respiratory Failure and with ARDS," *Intensive Care Medicine*, Vol. 19, No. 2, 1993, pp. 78-81.

[20] E.-J. D. Oudijk, W. B. M. Gerritsen, E. H. J. Nijhuis, D. Kanters, B. L. P. Maesen, J.-W. J. Lammers and L. Koenderman, "Expression of Priming-Associated Cellular Markers on Neutrophils during an Exacerbation of COPD," *Respiratory Medicine*, Vol. 100, No. 10, 2006, pp. 1791-1799.

[21] Q. Jöbsis, H. C. Raatgeep, S. L. Schellekens, A. Kroesbergen, W. C. Hop and J. C. de Jongste, "Hydrogen Peroxide and Nitric Oxide in Exhaled Air of Children with Cystic Fibrosis during Antibiotic Treatment," *European Respiratory Journal*, Vol. 16, No. 1, 2000, pp. 95-100.

[22] Q. Jöbsis, H. C. Raatgeep, S. L. Schellekens, W. C. Hop, P. W. Hermans and J. C. de Jongste, "Hydrogen Peroxide in Exhaled Air of Healthy Children: Reference Values," *European Respiratory Journal*, Vol. 12, No. 2, 1998, pp. 483-485.

[23] M. B. Schleiss, O. Holz, M. Behnke, K. Richter, H. Magnussen and R. A. Jorres, "The Concentration of Hydrogen Peroxide in Exhaled Air Depends on Expiratory Flow Rate," *European Respiratory Journal*, Vol. 16, No. 6, 2000, pp. 1115-1118.

[24] M. D. Davis and J. Hunt, "Exhaled Breath Condensate pH Assays," *Immunology and Allergy Clinics of North America*, Vol. 32, No. 3, 2012, pp. 377-386.

[25] J. Vaughan, L. Ngamtrakulpanit, T. N. Pajewski, R. Turner, T. A. Nguyen, A. Smith, P. Urban, S. Hom, B. Gaston and J. Hunt, "Exhaled Breath Condensate pH is a Robust and Reproducible Assay of Airway Acidity," *European Respiratory Journal*, Vol. 22, No. 6, 2003, pp. 889-894.

[26] M. M. Tomasiak-Lozowska, Z. Zietkowski, K. Przeslaw, M. Tomasiak, R. Skiepko and A. Bodzenta-Lukaszyk, "Inflammatory Markers and Acid-Base Equilibrium in Exhaled Breath Condensate of Stable and Unstable Asthma Patients," *International Archives of Allergy and Immunology*, Vol. 159, No. 2, 2012, pp. 121-129.

[27] A. I. Papaioannou, S. Loukides, M. Minas, K. Kontogianni, P. Bakakos, K. I. Gourgoulianis, M. Alchanatis, S. Papiris and K. Kostikas, "Exhaled Breath Condensate pH as a Biomarker of COPD Severity in Ex-Smokers," *Respiratory Research*, Vol. 12, 2011, p. 67.

[28] J. Hunt, Y. Yu, J. Burns, B. Gaston, L. Ngamtrakulpanit, D. Bunyan, B. K. Walsh, A. Smith and S. Hom, "Identification of Acid Reflux Cough Using Serial Assays of Exhaled Breath Condensate pH," *Cough*, Vol. 2, 2006, p. 3.

[29] B. K. Walsh, D. J. Mackey, T. Pajewski, Y. Yu, B. M. Gaston and J. F. Hunt, "Exhaled-Breath Condensate pH Can Be Safely and Continuously Monitored in Mechanically Ventilated Patients," *Respiratory Care*, Vol. 51, No. 10, 2006, pp. 1125-1131.

[30] J. P. Eiserich, R. P. Patel and V. B. O'Donnell, "Pathophysiology of Nitric Oxide and Related Species: Free Radical Reactions and Modification of Biomolecules," *Molecular Aspects of Medicine*, Vol. 19, No. 4-5, 1998, pp. 221-357.

[31] R. C. Strunk, S. J. Szefler, B. R. Phillips, R. S. Zeiger, V. M. Chinchilli, G. Larsen, K. Hodgdon, W. Morgan, C. A. Sorkness and R. F. Lemanske Jr., "Relationship of Exhaled Nitric Oxide to Clinical and Inflammatory Markers of Persistent Asthma in Children," *The Journal of Allergy and Clinical Immunology*, Vol. 112, No. 5, 2003, pp. 883-892.

[32] J. Liu, A. Sandrini, M. C. Thurston, D. H. Yates and P. S. Thomas, "Nitric Oxide and Exhaled Breath Nitrite/Nitrates in Chronic Obstructive Pulmonary Disease Patients," *Respiration*, Vol. 74, No. 6, 2007, pp. 617-623.

[33] A. van der Vliet, J. P. Eiserich, M. K. Shigenaga and C. E. Cross, "Reactive Nitrogen Species and Tyrosine Nitration in the Respiratory Tract: Epiphenomena or a Pathobiologic Mechanism of Disease?" *American Journal of Respiratory and Critical Care Medicine*, Vol. 160, No. 1, 1999, pp. 1-9.

[34] M. Corradi, P. Montuschi, L. E. Donnelly, A. Pesci, S. A. Kharitonov and P. J. Barnes, "Increased Nitrosothiols in Exhaled Breath Condensate in Inflammatory Airway Diseases," *American Journal of Respiratory and Critical Care Medicine*, Vol. 163, No. 4, 2001, pp. 854-858.

[35] E. Baraldi, G. Giordano, M. F. Pasquale, S. Carraro, A. Mardegan, G. Bonetto, C. Bastardo, F. Zacchello and S. Zanconato, "3-Nitrotyrosine, a Marker of Nitrosative Stress, Is Increased in Breath Condensate of Allergic Asthmatic Children," *Allergy*, Vol. 61, No. 1, 2006, pp. 90-96.

[36] B. Balint, S. A. Kharitonov, T. Hanazawa, L. E. Donnelly, P. L. Shah, M. E. Hodson and P. J. Barnes, "Increased Nitrotyrosine in Exhaled Breath Condensate in Cystic Fibrosis," *European Respiratory Journal*, Vol. 17, No. 6, 2001, pp. 1201-1207.

[37] J. D. Morrow, K. E. Hill, R. F. Burk, T. M. Nammour, K. F. Badr and L. J. Roberts 2nd, "A Series of Prostaglandin F2-Like Compounds are Produced *in Vivo* in Humans by a Non-Cyclooxygenase, Free Radical-Catalyzed Mechanism," *Proceedings of the National Academy Sciences of the USA*, Vol. 87, No. 23, 1990, pp. 9383-9387.

[38] P. Montuschi, P. J. Barnes and L. J. Roberts 2nd, "Isoprostanes: Markers and Mediators of Oxidative Stress," *FASEB Journal*, Vol. 18, No. 15, 2004, pp. 1791-1800.

[39] P. Montuschi, M. Corradi, G. Ciabattoni, J. Nightingale, S. A. Kharitonov and P. J. Barnes, "Increased 8-Isoprostane, a Marker of Oxidative Stress, in Exhaled Condensate of Asthma Patients," *American Journal of Respiratory and Critical Care Medicine*, Vol. 160, No. 1, 1999, pp. 216-220.

[40] P. Montuschi, J. V. Collins, G. Ciabattoni, N. Lazzeri, M. Corradi, S. A. Kharitonov and P. J. Barnes, "Exhaled 8-Isoprostane as An *in Vivo* Biomarker of Lung Oxidative Stress in Patients with COPD and Healthy Smokers," *American Journal of Respiratory and Critical Care Medicine*, Vol. 162, No. 3, 2000, pp. 1175-1177.

[41] P. Montuschi, S. A. Kharitonov, G. Ciabattoni, M. Corradi, L. van Rensen, D. M. Geddes, M. E. Hodson and P. J. Barnes, "Exhaled 8-Isoprostane as a New Non-Invasive Biomarker of Oxidative Stress in Cystic Fibrosis," *Thorax*, Vol. 55, No. 3, 2000, pp. 205-209.

[42] C. T. Carpenter, P. V. Price and B. W. Christman, "Exhaled Breath Condensate Isoprostanes Are Elevated in Patients with Acute Lung Injury or ARDS," *Chest*, Vol. 114, No. 6, 1998, pp. 1653-1659.

[43] M. Barreto, M. P. Villa, C. Olita, S. Martella, G. Ciabattoni and P. Montuschi, "8-Isoprostane in Exhaled Breath Condensate and Exercise-Induced Bronchoconstriction in Asthmatic Children and Adolescents," *Chest*, Vol. 135, No. 1, 2009, pp. 66-73.

[44] B. Samuelsson, "Leukotrienes: mediators of immediate hypersensitivity reactions and inflammation," *Science*, Vol. 220, No. 4597, 1983, pp. 568-575.

[45] P. Montuschi, S. A. Kharitonov, G. Ciabattoni and P. J. Barnes, "Exhaled Leukotrienes and Prostaglandins in COPD," *Thorax*, Vol. 58, No. 7, 2003, pp. 585-588.

[46] A. Bodini, C. D'Orazio, D. Peroni, M. Corradi, G. Folesani, E. Baraldi, B. M. Assael, A. Boner and G. L. Piacentini, "Biomarkers of Neutrophilic Inflammation in Exhaled Air of Cystic Fibrosis Children with Bacterial Airway Infections," *Pediatric Pulmonology*, Vol. 40, No. 6, 2005, pp. 494-499.

[47] S. Caballero Balanzá, A. Martorell Aragonés, J. C. Cerdá Mir, J. Belda Ramírez, J. B. Ramírez, R. Navarro Iváñez, A. Navarro Soriano, R. Félix Toledo and A. Escribano Montaner, "Leukotriene B4 and 8-Isoprostane in Exhaled Breath Condensate of Children with Episodic and Persistent Asthma," *Journal of Investigational Allergology and Clinical Immunology*, Vol. 20, No. 3, 2010, pp. 237-243.

[48] G. E. Carpagnano, G. P. Palladino, D. Lacedonia, A. Koutelou, S. Orlando and M. P. Foschino-Barbaro, "Neutrophilic Airways Inflammation in Lung Cancer: The Role of Exhaled LTB-4 and IL-8," *BMC Cancer*, Vol. 11, 2011, p. 226.

[49] E. Ono, H. Mita, M. Taniguchi, N. Higashi, T. Tsuburai, M. Hasegawa, E. Miyazaki, T. Kumamoto and K. Akiyama, "Increase in Inflammatory Mediator Concentrations in Exhaled Breath Condensate after Allergen Inhalation," *Journal of Allergy and Clinical Immunology*, Vol. 122, No. 4, 2008, pp. 768-773.

[50] S. Carraro, M. Corradi, S. Zanconato, R. Alinovi, M. F. Pasquale, F. Zacchello and E. Baraldi, "Exhaled Breath Condensate Cysteinyl Leukotrienes Are Increased in Children with Exercise-Induced Bronchoconstriction," *Journal of Allergy and Clinical Immunology*, Vol. 115, No. 4, 2005, pp. 764-770.

[51] L. Mastalerz, M. Sanak, J. Kumik, A. Gawlewicz-Mroczka, N. Celejewska-Wójcik, A. Cmiel and A. Szczeklik, "Exhaled Eicosanoids Following Bronchial Aspirin Challenge in Asthma Patients with and without Aspirin Hypersensitivity: The Pilot Study," *Journal of Allergy (Cairo)*, Vol. 2012, 2012, Article ID: 696792.

[52] M. Sanak, A. Gielicz, G. Bochenek, M. Kaszuba, E. Niżankowska-Mogilnicka and A. Szczeklik, "Targeted Eicosanoid Lipidomics of Exhaled Breath Condensate Provide a Distinct Pattern in the Aspirin-Intolerant Asthma Phenotype," *Journal of Allergy and Clinical Immunology*, Vol. 127, No. 5, 2011, pp. 1141-1147.

[53] I. D. Pavord and A. E. Tattersfield, "Bronchoprotective Role for Endogenous Prostaglandin E2," *Lancet*, Vol. 345, No. 8947, 1995, pp. 436-438.

[54] R. J. Church, L. A. Jania and B. H. Koller, "Prostaglandin E(2) Produced by the Lung Augments the Effector Phase of Allergic Inflammation," *Journal of Immunology*, Vol. 188, No. 8, 2012, pp. 4093-4102.

[55] V. Lucidi, G. Ciabattoni, S. Bella, P. J. Barnes and P. Montuschi, "Exhaled 8-Isoprostane and Prostaglandin E2 in Patients with Stable and Unstable Cystic Fibrosis," *Free Radical Biology and Medicine*, Vol. 45, No. 6, 2008, pp. 913-919.

[56] E. Huszár, Z. Szabó, A. Jakab, I. Barta, I. Herjavecz and I. Horváth, "Comparative Measurement of Thromboxane A2 Metabolites in Exhaled Breath Condensate by Different Immunoassays," *Inflammation Research*, Vol. 54, No. 8, 2005, pp. 350-355.

[57] M. G. Kiehl, H. Ostermann, M. Thomas, C. Müller, U. Cassens and J. Kienast, "Inflammatory Mediators in Bronchoalveolar Lavage Fluid and Plasma in Leukocytopenic Patients with Septic Shock-Induced Acute Respiratory Distress Syndrome," *Critical Care Medicine*, Vol. 26, No. 7, 1998, pp. 1194-1199.

[58] H. Schütte, J. Lohmeyer, S. Rosseau, S. Ziegler, C.

Siebert, H. Kielisch, H. Pralle, F. Grimminger, H. Morr and W. Seeger, "Bronchoalveolar and Systemic Cytokine Profiles in Patients with ARDS, Severe Pneumonia and Cardiogenic Pulmonary Oedema," *European Respiratory Journal*, Vol. 9, No. 9, 1996, pp. 1858-1867.

[59] C. E. Brightling, F. A. Symon, S. S. Birring, P. Bradding, I. D. Pavord and A. J. Wardlaw, "TH2 Cytokine Expression in Bronchoalveolar Lavage fluid T Lymphocytes and Bronchial Submucosa Is a Feature of Asthma and Eosinophilic Bronchitis," *Journal of Allergy and Clinical Immunology*, Vol. 110, No. 6, 2002, pp. 899-905.

[60] B. Barceló, J. Pons, A. Fuster, J. Sauleda, A. Noguera, J. M. Ferrer and A. G. N. Agustí, "Intracellular Cytokine Profile of T Lymphocytes in Patients with Chronic Obstructive Pulmonary Disease," *Clinical & Experimental Immunology*, Vol. 145, No. 3, 2006, pp. 474-479.

[61] T. L. Bonfield, J. R. Panuska, M. W. Konstan, K. A. Hilliard, J. B. Hilliard, H. Ghnaim and M. Berger, "Inflammatory Cytokines in Cystic Fibrosis Lungs," *American Journal of Respiratory and Critical Care Medicine*, Vol. 152, No. 6, 1995, pp. 2111-2118.

[62] S. S. Hacievliyagil, H. Gunen, L. C. Mutlu, A. B. Karabulut and İ. Temel, "Association between Cytokines in Induced Sputum and Severity of Chronic Obstructive Pulmonary Disease," *Respiratory Medicine*, Vol. 100, No. 5, 2006, pp. 846-854.

[63] W. A. Neveu, J. L. Allard, D. M. Raymond, L. M. Bourassa, S. M. Burns, J. Y. Bunn, C. G. Irvin, D. A. Kaminsky and M. Rincon, "Elevation of IL-6 in the Allergic Asthmatic Airway Is Independent of Inflammation But Associates with Loss of Central Airway Function," *Respiratory Research*, Vol. 11, No. 1, 2010, p. 28.

[64] S. K. Shahid, S. A. Kharitonov, N. M. Wilson, A. Bush and P. J. Barnes, "Increased Interleukin-4 and Decreased Interferon-γ in Exhaled Breath Condensate of Children with Asthma," *American Journal of Respiratory and Critical Care Medicine*, Vol. 165, No. 9, 2002, pp. 1290-1293.

[65] P. P. R. Rosias, E. Dompeling, M. A. Dentener, H. J. Pennings, H. J. E. Hendriks, M. P. A. Van Iersel and Q. Jöbsis, "Childhood Asthma: Exhaled Markers of Airway Inflammation, Asthma Control Score, and Lung Function Tests," *Pediatric Pulmonology*, Vol. 38, No. 2, 2004, pp. 107-114.

[66] C. M. H. H. T. Robroeks, Q. Jöbsis, J. G. M. C. Damoiseaux, P. H. M. Heijmans, P. P. R. Rosias, H. J. E. Hendriks and E. Dompeling, "Cytokines in Exhaled Breath Condensate of Children with Asthma and Cystic Fibrosis," *Annals of Allergy, Asthma & Immunology*, Vol. 96, No. 2, 2006, pp. 349-355.

[67] P. P. Rosias, C. M. Robroeks, A. Kester, G. J. den Hartog, W. K. Wodzig, G. T. Rijkers, L. J. Zimmermann, C. P. van Schayck, Q. Jöbsis and E. Dompeling, "Biomarker Reproducibility in Exhaled Breath Condensate Collected with Different Condensers," *European Respiratory Journal*, Vol. 31, No. 5, 2008, pp. 934-942.

[68] C. M. H. H. T. Robroeks, G. T. Rijkers, Q. Jöbsis, H. J. E. Hendriks, J. G. M. C. Damoiseaux, L. J. I. Zimmermann, O. P. van Schayck and E. Dompeling, "Increased Cytokines, Chemokines and Soluble Adhesion Molecules in Exhaled Breath Condensate of Asthmatic Children," *Clinical & Experimental Allergy*, Vol. 40, No. 1, 2010, pp. 77-84.

[69] C. Gessner, R. Scheibe, M. Wötzel, S. Hammerschmidt, H. Kuhn, L. Engelmann, G. Hoheisel, A. Gillissen, U. Sack and H. Wirtz, "Exhaled Breath Condensate Cytokine Patterns in Chronic Obstructive Pulmonary Disease," *Respiratory Medicine*, Vol. 99, No. 10, 2005, pp. 1229-1240.

[70] Z. Zietkowski, M. M. Tomasiak, R. Skiepko and A. Bodzenta-Lukaszyk, "RANTES in Exhaled Breath Condensate of Stable and Unstable Asthma Patients," *Respiratory Medicine*, Vol. 102, No. 8, 2008, pp. 1198-1202.

[71] Z. Zietkowski, R. Skiepko, M. M. Tomasiak-Lozowska, E. Zietkowska and A. Bodzenta-Lukaszyk, "Eotaxin in Exhaled Breath Condensate of Allergic Asthma Patients with Exercise-Induced Bronchoconstriction," *Respiration*, Vol. 82, No. 2, 2011, pp. 169-176.

[72] K. D. G. van de Kant, E. M. M. Klaassen, Q. Jöbsis, K. Koers, G. T. Rijkers, C. P. van der Grinten, O. C. P. van Schayck, V. L. Passos and E. Dompeling, "Wheezing in Preschool Children Is Associated with Increased Levels of Cytokines/Chemokines in Exhaled Breath Condensate," *The Journal of Allergy and Clinical Immunology*, Vol. 126, No. 3, 2010, pp. 669-671.

[73] T. Zakharkina, A.-R. Koczulla, O. Mardanova, A. Hattesohl and R. Bals, "Detection of Microorganisms in Exhaled Breath Condensate during Acute Exacerbations of COPD," *Respirology*, Vol. 16, No. 6, 2011, pp. 932-938.

[74] Z. Xu, F. Shen, X. Li, Y. Wu, Q. Chen, X. Jie and M. Yao, "Molecular and Microscopic Analysis of Bacteria and Viruses in Exhaled Breath Collected Using a Simple Impaction and Condensing Method," *PLoS ONE*, Vol. 7, No. 7, 2012, Article ID: e41137.

[75] G. E. Carpagnano, A. Koutelou, M. I. Natalicchio, D. Martinelli, C. Ruggieri, A. Di Taranto, R. Antonetti, F. Carpagnano and M. P. Foschino-Barbaro, "HPV in Exhaled Breath Condensate of Lung Cancer Patients," *British Journal of Cancer*, Vol. 105, No. 8, 2011, pp. 1183-1190.

[76] K. Chikasue, M. Kimura, K. Ikeda, T. Ohnishi, S. Kawanishi, T. Iio, M. Kataoka and Y. Arao, "Detection of Torque Teno Virus DNA in Exhaled Breath by Polymerase Chain Reaction," *Acta Medica Okayama*, Vol. 66, No. 5, 2012, pp. 387-397.

[77] E. Sonkoly, M. Ståhle and A. Pivarcsi, "MicroRNAs and Immunity: Novel Players in the Regulation of Normal Immune Function and Inflammation," *Seminars in Can-*

cer Biology, Vol. 18, No. 2, 2008, pp. 131-140.

[78] Y. Xie, N. W. Todd, Z. Liu, M. Zhan, H. Fang, H. Peng, M. Alattar, J. Deepak, S. A. Stass and F. Jiang, "Altered miRNA Expression in Sputum for Diagnosis of Non-Small Cell Lung Cancer," *Lung Cancer*, Vol. 67, No. 2, 2010, pp. 170-176.

[79] L. Yu, N. W. Todd, L. Xing, Y. Xie, H. Zhang, Z. Liu, H. Fang, J. Zhang, R. L. Katz and F. Jiang, "Early Detection of Lung Adenocarcinoma in Sputum by a Panel of MicroRNA Markers," *International Journal of Cancer*, Vol. 127, No. 12, 2010, pp. 2870-2878.

[80] R. M. Effros, J. Biller, B. Foss, K. Hoagland, M. B. Dunning, D. Castillo, M. Bosbous, F. Sun and R. Shaker, "A Simple Method for Estimating Respiratory Solute Dilution in Exhaled Breath Condensates," *American Journal of Respiratory and Critical Care Medicine*, Vol. 168, No. 12, 2003, pp. 1500-1505.

[81] P. Montuschi, "LC/MS/MS Analysis of Leukotriene B_4 and Other Eicosanoids in Exhaled Breath Condensate for Assessing Lung Inflammation," *Journal of Chromatography B*, Vol. 877, No. 13, 2009, pp. 1272-1280.

[82] G. de Laurentiis, D. Paris, D. Melck, M. Maniscalco, S. Marsico, G. Corso, A. Motta and M. Sofia, "Metabonomic Analysis of Exhaled Breath Condensate in Adults by Nuclear Magnetic Resonance Spectroscopy," *European Respiratory Journal*, Vol. 32, No. 5, 2008, pp. 1175-1183.

Abbreviations

COPD: Chronic obstructive pulmonary disease
EBC: Exhaled breath condensate
H_2O_2: Hydrogen peroxide
DNA: Deoxyribonucleic acid
RNA: Ribonucleic acid
MiRNAs: microRNAs
ROS: Reactive oxygen species
ALF: Airway lining fluid
BAL: Bronchoalveolar lavage
VOCs: Volatile organic compounds
O_2^- : superoxide anion
ARDS: Acute respiratory distress syndrome
FeNO: Fractional excretion of NO
RNS: Reactive nitrogen species
NO: Nitric oxide
CO_2: Carbon dioxide
COX: Cyclooxygenase
8: IP- 8-Isoprostane
5-LO: 5-Lipoxigenase
LT: Leukotriene
NSCLC: Non-small cell lung cancer
CysLTs: Cysteinyl-leukotrienes
EIB: Exercise induced bronchospasm
HETE: Hydroxyeicosatetraenoic acid
PGD2: Prostaglandin D2
EIA: Enzyme immunoassays
PGE2: Prostaglandin E2
TxA2: Thromboxane A2
RIA: Radioimmunoassay
Th0: Naïve T-cells
Th: T helper
IL: Interleukin
IFN-γ: Interferon-gamma
RANTES: Regulated and normal T cell expressed and secreted
MIP1α: Macrophage Inflammatory Proteins1α
PCR: Polymerase chain reaction
Treg: Regulatory T cells
NK cells: Natural killer cells
NMR: Nuclear magnetic resonance
HPCL: High performance liquid chromatography
GC/MS: Gas chromatography/mass spectrometry
LC/MS: Liquid chromatography/mass spectrometry

Development and Application of a Validated HPLC Method for the Determination of Clindamycin Palmitate Hydrochloride in Marketed Drug Products: An Optimization of the Current USP Methodology for Assay

Geoffrey K. Wu[1], Abhay Gupta[2], Mansoor A. Khan[2], Patrick J. Faustino[2]

[1]Office of Generic Drugs, Food and Drug Administration, Rockville, USA; [2]Division of Product Quality Research, Office of Pharmaceutical Science, Food and Drug Administration, Silver Spring, USA.

ABSTRACT

A simple efficient isocratic reversed-phase HPLC method was developed and validated for the determination of clindamycin palmitate hydrochloride (CPH) and its commercially available oral solution products. Separation was achieved on a Phenomenex Zorbax (Luna) cyano column (150 × 4.6 mm, 5 μm) with a Phenomenex cyano guard cartridge (4 × 3.0 mm) on Agilent 1050 series HPLC system. CPH and its resolution standard lincomycin were eluted isocratically at a flow rate of 1 mL/min with a simplified mobile phase (potassium phosphate buffer (5 mM, pH 3.0)—acetonitrile—tetrahydrofuran (20:75:5, v/v/v)) and detected at 210 nm. The column was maintained at 25°C. The method was validated according to USP category I requirements. Robustness and forced degradation studies were also conducted. CPH marketed drug products were obtained from a drug distributor and assayed for potency using the validated method. Validation acceptance criteria were met in all cases. The analytical range for CPH was 15 - 500 μg/mL and the linearity was $r^2 > 0.999$ over three days. The method was determined to be specific and robust. Both accuracy (92.0% - 103.8%) and precision (0.67% - 1.52%) were established across the analytical range for low, intermediate and high QC concentrations. Method applicability was demonstrated by analyzing two marketed products of CPH, in which results showed potency >98%. The method was determined to be an enhancement over the current USP methodology for assay as a result of increased efficiency, reduced organic solvents and the elimination of matrix modifiers. This method was successfully applied for the quality assessment of: 1) currently marketed drug products and 2) will in future assess the product quality of novel dosage forms of CPH for pediatric use.

Keywords: Clindamycin Palmitate Hydrochloride (CPH); HPLC; Method Validation; Pediatric Dosage Form

1. Introduction

Clindamycin palmitate hydrochloride (CPH, methyl 7-chloro-6,7,8-trideoxy-6-(1-methyl-*trans*-4-propyl-L-2-pyrrolidinecarboxamido)-1-thio-l-*threo*-*α*-d-*galacto*octo-pyranoside monohydrochloride, **Figure 1(a)**), is an ester of clindamycin and palmitic acid. Clindamycin (**Figure 1(b)**) is a lincosamide antibiotics first synthesized in 1967. It is active against gram-positive aerobes and highly active against both gram-positive and negative anaerobes [1]. CPH is a white to off-white crystalline solid with a melting point from 168°C - 178°C, CPH has a molecular formula of $C_{34}H_{63}ClN_2O_6S{\cdot}HCl$ and a molecular weight of 699.85. CPH is a white to off-white amorphous solid with a pKa of 7.6.

During the development of pediatric preparations, a series of clindamycin esters were synthesized to minimize its bitter taste. As a successful 2-palmitate ester derivative, CPH is microbiologically inactive; however, it can be readily cleaved to clindamycin by enzymes located in small-intestine and has been shown to demonstrate similar therapeutic strength [2]. It has been used to treat a wide spectrum of bacterial infections, for instance, infections of respiratory tract, skin and soft-tissue infections, osteomyelitis, gynecological infections, as well as

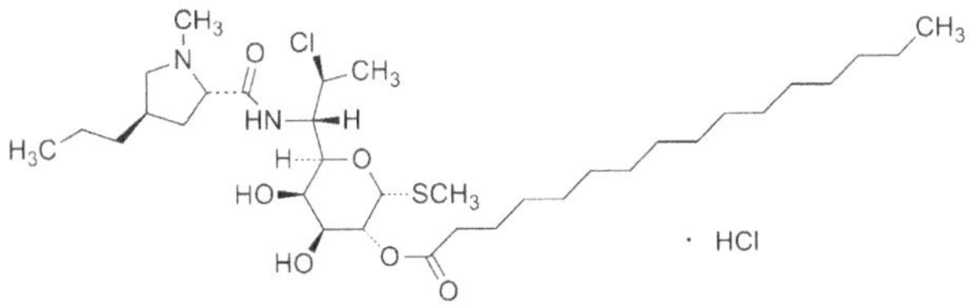

(a)

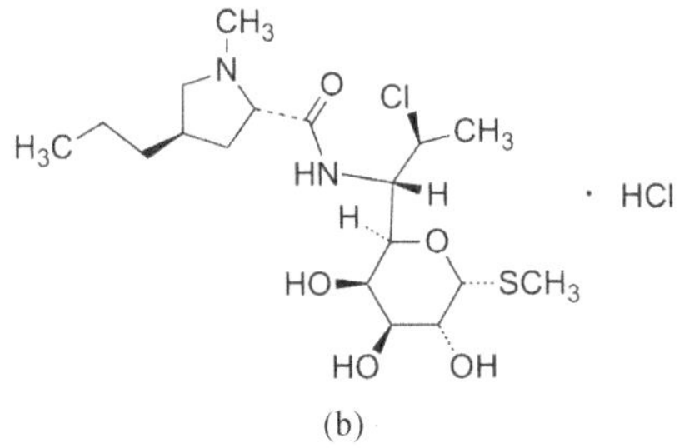

(b)

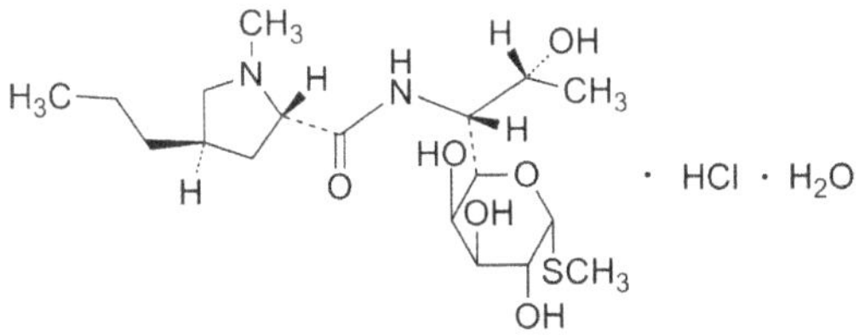

(c)

Figure 1. Chemical structures of clindamycin palmitate hydrochloride (a), clindamycin hydrochloride (b) and lincomycin hydrochloride (c). (a) Clindamycin palmitate hydrochloride (CPH). (b) Clindamycin hydrochloride. (c) Lincomycin hydrochloride monohydrate.

bacterial endocarditis.

In contrast to the other two clinically available clindamycin salts (*i.e.*, clindamycin hydrochloride and clindamycin phosphate), there are only two HPLC analytical methods reported for CPH analysis and neither of them is ideal to implement or simple in their execution. Both methods required complex mobile phase compositions, with high organic content while both yielding longer retention times. The first method is a compendial method from the United State Pharmacopoeia using a C18 reversed-phase chromatographic column with refractive index (RI) detector, and with a mobile phase containing ~92% of methanol with a total run time of 30 min [3]. This is an adaptation from a method first published in 1978 by Brown [4]. The method is still being widely used by other researchers with minor modifications [5]. However, the use of RI detector can significantly compromise method robustness since the detector is sensitive to various environmental factors, such as small changes to ambient temperature, and operational factors such as flow rate, column pressure and non-compatibility with gradient elution and ultimately difficult to implement. The second method reported by Raju and colleagues [6] was an isocratic HPLC method using UV detector (210 nm) also using methanol as a major (90%, v/v%) component in the mobile phase. However, the mobile phase did not have buffering capacity and triethylamine (TEA) was used as ion pairing agent complicating the mobile phase preparation. TEA acts as a base causing the mobile phase pH to increasee, which negatively impacts the robustness of non-buffer mobile phase systems. Additionally TEA can not be easily washed out of reverse phase columns typically requiring 100 - 200 column volumes, thereby limiting column use and ultimately column lifetime.

Developing a practical HPLC method for CPH is challenging due to its amphiphlic nature and highly pH-dependent solubility in aqueous medium [7]. In its free base form, when pH is close to or above its pKa (7.6), CPH is extremely insoluble. For instance at pH 7.4 it has an aqueous solubility of <0.0002 mg/mL. Once the pH drops to around 4, its solubility increases to 0.238 mg/mL; if micelle formation is considered, its solubility can increase to 50 mg/mL at pH 3.6. Such a solubility variation was observed in our laboratory studies. This solubility phenomenon is significantly different from the widely used hydrochloride salt where solubility is reported to be greater than 500 mg/mL in water [8]. Additionally the ionic strength and the presence of a high concentration of inorganic salts also have a significant impact on reducing CPH aqueous solubility. However, CPH is freely soluble in methanol and acetonitrile [5,9].

In order to optimize the USP method and provide a more robust means for assessing product quality (e.g., content uniformity and dissolution) control, a novel HPLC-UV method with well controlled mobile phase pH, low ionic strength, robust performance, and ease of system setup was developed and validated here. To demonstrate the system suitability, lincomycin, the synthesis starting material and major impurity, was used in our study as a resolution and impurity standard [5,9]. Other possible impurities in CPH drug substance include clindamycin hydrochloride which was assayed in our system. Sample detection was performed by a UV detector at 210 nm wavelength; phosphate buffer (pH 3.0) was employed to precisely control the acidity of the mobile phase; and no ion pairing agent was required for chromatographic selectivity or resolution. The method can selectively detect CPH and its related impurities with acceptable resolution. This simple and robust method will be a substantial improvement to the current USP method to evaluate product quality and drug stability.

2. Experimental

2.1. Materials

Clindamycin palmitate hydrochloride and lincomycin (resolution standard, **Figure 1(c)**) certified reference standards were purchased from the United States Pharmacopeia (Rockville, MD). Clindamycin palmitate hydrochloride drug substance was purchased from RIA International LLC (East Hanover, NJ). Clindamycin HCl was purchased from TCI (Tokyo, Japan). Clindamycin palmitate hydrochloride drug products were purchased from Bradley Drug-Washington Wholesale Drug Exchange (Bethesda, MD). Nylon syringe filters were purchased from Sun SRI (Rockwood, TN). HPLC grade monobasic potassium phosphate and ACS grade phosphoric acid were purchased from Fisher Scientific (Fairlawn, NJ). HPLC grade acetonitrile was purchased from Burdick and Jackson (Muskegon, MI). HPLC ready 18 MOhm water was obtained, in-house, from a Milli-Q Gradient A-10 water purification system, Millipore Corp. (Bedford, MA).

2.2. Instrumentation and Chromatographic Conditions

Hewlett Packard 1050 series (Agilent Technologies, Wilmington, DE) HPLC system was used for the analysis. It is equipped with a quaternary pump, online degasser, column heater, autosampler, and diode array detector (DAD). ChemStation software (Agilent Technologies, Wilmington, DE) was used to perform data collection and analysis. Separation was achieved on a Phenomenex Luna cyano column (150 × 4.6 mm, 5 μm) in series with a Phenomenex cyano SecurityGuard column (4.0 × 3.0 mm, Phenomenex, Torrance, CA). The elution was isocratic with a mobile phase of 5 mM KH_2PO_4 (pH 3.0)—acetonitrile—tetrahydrofuran (20:75:5, v/v/v). The column temperature was maintained at 25°C. The injection volume was 30 μL with UV detection at 210 nm.

2.3. Preparation of Standard Solutions

CPH stock solution I of 10 mg/mL was prepared in phosphate buffer (5 mM, pH 3.0) using the USP reference standard. Calibration standard solutions at seven levels were prepared by serially diluting the stock solution I with the HPLC mobile phase to concentrations of 15, 25, 50, 100, 250, 375, and 500 μg/mL over the analytical range of 15 - 500 μg/mL.

CPH stock solution II of 10 mg/mL was prepared in phosphate buffer (5 mM, pH 3.0) using USP reference standard. Quality control (QC) standard solutions were prepared daily by diluting the stock solution II with the HPLC mobile phase to the final QC concentrations of 15, 100 and 500 μg/mL.

Lincomycin stock solution I of 10 mg/mL was prepared in phosphate buffer (5 mM, pH 3.0) using the USP reference standard. This solution was later used to prepare a mixture of CPH and lincomycin for daily system suitability tests. Clindamycin hydrochloride stock solution of 10 mg/mL was prepared in the same way.

2.4. Sample Preparation from the Marketed Products

Two CPH products (brand-name and generic), both in a powder form for oral solution for pediatric patients, were designated as A- and B-powder respectively and assayed for potency. For each product, powder equivalent to about 5 mg of CPH was accurately weighed on an analytical balance and transferred to a 10 mL volumetric flask. About 8 mL of HPLC mobile phase was added to the flask and the content was sonicated for 5 min. Additional mobile phase was added to adjust the volume to 10 mL. The resulting solution was filtered via a 0.20 μm nylon filter into standard analytical glass vials. The filtrate, containing ~500 μg/mL CPH, was further diluted in mobile phase to 250 and 125 μg/mL. All three concentrations (125, 250 and 500 μg/mL) were injected into the HPLC for potency measurements.

2.5. Method Validation

The method was validated according to the United States Pharmacopeia requirements for Category I for assay [10]. The following validation characteristics were evaluated: linearity, range, accuracy, precision, specificity, limit of quantitation and robustness.

System suitability standard solution which contained 100 μg/mL CPH and 10 μg/mL lincomycin as a resolution standard was prepared by diluting and mixing CPH and lincomycin stock solutions with mobile phase. System suitability was determined from six replicate injections of the system suitability standard before sample analysis. The acceptance criteria were: relative standard deviation (RSD) for peak area less than 2.0%, greater than 2000 plates per column (13,333 plates per meter), a peak width at half height of less than 0.25 minute, a resolution for CPH and lincomycin greater than 1.5, a USP tailing factor less than 2.0. Equation for calculating resolution (R) is list below:

$$R = 1.18 \times \left[\frac{(t_2 - t_1)}{(W_2 + W_1)} \right]$$

where t_2 and t_1 were respective retention times for CPH and lincomycin, and W_2 and W_1 were the peak widths at half height. Acquired results were used to evaluate critical operational parameters of the chromatographic system to ensure adequate resolution and reproducibility immediately prior to sample analysis.

Standard calibration curves were prepared with seven calibrators over a concentration range of 15 - 500 μg/mL for CPH. Linear least square regression analysis was then employed to correlate peak areas and drug concentrations.

The resultant standard curves were evaluated for intra- and inter-day reproducibility.

Accuracy and precision of the method were determined for the drug substance by analyzing QC standard samples at three concentrations levels of CPH (15, 100, and 500 μg/mL) over the analytical range. The method precision was established by injecting five standard QC samples at each concentration level for the intra-day precision and on three days for the intermediate precision. Precision was expressed by the relative standard deviation (RSD, %) of the analyte peak areas. Accuracy was expressed as percent recovery (%) by comparing the amount determined from the quality control standards against the respective nominal values. Accuracy of the method was also tested on two drug products at three concentrations with 3 replicates at each level, using the ICH Q2R1 approach of standard addition [11]. In this approach the following changes were made to the procedures described in section 2.4: only 50% of the nominal amount of drug product was placed in the flask and then the drug product was spiked with CPH stock solution III up to the target concentrations. The target concentrations were 125, 250 and 500 μg/mL of CPH in total; in addition, samples containing only 50% of the nominal amount of drug product without spiking were prepared and analyzed at three corresponding levels. Percent recovery (%) of CPH was calculated by comparing the known spiked amount of CPH to the amount CPH detected after subtracting the un-spiked (50% product) samples.

Method robustness was evaluated by analyzing the system suitability standard and evaluating system suitability parameters after varying the HPLC pump flow rate (± 2.5%), auto-sampler injector volume (± 10%), column compartment temperature (± 2˚C), mobile phase pH (± 0.1) and organic composition (20% regarding tetrahydrofuran content), respectively.

Method specificity was evaluated by analyzing a mixture of CPH drug substance and lincomycin (a major synthetic precursor of CPH), and two drug products versus sample blank. In addition, CPH standards were subjected to forced degradation under different conditions: elevated temperature (55˚C for 24 hours), light (365 nm for 24 hours at room temperature), 10% peroxide (for 24 hours at room temperature), acidic (0.1 M HCl for 4 hours at room temperature), and basic (0.1 M NaOH for 4 hours, room temperature). Resulting HPLC chromatograms were compared to determine if CPH was co-eluting with lincomycin or other degradation products or excipient. Full spectrum (200 - 560 nm) chromatograms were acquired and Agilent ChemStation peak purity algorithms were also used to evaluate specificity.

3. Results and Discussion

3.1. Optimization of HPLC Method

Similar to other derivatives of lincomycin, CPH has a relatively weak UV absorption with an absorbance peak maxima around 206 nm. Different from the other lincomycin derivatives, CPH is an amphiphilic molecule due to the hydrophobicity of palmitate (14 carbon chain) and the basic nature of clindamycin. Its amphililicity leads to an anomalous solubility feature: it can only be reasonably dissolved in aqueous medium at a pH below 3.7 (53.2 mg/mL) because of micelle formation; above this pH, it is practically insoluble (e.g., 0.001 mg/mL at pH 5.8) [7]. During solubility studies there was difficulty dissolving CPH in aqueous medium and micelle formation was observed. These observations were contradictory to most of the publications and even USP material safety data sheet (MSDS), which claimed that CPH is freely soluble [12]. Both matters complicate developing a simple and robust HPLC method for CPH. This may explain why there are only two major groups of analytical methods available for CPH in the scientific literature.

The first group of methods is represented by the compendial method recorded in USP36/NF31 (CPH monograph); other similar methods were developed following minor modifications to this USP method. Acetate organic salts were used to constitute the mobile phases in these methods and this rendered RI detector as the only choice due to the high UV cutoff (~230 nm) of these salts. However, the well-known disadvantages of RI detector are its sensitivity to environmental influence, and long equilibration periods which may negatively impact routine operations. Recently another HPLC method was reported using UV detector at 210 nm for CPH analysis, in which the mobile phase did not possess any buffering capacity [6]. A common feature for all the aforementioned methods is that a high percentage (>90%) of methanol was used in mobile phase. Unfortunately like acetate salts, methanol has a UV cutoff around 210 nm, which is the chosen detection wavelength in the last method. Therefore, an HPLC method void of these problems is necessary to best assess the product quality of CPH. To this end, our goal was to develop a simple and robust HPLC method, ideally possessing buffering capacity and using UV detection at the absorbance maxima.

Initially a C18 column, used in all the published methods, was chosen and a series of mobile phase screening was conducted (**Table 1**). The mobile phase (MP1), composed of H_2O/methanol (10:90) supplemented with 0.5% triethylamine (TEA), reported by Raju *et al.* was first evaluated [6]. CPH could be eluted with acceptable symmetry, but the CPH retention was prohibitively long (~33 min) on a Phenomenex Luna C18 (2) column (250

Table 1. List of major mobile phases and HPLC columns screened.

Name	Composition	HPLC Column	Notes
MP1	H_2O—Methanol (10:90, v/v) supplemented with 0.5% (v%) TEA, pH adjusted to 5 with H_3PO_4	Phenomenex Luna C18 (2) column (250 × 4.6 mm, 5 μm, 100 Å)	1. Long retention time (~33 min) 2. Presence of ghost peak
MP2	Ammonium acetate (10 mM, pH 4.5)—acetonitrile-methanol (30:65:5, v/v/v)	Phenomenex Luna C18 (2) column (250 × 4.6 mm, 5 μm, 100 Å)	1. No CPH peak observed due to a high UV cutoff of ammonium acetate at 230 nm
MP3	Ammonium formate (5 mM, pH 3.8)—acetonitrile-methanol (30:65:5, v/v/v)	Phenomenex Luna C18 (2) column (250 × 4.6 mm, 5 μm, 100 Å)	1. No CPH peak observed due to a high UV cutoff of ammonium formate at 210 nm
MP4	Phosphate buffer (10 mM, pH 4.5)—acetonitrile-methanol (30:65:5, v/v/v)	Phenomenex Luna C18 (2) column (250 × 4.6 mm, 5 μm, 100 Å)	1. No CPH peak observed
MP5	H_2O—acetonitrile (10:90, v/v) supplemented with 0.5% TEA, pH adjusted to 5.0 with H_3PO_4	Phenomenex Luna C18 (2) column (250 × 4.6 mm, 5 μm, 100 Å)	1. Long retention time (~41 min) 2. No ghost peak observed
MP6	H_2O—acetonitrile-THF-TEA (100:900:8:0.5, v/v/v/v), pH adjusted to 5.0 with H_3PO_4	Phenomenex Luna C18 (2) column (250 × 4.6 mm, 5 μm, 100 Å)	1. Long CPH retention time > 10 min
MP7	H_2O—acetonitrile-THF-TEA (200:800:8:0.5, v/v/v/v), pH adjusted to 5.0 with H_3PO_4	Phenomenex Luna C18 column (1000 × 4.6 mm, 3 μm, 100 Å)	1. CPH retention time reduced to under 10 min
MP8	Potassium phosphate (5 mM, pH 3.)—acetonitrile- THF (200:800: 8, v/v/v)	Phenomenex Luna CN column (150 × 4.6 mm, 5 μm, 100 Å)	1. Short CPH retention time 2. No ghost peaks 3. Presence of buffering capacity

× 4.6 mm, 5 um, 100 Å) with a flow rate of 1 mL/min at ambient temperature. More importantly, ghost peaks had been observed and the source of these peaks was highly possible the mobile phase components, where the sample elution may have been incomplete. Following the reported preparation, pH (5.0) of the mobile phase was simply adjusted by titrating the mixture using phosphoric acid, although this procedure is not recommended since a pH meter generally does not measure properly in the presence of a large amount of organic solvent. This is also important, suggesting that ionization from TEA can significantly increase the pH of the MP and the pKa of CHP is 7.6.

In order to introduce buffering capacity and eliminate the ghost peaks, aqueous buffers were used to replace DI water and acetonitrile was employed to substitute methanol. The additional mobile phases that were screened include: ammonium acetate (10 mM, pH 4.5)—acetonitrile—methanol (30:65:5, v/v/v, MP2) and ammonium formate (5 mM, pH 3.8)—acetonitrile—methanol (30:65: 5, v/v/v, MP3); no major peak of CPH could be observed since these two buffers have high UV cutoffs at 230 and 210 nm, respectively. Phosphate buffer is generally recognized as the most ideal buffer with a low UV cutoff < 200 nm. Therefore a mobile phase (MP4), composed of phosphate buffer (pH 4.5, 10 mM)—acetonitrile—methanol (30:65:5, v/v/v) was examined; however, no CPH peak was observed either. These mobile phases were not supplemented with triethylamine (TEA), which could be critical to the successful elution of CPH on C18 columns.

So TEA was reintroduced and the following mobile phase combinations were tested. The first one (MP5) consisted of H_2O—acetonitrile (10:90, v/v) supplemented with 0.5% TEA with pH 5.0 adjusted by phosphoric acid. Comparing against the published mobile phase, the only difference was that acetonitrile was used to replace methanol. Results showed that a major peak of CPH could be observed around 41 min when isocratically eluted at a flow rate of 1 mL/min at 30°C; more importantly, no ghost peaks were observed since acetonitrile has low UV cutoff (~190 nm). Using MP5 while increasing column temperature to 35°C; surprisingly this only shortened the retention time from 41 to 40 min. In order to further reduce the retention time, tetryhydrofuran (THF), an organic modifier, was added to the current mobile phase. The influence of THF content was examined and the volume percentages screened of were 1%, 2%, 5%, 8% and 10%. And the CPH retention time was reduced to 37, 35, 26, 17 and 19 min, respectively. Therefore, 8% of THF supplement was found appropriate; however, TEA seemed indispensible for efficient elution.

Thus a more promising mobile phase (H_2O—acetonitrile-THF-TEA, 100:900:8:0.5, v/v/v/v, MP6) was identified; but the percentage of acetonitrile was too high. So the ratio was adjusted to 200:800 (H_2O—acetonitrile, v/v, MP7); and this new mobile phase combination resulted in a more efficient CPH retention time around 22 min. Using this new mobile phase as a working reference, we migrated to a shorter Phenomenex Luna C18 column (100 × 4.6 mm, 3 μm, 100 Å) aiming to reducing CPH retention time to under 10 min. With this HPLC condition, CPH was eluted out at 8.6 min. Further adjustment of H_2O—acetonitrile to 100:900 and 300:700 was conducted; the CPH retention time shifted to 6.7 and 17 min, respectively. These results were satisfactory; however, the mobile phase composition was still complex without a buffering capacity.

To better address these issues, the optimization of the stationary phase and HPLC column selection were performed. As observed, CPH bound to C18 stationary phase tightly so that TEA was necessary to form an ion pair to achieve efficient elution. A major factor contributing to this strong interaction is the fourteen carbon chain of palmitate. In order to reduce the affinity of CPH to the column, a cyano stationary phase with lower hydrophobicity was selected since it is more suitable for efficient elution of basic amines. To ensure the consistency, a Phenomena Luna CN column (150 × 4.6 mm, 5 μm, 100 Å) was selected. Stationary phase change allowed for significant modifications to the mobile phase composition: DI water was replaced by buffer with a ratio to acetonitrile as 200:800 (v/v) and TEA was eliminated; while 8% of THF was kept as an organic modifier. Given the low UV-cutoff at 200 nm, potassium phosphate buffer (5 mM, pH 3.0) was employed and results showed that this mobile phase could provide better peak symmetry and less baseline noise. Under the current chromatographic condition, CPH had a retention time of 5.5 min when isocratically eluted with a flow rate of 1 mL/min at 25˚C.

Therefore, to decrease the chromatographic elution time we chose to analyze CPH at an acidic pH (pH = 3.0) due to the hydrophobicity of palmitate and the basic nature of clindamycin. At this pH, CPH is positively charged and relatively easier to be eluted off a reverse phase column. Additionally, this pH is >4 units away from the pKa of CPH ensuring the constant composition of the ionic species. The selection of a cyano stationary phase further facilitated the efficient elution of CPH with an ideal elution time and a lower organic content.

In summary, a novel HPLC analysis method for CPH is established on a Phenomenex Luna CN column (150 × 4.6 mm, 5 μm, 100 Å) performed at 25˚C. After minor adjustment, the mobile phase was composed of phosphate buffer (5 mM, pH 3.0)—acetonitrile—tetrahydrofuran (20:75:5, v/v/v). CPH was eluted isocratically at a flow rate of 1 mL/min with a run time of 10 min; its retention time was around 5.6 min, detected at 210 nm (**Figure 2(a)**). This method was a significant improvement over the previous methods. First, a simple mobile phase was formulated without using ion pairing agent (e.g., triethylamine). Second, it possessed a buffering capacity to ensure the ionized state of CPH and retention time reproducibility. Third, the commonly available UV detector, instead of RI detector, was adopted to ensure the robustness of the analysis. Fourth, acetonitrile with a low UV cutoff (~190 nm) was employed to replace methanol (UV cutoff @ 240 nm) so that the interference from the mobile phase was minimized. Lastly, this method had enhanced selectivity over CPH and its related synthetic impurity and is a potential analytical tool to evaluate product stability.

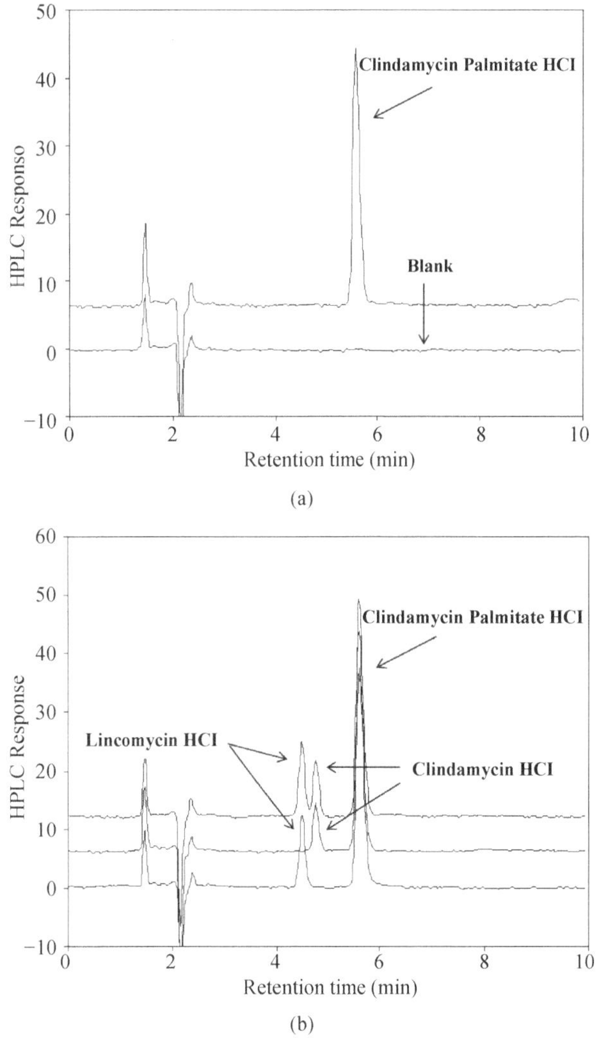

Figure 2. HPLC chromatograms of blank mobile phase and clindamycin palmitate hydrochloride USP reference standard (a), and system suitability samples (b). Retention times for lincomycin, clindamycin HCl and clindamycin palmitate hydrochloride were 4.47, 4.74 and 5.60 min, respectively.

3.2. Method Validation

The following method validation characteristics were addressed for CPH: accuracy, precision, specificity, linearity, range, and robustness. The method validation for CPH meets the requirements for USP Category I [10].

3.3. System Suitability

The system suitability test is designed to ensure the validity of the analytical procedure as well as the conformation of the resolution between different peaks of interest. All critical parameters tested met the acceptance criterion on all days, as listed in **Table 2**. Adequate resolution of >1.5 between CPH and lincomycin demonstrated the method specificity in the presence of its synthetic precursor, as shown in **Figure 2(b)**. Clindamycin HCl, another possible impurity, was also well separated from CPH.

3.4. Linearity and Range

Linearity of the method was confirmed by preparing standard curves for the analytical range of 15 - 500 μg/mL for CPH for three days. The results, summarized in **Table 3**, demonstrated an excellent correlation between analyte peak area and concentration over the analytical range with $r^2 \geq 0.999$ using a least squares linear regression function.

3.5. Accuracy and Precision

Accuracy and precision was established across the analytical range for CPH. The intra- and inter-day accuracy and precision were calculated from the QC samples for CPH. Results for the intra-day accuracy and precision of CPH were summarized in **Table 4**. In order to meet the experimental requirements, the acceptance criterion for low concentrations was set to be ± 10%, and ± 5% for medium and high concentrations. In the future use of this method, this design would allow us to capture the very early dissolution time points for pediatric dosage forms. The above analytical range will be necessary to achieve our analytical objectives and able to successfully and accurately analyze these dosage forms to best evaluate their release characteristics. Moreover, we were aware that even small increase of the lower limit of quantitation would significantly impact the accuracy in a positive manner.

The accuracy results for CPH in the two drug products were summarized in **Table 5**. Results for the accuracy of CPH tested in drug products A- and B- powder at three concentrations levels (125, 250 and 500 μg/mL) by the technique of standard addition ranged from 100.62% to 104.82%. The recovery was within 100 ± 5% for all samples.

3.6. Robustness

To ensure the HPLC method insensitive to minor changes in the experimental conditions, it is important to demonstrate robustness of the method. As pointed out by Dejaegher and colleagues [13], the following responses were evaluated: CPH retention time, peak area, capacity factor, USP tailing, resolution between CPH and lincomycin, and theoretical plate numbers. Results were summarized in **Table 6** and none of the alterations caused a significant change in these responses and all responses met the acceptance criteria for system suitability. Therefore the method is robust and reliable to quantitate CPH.

3.7. Specificity and Forced Degradation

Full spectrum data acquisition was performed for system suitability standard (**Figure 2**) and two products as well to examine for co-eluting peaks (**Figure 3**). After peak

Table 2. System suitability test (n = 6).

USP Criteria	Specification	Day 1	Day 2	Day 3	Pass/Fail
Retention time	RSD ≤ 2.0%	0.047%	0.113%	0.130%	Pass
Capacity factor k'	> 1.0	2.674	2.694	2.686	Pass
Symmetry	> 0.50	0.659	0.827	0.626	Pass
Area	RSD ≤ 2.0%	1.53%	1.23%	1.51%	Pass
Theoretical plates/meter	> 13,333 (2000/column)	21,501	21,461	21,909	Pass
USP tailing	< 2.0	1.384	1.109	1.293	Pass
Resolution	> 1.5	2.39	2.27	2.37	Pass

Table 3. Parameters and linearity of clindamycin palmitate hydrochloride calibration curves.

Standard curve	Analytical range (μg/mL)	Calibrators	Slope	y-intercept	R^2 value
Day 1	15 - 500	7	3.89	8.87	0.9996
Day 2	15 - 500	7	3.83	8.74	0.9999
Day 3	15 - 500	7	3.89	9.53	0.9997

Table 4. Clindamycin palmitate hydrochloride accuracy and precision: drug substance (% recovery, n = 5).

		Validation solutions (µg/mL)		
		15	100	500
Accuracy (% recovery)	Day 1	92.3	100.5	99.5
	Day 2	92.0	103.3	101.8
	Day 3	94.0	103.8	101.3
Precision (% RSD)	Day 1	0.67	1.52	0.73
	Day 2	1.07	0.83	0.91
	Day 3	0.97	0.89	0.99
	Intermediate	1.18	1.18	1.21

Note: The acceptance criteria for accuracy: not more than 10% at the low QC levels, and 5% at the medium and high levels.

Table 5. Clindamycin palmitate hydrochloride recovery: drug product (% recovery, n = 3) via standard addition test.

Product	50% weight product spiked to		
	125 µg/mL	250 µg/mL	500 µg/mL
A-powder	104.8	103.5	103.2
B-powder	100.8	100.6	104.4

purity analysis, results demonstrated that no co-eluting peaks of CPH and lincomycin was observed; no impurities were detected in CPH peak; and excipients from the products were determined not co-eluted with CPH by observation following a spectral peak purity CPH evaluation. Under the current stress-test conditions, CPH was stable under the stress conditions of heat, light and oxidation but labile in presence of acid and base. However, no co-eluting peaks were observed from these stressed samples. Together with all the aforementioned observations, we determined that this method is specific for CPH analysis.

3.8. Analysis of the Marketed Products

In the market there is only one solid dosage form available for CPH, flavored granules for oral solution with a single dose strength of 75 mg/5 mL upon reconstitution. Currently, there are one brand-name and two generic products being marketed. To ensure the representativeness, both brand-name (one product) and generic product (one of the two) were purchased and analyzed by this validated method. Representative chromatograms were shown in **Figure 3**. Results for CPH potency are summarized in **Table 7**. All products had a potency of >98%. Although only two drug products were tested due to the limited number of products available, this method demonstrated strong applicability for product potency analysis.

In addition, this method has been proved applicable for potency analysis of novel CPH formulations prepared in house. It also demonstrated an applicability to quantitate CPH release from these novel formulations during dissolution testing (data not published).

4. Conclusion

A simple and efficient and robust reverse-phase HPLC method for clindamycin palmitate hydrochloride was found to be accurate, precise, specific, and linear across

Table 6. Robustness test results for the HPLC method for clindamycin palmitate hydrochloride analysis.

Parameters	Change	Retention time, min (RSD %)	Peak area (RSD%)	Capacity factor, k' (RSD%)	USP Tailing (RSD%)	Resolution (RSD%)	Plates/meter (half width, RSD%)
Nominal condition		5.570 (0.062)	347.6 (1.654)	2.713 (0.080)	1.455 (1.215)	2.338 (1.507)	20,169 (2.586)
Flow Rate, mL/min	0.975 (Low)	5.585 (0.065)	350.9 (1.713)	2.630 (0.094)	1.355 (1.453)	2.333 (1.155)	22,179 (1.598)
	1.025 (High)	5.584 (0.050)	343.5 (0.856)	2.815 (0.077)	1.339 (1.032)	2.332 (0.550)	22,559 (1.078)
Injection volume, µL	27 (Low)	5.578 (0.038)	319 (0.748)	2.719 (0.055)	1.372 (0.788)	2.332 (1.626)	21,128 (2.183)
	33 (High)	5.581 (0.058)	384.6 (1.449)	2.720 (0.076)	1.369 (0.994)	2.351 (1.110)	22,104 (2.188)
Column temperature, °C	23 (Low)	5.579 (0.080)	357.4 (1.586)	2.719 (0.108)	1.443 (1.041)	2.421 (1.527)	19,764 (1.148)
	27 (High)	5.586 (0.057)	344.9 (1.739)	2.724 (0.088)	1.359 (1.190)	2.339 (0.559)	22,730 (1.901)
Organic content, %	74/6 (Low)	5.609 (0.155)	345.0 (1.358)	2.739 (0.201)	1.443 (1.429)	2.755 (0.345)	20,766 (2.068)
	76/4 (High)	5.594 (0.063)	352.5 (1.218)	2.729 (0.082)	1.452 (1.415)	2.368 (0.496)	20,216 (1.111)
pH	2.9 (Low)	5.588 (0.064)	355.6 (1.588)	2.726 (0.096)	1.471 (1.269)	2.445 (2.080)	21,644 (2.648)
	3.1 (High)	5.598 (0.124)	353.7 (1.710)	2.732 (0.176)	1.374 (1.455)	2.606 (0.807)	20,076 (1.455)

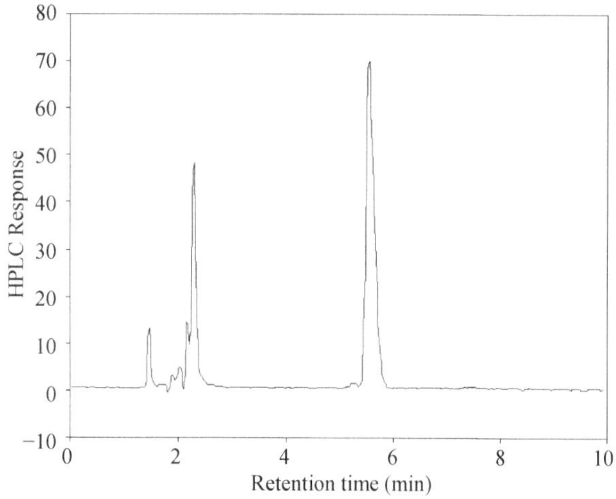

(a)

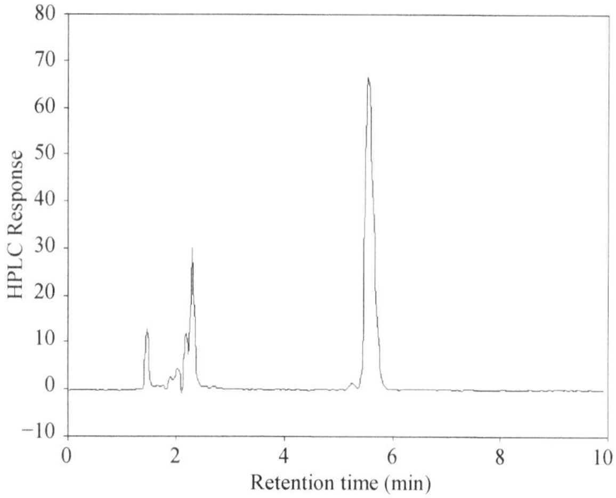

(b)

Figure 3. HPLC chromatograms of clindamycin palmitate hydrochloride products: A-powder (a) and B-powder (b). CPH had a retention time of 5.60 min.

Table 7. HPLC results of clindamycin palmitate hydrochloride products (n = 3).

Product	Dose strength	CPH potency, % (RSD %)
A-powder	75 mg	103.5 (0.85)
B-powder	75 mg	101.9 (1.12)

the analytical range. The significant advantages of this method, over existing methods including the current USP methodology, were the simplicity of mobile phase preparation, buffering capacity, reduced consumption of organic solvents and avoidance of ion pairing agents. The method was robust and demonstrated to be selective for the determination of CPH in the presence of its primary synthetic impurity in drug substance and products. The current method represents a significant improvement over the current USP methodology for CPH (USP CPH monograph). The method was successfully applied for the assessment of product quality of various marketed CPH dosage forms. It is also promising to be further developed into a selective stability indicating method if impurities standards are readily available [14].

5. Acknowledgements

This study was supported in part by the NIH pediatric formulation initiative and intramural FDA research funds. The authors would like to acknowledge Gretchen Whitesell's editorial assistance.

REFERENCES

[1] A. A. Forist, R. M. DeHaan and C. M. Metzler, "Clindamycin Bioavailability from Clindamycin-2-Palmitate and Clindamycin-2-Hexadecylcarbonate in Man," *Journal of Pharmacokinetics and Biopharmaceutics*, Vol. 1, No. 2, 1973, pp. 89-98.

[2] M. Cimbollek, N. Nies, A. Liebendorfer, *et al.*, "The Potential of the Prodrug Clindamycin Palmitate as an Implantable Slow Release form of the Antibotic Clindamycin for Heart Valves," *Journal of Controlled Release*, Vol. 33, No. 1, 1995, pp. 47-53.

[3] United States Pharmacopeia, "Clindamycin Palmitate Hydrochloride," United States Pharmacopeia (36) and National Formulary (31), 2013, pp. 3028-3029.

[4] L. W. Brown, "High-Pressure Liquid Chromatographic Assays for Clindamycin, Clindamycin Phosphate, and Clindamycin Palmitate," *Journal of Pharmaceutical Sciences*, Vol. 67, No. 9, 1978, pp. 1254-1257.

[5] C. Bharathi, P. Jayaram, J. S. Raj, *et al.*, "Identification, Isolation and Characterization of Impurities of Clindamycin Palmitate Hydrochloride," *Journal of Pharmaceutical and Biomedical Analysis*, Vol. 48, No. 4, 2008, pp. 1211-1218.

[6] C. B. V. N. Raju, G. Panda, G. N. Rao, *et al.*, "HPLC-UV Assay Method for Clindamycin Palmitate Hydrochloride as Drug Substance and Oral Solution," *Analytical Letters*, Vol. 41, No. 11, 2008, pp. 2033-2043.

[7] E. L. Rowe, "Anomalous Solution Behavior of 2-Palmitate Esters of Lincomycin and Clindamycin," *Journal of Pharmaceutical Sciences*, Vol. 68, No. 10, 1979, pp. 1292-1296.

[8] L. W. Brown and W.F. Beyer, "Clindamycin Hydrochloride," In: K. Florey and G. Rafik Bishara, Ed., *Analytical Profiles of Drug Substances*, Acedemic Press, Inc., London, 1981, pp. 75-91.

[9] S. M. Wang, S. S. Bu, H. M. Liu, *et al.*, "Separation and Characterization of Clindamycin Phosphate and Related Impurities in Injection by Liquid Chromatography/Electrospray Ionization Mass Spectrometry," *Rapid Communications in Mass Spectrometry*, Vol. 23, No. 6, 2009, pp.

899-906.

[10] United States Pharmacopeia, "General Information <1225> Validation of Compendial Procedures," *United States Pharmacopeia* (36) *and National Formulary* (31), 2013, pp 983-988.

[11] "Guidance for Industry Q2B Validation of Analytical Procedures: Methodology," International Conference on Harmonization of Technical Requirements for Registration of Pharmaceuticals for Human Use (ICH), 1996.

[12] United States Pharmacopcia, "Materials Safety Data Sheet: Clindamycin Palmitate Hydrochloride," 2007. http://www.usp.org/pdf/EN/referenceStandards/msds/1137005.pdf

[13] B. Dejaegher and Y.V. Heyden, "Ruggedness and Robustness Testing," *Journal of Chromatography A*, Vol. 1158, No. 1-2, 2007, pp. 138-157.

[14] M. Bakshi and S. Singh, "Development of Validated Stability-Indicating Assay Methods—Critical Review," *Journal of Pharmaceutical and Biomedical Analysis*, Vol. 28, No. 6, 2002, pp. 1011-1040.

Novel Approach for Quantitative Measurement of Matrix Metalloprotease-1 (MMP1) in Human Breast Cancer Cells Using Mass Spectrometry*

Ming Fa[1], Kathleen Hoch[2], Xuejun Fan[1], William P. Dubinsky[2], Zhiqiang An[1#], Ningyan Zhang[1#]

[1]Texas Therapeutics Institute, Houston, USA; [2]Proteomics Center, Brown Foundation Institute of Molecular Medicine, The University of Texas Health Science Center at Houston, Houston, USA.

ABSTRACT

Identification and quantification of low abundance growth factors and regulators in complex biological samples still present a challenging task in analytical biochemistry. Immunoassays are often used for such purpose but immunoassays face limitation of both availability and qualities of antibody reagents that are necessary for development of immune assays. With genomics data base available, mass spectrometry (MS) can analyze protein tryptic peptides directly for quantitative determination of proteins. In this study, we report a method for detection of matrix metalloproteinase 1 (MMP1), an important extracellular matrix modulator, in human breast cancer cells by quadrupole time-of-flight (Q-TOF) MS. Absolute quantification of MMP1 was conducted using the selected reaction monitoring (SRM) on a triple quadrupole (Triple-Quad) MS via transitions selected from MMP1 tryptic peptides using non isotope labeled MMP1 protein as a titration standard. In comparison with immune based assay, this MS method showed picogram level sensitivity for quantitative determination of MMP1 in total cell lysates. Our results demonstrated the feasibility of absolute quantification of low abundance proteins using label-free protein standard by mass spectrometry. Therefore, this method provides not only advantages of high sensitivity but also cost saving in comparison with the commonly used mass spectrometry that currently employs isotype labeled proteins for quantitative analysis.

Keywords: Matrix Metalloprotease-1 (MMP1); SKBR3 Breast Cancer Cells; Mass Spectrometry; Selected Reaction Monitoring; Absolute Quantification

1. Introduction

Liquid chromatography-tandem mass spectrometry (LC-MS/MS) has been widely used for protein analysis, ranging from protein expression profiling to specific post-translational modifications. However, quantitative detection of low abundance proteins in complex biological samples is still difficult and often requires multiple enrichment steps before analysis using mass spectrometry (MS). To increase sensitivity of MS analysis, a quadrupole time-of-flight (Q-TOF) MS [1] has been used for shotgun analysis of complex biological samples. In recent years, a more sensitive and selective approach named selected reaction monitoring (SRM), also called multiple reaction monitoring (MRM), has become increasingly popular in targeted and quantitative proteomics studies [2]. Triple quadrupole (Triple-Quad) mass analyzer enables targeting specific proteins of interest from a complex biological mixture based on precursor ions (charged peptide) and collision-induced product ions (fragmented peptide), and the SRM method provides both high sensitivity and selectivity for quantitative analysis of low abundance proteins in complex biological samples [3-5].

Matrix metalloproteinases (MMPs) are a family of zinc-dependent endopeptidases that modulate extracellular matrix proteins as well as activate or process protein receptors and regulators [6]. MMPs have been strongly implicated in multiple stages of cancer progression and promoting invasive and metastatic properties. Our recent studies suggest that MMPs in the tumor microenvironment may compromise the efficacy of therapeutic antibodies by proteolytic cleavage of antibody hinge region which results in the loss of Fc-mediated function such as antibody dependent cell mediated cytotoxicity (ADCC) [7]. Therefore, analysis of MMP expression in tumor samples is important for understanding

*Funding source: This work was supported in part by a grant from Janssen Research & Development, L.L.C and the Welch Foundation Grant AU00024.

#Corresponding authors.

their roles in cancer progression. Commonly used methods for studying MMP protein expression include immunoassays and zymograms using unnatural protease substrate for detection of proteolytic activities after separation on gel electrophoresis. LC-MS/MS detection method can provide high sensitivity for identification of specific proteins based on their amino acid sequences and SRM-MS can provide absolute quantitation through a purified protein standard. As expression levels of MMPs are generally in the nanogram to picogram (ng-pg)/µg of cellular protein [8], method for detection of such low abundance of proteins by MS in complex biological samples is not well established. In the present work, we report a new mass spectrometry based method for identification and absolute quantification of a low abundance protein (MMP1) in cancer cell lysates using label free protein as standard and the method should applicable to detection of other low abundance proteins in complex biological samples.

2. Material and Method

2.1. Chemicals and Reagents

Urea (99.5%), dithiothreitol (DTT), iodoacetamide, acetonitrile, and ammonium bicarbonate were from Sigma-Aldrich (St. Louis, MO). Tris (2-carboxyethyl) phosphine (TECP) 0.5 M solution was from Thermo Scientific (Rockford, IL). Bradford Protein Assay Kit was from Bio-Rad (Hercules, CA). TPCK (L-1-tosylamido-2-phenylethyl chloromethyl ketone) treated trypsin was from AB SCIEX (Foster City, CA). Strong cation exchange (SCX) column-ICAT® and buffer kit were from Applied Biosystems (Foster City, CA). Oasis HLB 1ml (10 mg) reverse phase cartridges were from Waters (Milford, MA). Stable isotope labeled peptides were synthesized by JPT (Berlin, Germany). Recombinant human MMP-1 (produced in murine myeloma cell line) was from R&D Systems (Minneapolis, MN). Breast cancer cells lines SKBR3 and MCF-7 were obtained from American Type Culture Collection (ATCC).

2.2. Sample Preparation

Breast cancer cells SKBR3 and MCF-7 were cultured in RPMI media at 37˚C incubator, 5% CO_2 as recommended by ATCC and cell extracts were prepared as previously reported [7]. Briefly, breast cancer cells were cultured to 80% - 90% confluency and cell lysates were made with RIPA buffer with EDTA (Boston BioProducts, Ashland, MA) plus protease inhibitor cocktail (EMD Millipore, Billerica, MA). Total lysate proteins were measured by Bradford protein assay kit according to manufacturer's suggestion. One hundred microgram (100 µg) of cell lysate was dried in a SpeedVac concentrator (Thermo Electron Corporation Pty. Ltd., Elandsfontein, South Africa) to total dryness and reconstituted in 40 µl freshly prepared 8M Urea in 150 mM Tris-HCl, pH 8.5. The denatured lysate proteins were reduced by 5 mM TCEP (15 minutes at room temperature) and subsequently alkylated using 10 mM iodoacetamide for 30 minutes at room temperature in the dark. Then 240 µL of 150 mM Tris-HCl, pH 8.5 was added to dilute the urea to less than 2 M. For the tryptic protein digestion, trypsin was added to the solution at enzyme/protein ratio of 1:20 and was incubated at 37˚C for 12 hours. The digests were dried in a SpeedVac, resuspended in 0.5 ml of 10 mM KH_2PO_4, pH 8.0 containing 25% acetonitrile, and loaded onto a SCX column. The SCX column was washed with the same buffer and stepwise elution was performed with increasing concentrations of KCl (52.5, 105, 157.5, 253.75 and 350 mM) in the same buffer. The eluted fractions were dried again and reconstituted in 0.5 ml of 0.1% trifluoroacetic acid (TFA) in water before applying on Oasis® HLB 1ml C_{18} cartridges (Waters Co., Milford, MA) for desalting. Eluted fractions from SCX were dried and reconstituted in solvent containing 2% acetonitrile/0.1% formic acid (solvent A) for MS analysis.

2.3. Mass Spectrometry Instrumentation

Mass spectrometric analyses were performed on 6538 UHD Accurate-Mass Q-TOF LC/MS and 6430 Triple-Quad LC/MS (Agilent Technologies, Santa Clara CA). Liquid chromatography (LC) was performed on Agilent HPLC-Chip II: G4240-62010 using a 7.5 × 150 mm C - 18 analytical column with 1 µl of sample in each injection. The mobile phase buffer was solvent A and the flow rate was 0.3 µL/minutes. Tryptic peptides were eluted with a linear gradient from 5% to 30% of Solvent B (90% acetonitrile/0.1% formic acid) over 100 minutes for Q-TOF MS and 30 minutes for Triple Quad MS. For Q-TOF experiment, electrospray voltage (V) was set at 1850 V and collision energy was set based on the equation: (slope)*(m/z)/100 + offset where slope was equal to 3.6 and offset was –4.8. For SRM analysis, collision energy was set at 100 constant value. Data acquisition and analysis were carried out using Agilent's Workstation Software-MassHunter Qualitative Analysis (version B.04.00).

2.4. Proteomic Data Analysis

Database search for proteins and peptides matching tandem spectra of lysate sample was performed using Mascot software (Matrix Science, Boston, MA). Data files from Agilent MS spectrometers were converted to Mascot Generic Format (MGF) and searched in protein sequence database, SwissProt. The peptide mass tolerance was set at 10 ppm and the fragment ion mass tolerance was at 0.05 Da.

2.5. MMP1 Determination Using Reverse Phase Protein Array

Fifty microgram (50 μg) of protein lysate of SKBR3 cancer cells was used for MMP1 determination with the Quantibody human MMP array kit (RayBiotech Inc., Norcross, GA) according to manufacturer's instruction. Briefly, the glass chip was blocked with BSA containing blocking solution and cancer cell lysates were added into wells in the glass slide and incubated overnight at 4°C. The next day, the glass chip was washed 5 times (5 minutes each time) with wash buffer and detection antibody with fluorescent dye-conjugate (Cyanine 3) was added and incubated for 2 hours at 25°C. After washing step, fluorescence signal was imaged using a GenePix 4100A (Molecular Devices, Sunnyvale, CA) and the data was analyzed using QAH-MMP1 GAL software. The concentration of MMP1 was determined based on a titration curve made from different concentrations of MMP1 standard from the array kit.

3. Results

3.1. Identification of MMP1 Using Q-TOF MS

To improve the MS sensitivity for low abundance proteins, tryptic peptides from cancer cell lysates were applied to a strong cation exchange (SCX) column for further partitioning into 5 fractions with increasing concentrations of KCl. All five fractionated lysates were analyzed by Q-TOF MS. The most abundant proteins (214 ions based on the preliminary run) were preset to be excluded during the acquisition. Mascot search picked up several MMPs among which MMP1 appeared to be dominant based on its quantitation index (score = 52; emPAI = 0.13). **Table 1** presents the precursor ions from SKBR3 lysate matching a number of peptides based on MMP1 amino acid sequences (MMP1_HUMAN, P03956). The MMP1 peptides covering different regions were detected in four (#1, 2, 4, and 5) out of five fractions from the SCX separation (**Table 1**) and confirmed presence of MMP1 in the tryptic lysates.

3.2. Detection of MMP1 by SRM on Triple-Quad MS

Analysis of MS spectra of MMP1 digests indicated that a doubly charged peptide LTFDAITTIR showed the strongest ion intensity as well as defined tandem mass spectrum (**Figure 1**). Therefore a number of product ions of this peptide were chosen as transitions for SRM analysis of MMP1 on Triple-Quad MS. BLAST search (NCBI) of protein database indicates that LTFDAITTIR is a unique tryptic peptide for MMP1 protein. The selected transitions for this signature peptide LTFDAITTIR were 575.8→603.4; 575.8→674.4; 575.8→789.4; 575.8 →936.5 m/z, based on their strong ion abundances. Stable isotope labeled peptide LTFDAITTIR with [^{13}C, ^{15}N]Arginine was spiked at 0.2 pmol/μl into samples as internal reference. Consistent with Q-TOF MS data, all four transitions were detected by SRM on Triple Quad MS and the fraction 5 from SCX separation showed significant stronger ion intensities. **Figure 2** indicates the SRM chromatogram for MMP1 transitions in fraction 5 of SCX and endogenous MMP1 transition peaks are aligned well with those corresponding heavy peptides labeled with stable isotopes at the same retention time.

3.3. Quantification of MMP1 by SRM Using Non-Isotope Labeled MMP1 Protein Standard

In our effort to quantify MMP1 concentration in SKBR3 cancer cell lysate, we tried to use non-isotope labeled MMP1 protein as a standard and to correlate MMP1 concentration with its transitional signals. Various amounts of purified human MMP1 protein (0, 50, 100 and 200 ng)

Table 1. MMP1 peptides discovered by MASCOT search in fractionated SKBR3 cell lysates.

Peptide sequence	Mass	Precursor ion (m/z)	Z	Fraction #
ADVDHAIEK	996.4876	499.2462	2	1
LTFDAITTIR	1149.6394	575.8262	2	5
LTFDAITTIRGEVMFFK	2147.9768	717.0029	3	4
SQNPVQPIGPQTPK	1489.7889	497.6057	3	1
VDAVFMK	808.4153	405.2180	2	4
VTGKPDAETLKVMK	1611.7943	538.2728	3	1,2
YLEKYYNLK	1232.6441	617.3310	2	1

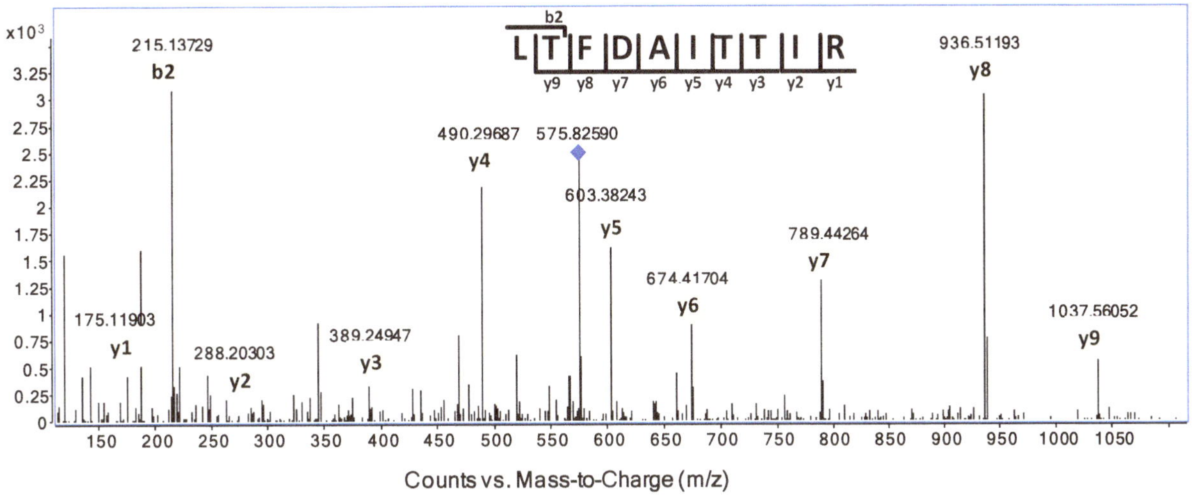

Figure 1. Tandem MS Spectrum of MMP1 peptide LTFDAITTIR (575.8 m/z, Z=2) showing product ions used as transitions for SRM. Diamond indicates precursor ion 575.8 m/z, Z=2. All the y-ions (y1-y9) and one b-ion (b2) were marked on the peptide sequence. The experiments were repeated 3 times and representitive of mass spectra is shown.

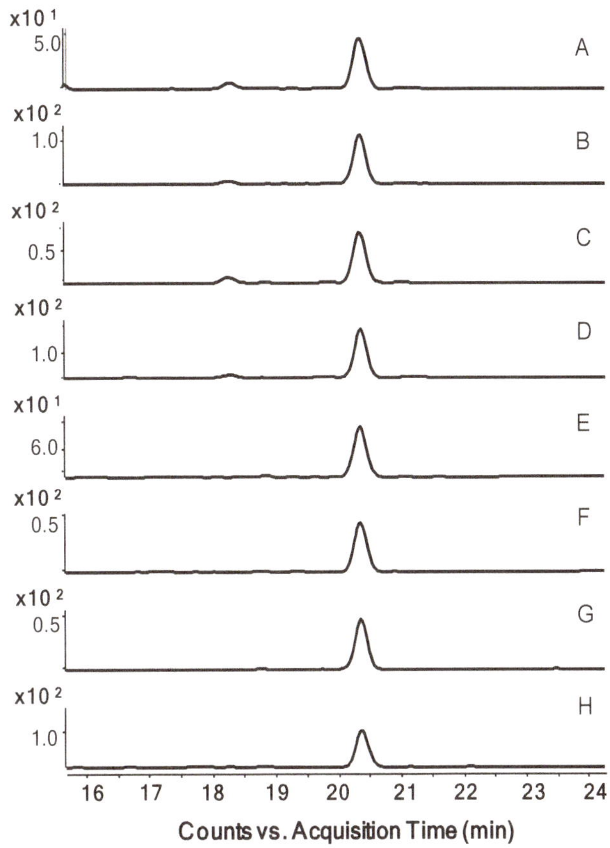

Figure 2. Extracted SRM chromatograms of MMP1 transitions detected in fractionated SKBR3 cell lysate on Triple-Quad MS. Stable isotope labeled peptide LTFDAITTIR (Label:13C(6)15N(4)) was spiked at 0.2 pmol/μl into SCX partitioned SKBR3 cell lysate (fraction #5) and measured on Triple-Quad LC/MS (Agilent Technologies) coupled to Agilent HPLC-Chip II: G4240-62010 as described in the Method section. SRM chromatograms (one from triplicate runs) for all LTFDAITTIR transitions were extracted using Agilent's MassHunter Qualitative Analysis Software. Panels A through D show transitions of spiked heavy peptide (A: 580.8→613.4 m/z; B: 580.8→684.4 m/z; C: 580.8→799.4 m/z; D: 580.8→946.5 m/z). Panels E through H show transitions of natural peptide from lysate sample (E: 575.8→603.4 m/z; F: 575.8→674.4 m/z; G: 575.8→789.4 m/z; H: 575.8→936.5 m/z).

were spiked individually into an aliquot of 100 μg cell lysate made from MCF-7 cancer cells that have no detectable level of MMP1 based on data from antibody array analysis (**Figure 3**). The MMP1 protein standard was spiked in MCF-7 cancer cell lysates and analyzed using the same procedures as that for SKBR3 cell lysates on Triple Quad LC/MS. For quantification, the signal from transition 575.8→789.4 m/z in peptide LTFDAITTIR was used for developing standard curve and calculating the amounts of MMP1 protein. The transition 575.8→789.4 m/z in MCF-7 lysate without spiked standard was at background level, suggesting that endogenous MMP1 in MCF-7 cells had little contribution to the ion intensity. Ion intensities (integrated peak area) of the transition 575.8→789.4 m/z were plotted against the corresponding concentrations of spiked MMP1 standard in MCF-7 lysate and the ion intensities showed linear correlation with MMP1 concentrations (**Figure 4**). MMP1 protein concentration in SKBR3 cell lysates (0.85ng/μg total lysate protein) was calculated using the standard curve. In order to evaluate the sensitivity and accuracy of the method, the same lysates were also assayed using immune based protein array method. The MMP1 concentration obtained by SRM approach in SKBR3 cell lysates was comparable to that determined by the protein array assay at 0.85 ng/μg and 0.99 ng/μg protein (**Figure 3**), respectively.

4. Discussion

Antibody based immuno assays are commonly used for detecting low abundance protein from complex biological samples. Since immune assays depend on specific antibodies for detection, lack of available antibody reagents or poor quality of reagents often pose limitation on reliability of immune assays. With increased resolution and sensitivity, MS technology is increasingly used for quantitative protein analysis and biomarker discovery. In comparison with immune assays, sequence based MS analysis provides advantage of unambiguous sequence identification of target proteins and has no need for expensive antibody reagents that are necessary in immune based assays.

LC-MS/MS applications can be classified into two general categories: qualitative and quantitative analysis [9]. Qualitative analysis typically is used for identification of unknown target compounds in applications such as

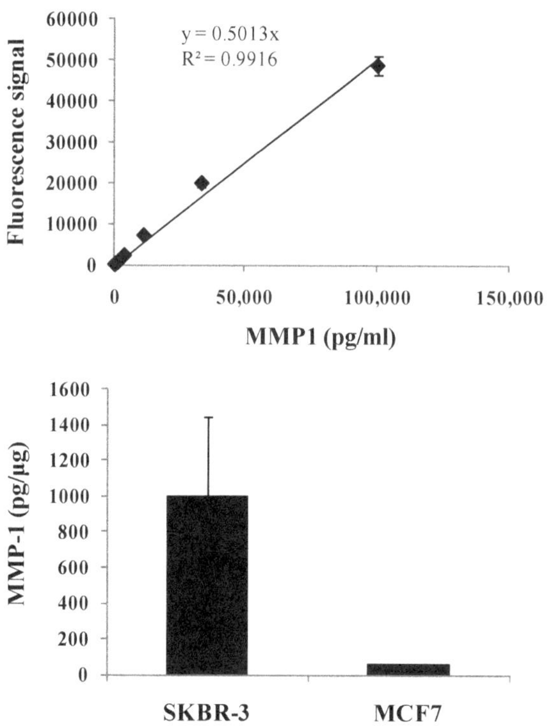

Figure 3. MMP1 detection using an antibody array method. The concentrations of MMP1 from SKBR3 and MCF-7 cell lysates (lower bar graph) were determined based on a titration curve (upper plot) made from various quantities of standard MMP1 antigen (137 - 100,000 pg/ml, in 3-fold titration) supplied in the array kit. Each assay points have 4 replications (n = 4) and error bars show the standard deviations. Formula derived from the linear slope and correlation coefficient (shown in the graph insert) were generated using linear fitting with GraphPad Prism software.

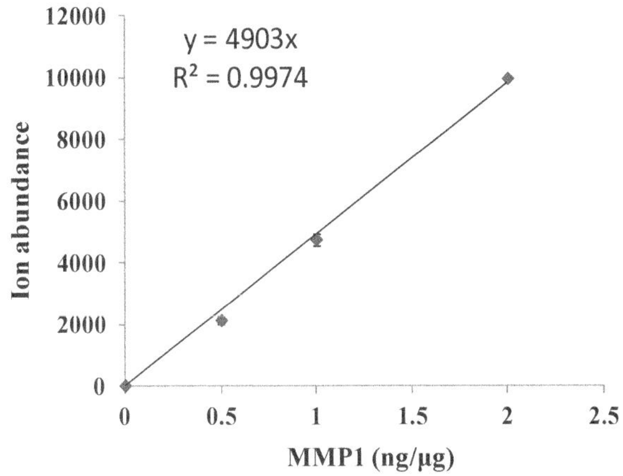

Figure 4. MMP1 protein standard curve based on transition (575.8→789.4 m/z). MCF-7 cell lysate spiked with different amount of MMP1 (0, 50, 100 and 200 ng) were trypsin digested and processed in the same fashion as that for SKBR3 cell lysate. The ion abundance signals (Y-Axis) from MMP1 peptide transition 575.8→789.4 m/z on Triple Quad LC/MS were plotted against the amounts of MMP1 protein spiked in MCF-7 lysates (X-axis). The ion intensity of transition represents average of integrated peak area (n = 3) and error bar indicates the standard deviations. Formula derived from the linear slope and correlation coefficient (shown in the graph insert) were generated using linear fitting with GraphPad Prism software.

proteomic profiling. This type of analysis requires the acquisition of a complete mass spectrum for each chromatographic data point. The TOF (as well as Fourier transform) mass analyzer can obtain high-resolution mass spectra throughout a broader mass range with great sensitivity and is well suited for qualitative analysis. Quantitative applications, on the other hand, often target specific known compounds, which makes the acquisition of a complete mass spectrum unnecessary. However, it demands the lowest detection limit. Despite somewhat lower mass resolution, the quadrupole mass analyzer generally achieves lower detection limit by operating under selected-ion monitoring, therefore better suited for quantitative analysis. SRM approach used in this study which is primarily performed on triple quadrupole mass spectrometer maximizes the selection and quantification power of quadrupole mass analyzer for targeted proteomics. In addition, proper sample purification is often required to achieve best sensitivity for both types of analyses, especially in detecting low abundance proteins. Nevertheless, it is still very challenging to have high sensitivity and quantitative MS analysis to detect low abundance proteins such as growth factors and regulators like MMP1 in complex biological samples. The reported lowest detection limit in complex biological samples currently ranges from 1 - 10 ng/ml in plasma [5] and sensitivity of MS detection in cell lysates is not well established. This study demonstrated that combination use of Q-TOF and SRM mass spectrometry can provide detection at nanogram to picogram level in cell lysates and the sensitivity is comparable to that by immune assays.

In this report we presented an example of detecting a low abundance protein MMP1 in cancer cell lysates by Q-TOF and SRM on Triple Quad MS. The overall working flow is summarized in **Figure 5**. To facilitate the detection of low abundance proteins, this study used sample enrichment through ion exchange chromatography and setting exclusion list for most abundant ions to lower the proteomic complexity in cell lysate mixture. Moreover, the MS/MS spectra of MMP1 peptides directly provided the chance to select transitions that would be used in SRM analysis. By employing both enrichment through SCX separation and SRM quantitation using selective peptide sequences from MMP1, we were able to determine MMP1, a low abundant protein, in complex cancer cell lysates with assay sensitivity comparable to that using immune based reverse phase protein array method in low picogram per microgram of cell lysate. Although proteomic tools (*e.g.* Skyline, MaxQaunt) are available to obtain transitions based on protein sequences, the best way of obtaining sensitive and reliable transitions is still through shotgun proteomic data using protein of interest [10]. This approach is particularly effective when the instrument for qualitative analysis is compatible to that for SRM based quantification. In this study, we used Agilent's 6538 Q-TOF LC/MS and 6430 Triple-Quad LC/MS that allows instrumental settings from Q-TOF LC/MS directly applicable to triple-quad LC/MS for SRM detection.

Based on MS/MS spectra of MMP1 peptides, we selected four product ions from a MMP1 signature peptide LTFDAITTIR (575.8 m/z) as transitions for SRM measurement because they were the strongest product ions shown on the spectra of all MMP1 precursor ions detected on Q-TOF MS. All four transitions were detectable by SRM on triple-quad MS for fractionated SKBR3 lysates.

SRM based quantification is usually conducted based on stable isotope-labeled standard (SIS) peptides [11]. Absolute quantification can be achieved by comparing the transition signal of the "heavy" reference peptide with that of a "light" endogenous peptide that results from proteolytic cleavage of the target protein. However, there is always a bias for the quantity of target protein to be represented by that of its proteolytic digests due to the fact that both the degree of protease digestion of any given protein and the ionization efficiency of its proteolytic peptides are not known. To overcome the limitation of using SIS peptide as titration standard, a new strategy (Protein Standard Absolute Quantification) that uses *in*

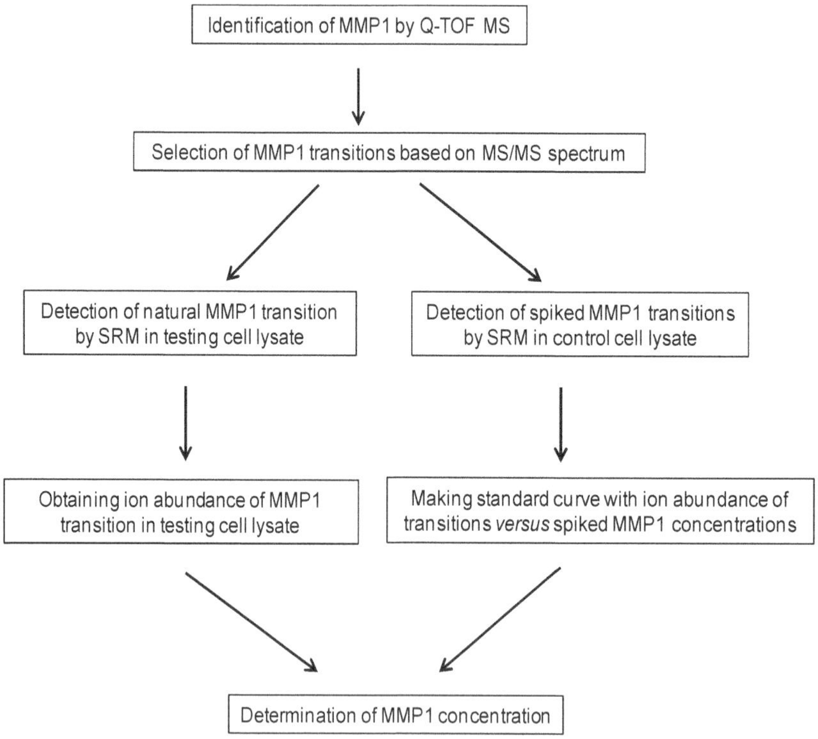

Figure 5. Diagram of key steps used for the quantitative measurement of MMP1.

vitro synthesized isotope-labeled full-length proteins as standards is merging as a powerful tool for absolute quantification [12]. As those protein standards perfectly match the biochemical properties of the target proteins, they can be directly added into the samples to be analyzed, allowing a highly accurate quantification of proteins even in complex samples. However, this approach can be expensive and time consuming. Although some efforts have been made to lower the production cost of isotope labeled protein such as implanting multiple transitions into a single hybrid protein [13], spiking entire labeled protein as quantification reference is still necessary if the complex biological sample needs to go through purification steps for target protein enrichment. Here we demonstrated the feasibility of protein quantification by SRM using non-labeled target protein as titration standard. Major advantages of using non-isotope protein as standard are simple and cost effective, as compared to making its full length isotope labeled counterpart.

Although the quantitative determination using the non-labeled protein standard requires analysis of the purified protein in a background negative lysate, the approach can work effectively by selecting the "control lysate" for spiking pure target protein with low interference. As both control lysate with spiked standard and sample lysates were processed under identical condition, experimental variation in the assay should be cancelled when calculating the testing samples using the control spiked standard curve. The control should have similar proteomic background to the testing sample with undetectable amount of target protein. Shotgun proteomic analyses of cell lysates from both SKBR3 and MCF-7 cells on Q-TOF LC/MS showed that their base peak chromatograms (BPC) were well overlapped, suggesting very similar proteomic background for the two cell lines. In addition, we picked MCF-7 cells because it had undetectable target protein MMP1 by MS method. If fractionation of lysates is necessary, it is also important for both target and standard proteins to undergo identical treatment. Overall, the method described in this study showed clear advantage over peptide based quantification in its accuracy. It can provide a good alternative when the production of isotope labeled full length protein is not available.

5. Conclusion

This study demonstrated the possibility of using SRM based mass spectrometry for quantitative measurements of low abundance proteins with sensitivity comparable to immune assays. This method can provide valuable alternative to immune assays for quantitative analysis of low expressing cellular proteins without needs of using specific antibodies for detection. This method also employed non-isotope labeled protein standard and provided a cost saving approach in comparison with commonly reported mass spectrometry for quantitative determination of proteins using isotope-labeled internal standard.

6. Acknowledgements

We thank Dr Nataliya Bulayeva for her technical assistance.

REFERENCES

[1] I. V. Chernushevich, A. V. Loboda and B. A. Thomson, "An Introduction to Quadrupole-Time-of-Flight Mass Spectrometry," *Journal of Mass Spectrometry*, Vol. 36, No. 8, 2001, pp. 849-865.

[2] V. Lange, P. Picotti, B. Domon and R. Aebersold, "Selected Reaction Monitoring for Quantitative Proteomics: A Tutorial," *Molecular Systems Biology*, Vol. 4, No. 222, 2008, pp. 1-14.

[3] L. Anderson and C. L. Hunter, "Quantitative Mass Spectrometric Multiple Reaction Monitoring Assays for Major Plasma Proteins," *Molecular & Cellular Proteomics*, Vol. 5, No. 4, 2006, pp. 573-588.

[4] P. Mitchell, "Proteomics Retrenches," *Nature Biotechnology*, Vol. 28, No. 7, 2010, pp. 665-670.

[5] H. Keshishian, T. Addona, M. Burgess, E. Kuhn and S. A. Carr, "Quantitative, Multiplexed Assays for Low Abundance Proteins in Plasma by Targeted Mass Spectrometry and Stable Isotope Dilution," *Molecular & Cellular Proteomics*, Vol. 6, No. 12, 2007, pp. 2212-2229.

[6] G. Murphy and H. Nagase, "Progress in Matrix Metalloproteinase Research," *Molecular Aspects of Medicine*, Vol. 29, No. 5, 2008, pp. 290-308.

[7] X. Fan, R. J. Brezski, M. Fa, H. Deng, A. Oberholtzer, A. Gonzalez, W. P. Dubinsky, W. R. Strohl, R. E. Jordan, N. Zhang and Z. An, "A Single Proteolytic Cleavage within the Lower Hinge of Trastuzumab Reduces Immune Effector Function and *in Vivo* Efficacy," *Breast Cancer Research*, Vol. 14, No. 4, 2012, p. R116.

[8] A. Kohrmann, U. Kammerer, M. Kapp, J. Dietl and J. Anacker, "Expression of Matrix Metalloproteinases (MMPs) in Primary Human Breast Cancer and Breast Cancer Cell Lines: New Findings and Review of the Literature," *BMC Cancer*, Vol. 9, No. 188, 2009, pp. 1-20.

[9] B. K. Choi, D. M. Hercules, T. Zhang and A. I. Gusev, "Comparison of Quadrupole, Time-of-Flight, and Fourier Transform Mass Analyzers for LC-MS Applications," *Current Trends in Mass Spectrometry*, Vol. 18, No. 5S, 2003, pp. 524-531.

[10] A. Prakash, D. M. Tomazela, B. Frewen, B. Maclean, G. Merrihew, S. Peterman and M. J. Maccoss, "Expediting the Development of Targeted SRM Assays: Using Data from Shotgun Proteomics to Automate Method Development," *Journal of Proteome Research*, Vol. 8, No. 6, 2009, pp. 2733-2739.

[11] D. S. Kirkpatrick, S. A. Gerber and S. P. Gygi, "The Absolute Quantification Strategy: A General Procedure for the Quantification of Proteins and Post-Translational Modifications," *Methods*, Vol. 35, No. 3, 2005, pp. 265-273.

[12] V. Brun, A. Dupuis, A. Adrait, M. Marcellin, D. Thomas, M. Court, F. Vandenesch and J. Garin, "Isotope-Labeled Protein Standards: Toward Absolute Quantitative Proteomics," *Molecular & Cellular Proteomics*, Vol. 6, No. 12, 2007, pp. 2139-2149.

[13] R. J. Beynon, M. K. Doherty, J. M. Pratt and S. J. Gaskell, "Multiplexed Absolute Quantification in Proteomics Using Artificial QCAT Proteins of Concatenated Signature Peptides," *Nature Methods*, Vol. 2, No. 8, 2005, pp. 587-589.

A Study on Measurement of Forest Soil Redox Potential

Mu Liu[1,2], Chengde Luo[1], Xiaoyun Liu[2], Gang Wan[2], Bin Dai[1]

[1]College of Forestry, Sichuan Agricultural University, Ya'an, China; [2]Promotion Center of Forestry Technology, Forestry Bureau of Fushun County, Fushun, China.

ABSTRACT

Redox potential has principal influences on formation and attribute of soils, so it is necessary to research the measurement of soil redox potential. The self-made potentiometer and two indigenous FJA-5 potentiometers were employed to measure the Eh values of fir forest soils *in situ*, which lay in Savage Sea Beauty Spot in Sichuan Province. Simultaneously fresh soils *in situ* were collected and made into suspensions with various soil-water ratios and standing time to measure Eh values. The results demonstrated that the total tendency of Eh values measured by the self-made potentiometer was gradually falling with soil-water ratios increasing and standing time prolonging. There is a great difference between Eh values measured in laboratory and *in situ*. Only *in situ* may the measured Eh values show realistic oxidation-reduction conditions and formation characteristics of forest soil. The self-made potentiometer has a more preferable performance than indigenous FJA-5 potentiometers and it can be applied to measuring forest soil Eh *in situ*.

Keywords: The Self-Made Potentiometer; Redox Conditions; Suspensions; Soil-Water Ratios and Standing Time

1. Introduction

Redox reaction is one of the most basic forms of material motion in nature and a vital dynamic phase in soil formation. Redox potential is an intensity index, which decides the direction and extent of redox reaction, and is used for estimating soil's redox ability and showing clearly morphological characteristics and mobile performance of mineral elements in soil profile [1]. So far, Eh value has been mostly measured in soft hydromorphic soils, such as paddy soil [2-11]. For unhydromorphic soils, their Eh values were usually measured after the soil was made into suspensions or disposed else [12-21]. However there are fewer researches on measurement of the forest soil redox potential. The environment and substance composition of soil itself have an extremely profound impact on redox potential, so it is important to research the methods and technologies about measurement of forest soil Eh.

2. Materials and Methods

2.1. Developing Redox Potentiometer Alone

At present most of redox potentiometers made in China are matched with acidimeters. They are usually used to measuring pH and Eh in soil solution. Sometimes they contribute to detect Eh values in paddy soils or other soft sediment. Hence the special redox potentiometer for the forest soil Eh was made by ourselves and its measured results would be compared with the indigenous FJA-5 potentiometers' results.

A self-developed platinum electrode, a commercially available saturated calomel electrode and a DT-9205B multimeter were combinated into the self-made potentiometer (**Figure 1**). It is necessary to adjust the potential of platinum electrode before measurement so that exact and stable Eh values can be obtained [22-25]. In addition the sand core lies in bottom of saturated calomel electrode should be slightly immersed in saturated KCl to decrease the disturbance of liquid junction potential as far as possible [26,27].

2.2. Experimental Design

There were two experiments, namely, *in situ* test and laboratory test. *In situ* test was achieved in Savage Sea Beauty Spot in Sichuan Province, where the soil under fir forest was tested. The soil layer was respectively Layer A_d, the biggest eluvial horizon, and Layer B_2, the biggest illuvium. Eh values were measured in February 2011. Electrodes of the self-made potentiometer and FJA-5 potentiometers were inserted in the central section of Layer A_d and the electrodes of self-made potentiometer

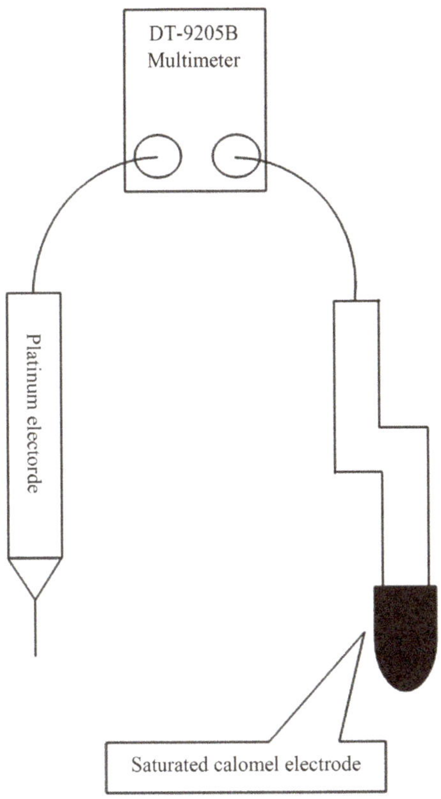

Figure 1. The schematic diagram of the self-made redox potentiometer.

were between that of two FJA-5 potentiometers. And what's more, the distance between platinum electrode and saturated calomel electrode ought to be near (1 - 3 cm), which could lower the resistance and determinate error.

Opened the power supply of DT-9205B multimeter linked with the self-made potentiometer and record the numerical value in display screen after it was stable. When the amplitude of fluctuation of the numerical value was less than 1 mv in five minutes, the value was deemed to be stable. Here it was emphasized that the numerical value in display screen only meaned the potential difference between platinum electrode and saturated calomel electrode instead of Eh value. The Eh value must be the summation of this potential difference and the constant potential provided by saturated calomel electrode at corresponding temperature. Simultaneously opened the power supply of FJA-5 potentiometers and the numerical value in display screen was exactly the Eh value. After finishment of Eh measurement in Layer A_d, all electrodes were taken out and washed by distilled water, then they were cleaned by dry filter papers to continue measuring Eh values of Layer B_2 with the same method. Eh values of each layer were repeatedly measured three times. Furthermore fresh soil of both Layer A_d and Layer B_2 were collected, preserved with fresh-keeping packaging and brought back to laboratory.

10 g fresh soil samples were weighed and put in 50-ml beakers with high pattern in which additional distilled water with different volumes were mixed and suspendsions with different soil-water ratios were formed. These suspensions had five soil-water ratios together, that was, 1:1, 1:2, 1:3, 1:4 and 1:5. As far as they were formed, the standing time was began to be recorded, which was respectively 0, 1, 2, 6, 12, 24, 48, 96, 144, 192 and 240 hours. Then Eh values were measured under varied soil-water ratio and standing time by the self-made potentiometer and FJA-5 potentiometers. As same as *in situ* test, Eh values were repeatedly measured three times. At last average values were worked out to draw the curve graph where Eh values were changing with changed soil-water ratios and standing time.

3. Results and Discussion

3.1. Laboratory Analysis

The Eh values measured by the self-made potentiometer were gradually falling with soil-water ratios increasing and standing time prolonging (**Figure 2**), whereas the Eh values measured by FJA-5 potentiometers were sometimes high and sometimes low, and irregular (**Figures 3** and **4**). The approximate tendency of Eh values of various suspensions with incremental soil-water ratios and prolonged standing time could be analysed according to Nernst Equation

$$Eh = E^0 + \frac{RT}{nF} LOG \frac{OX}{RED} \quad [28].$$

If soil-water ratio is smaller, the concentration of redox systems in soil is higher, the oxidizing matters' oxidation capability is more powerful, even the concentration of oxidizing matters (OX) is greater than that of re-

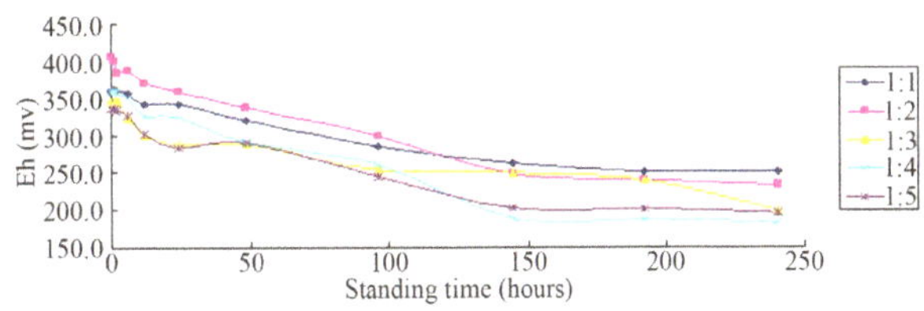

(a)

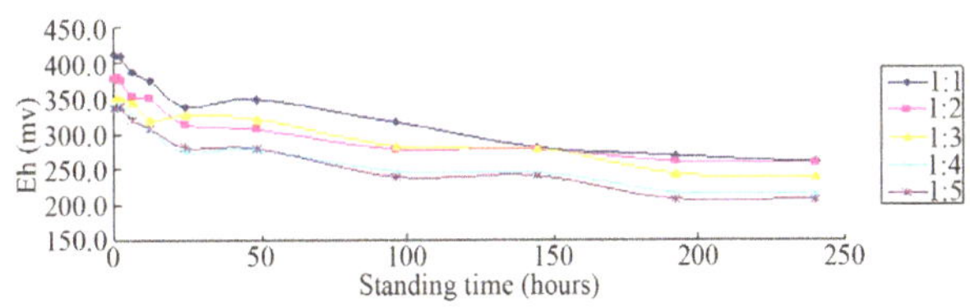

(b)

Figure 2. The variational curves of Eh values in Layer A_d (a) and Layer B_2 (b) measured by the self-made potentiometer. 1:1, 1:2, 1:3, 1:4 and 1:5 were the soil-water ratios.

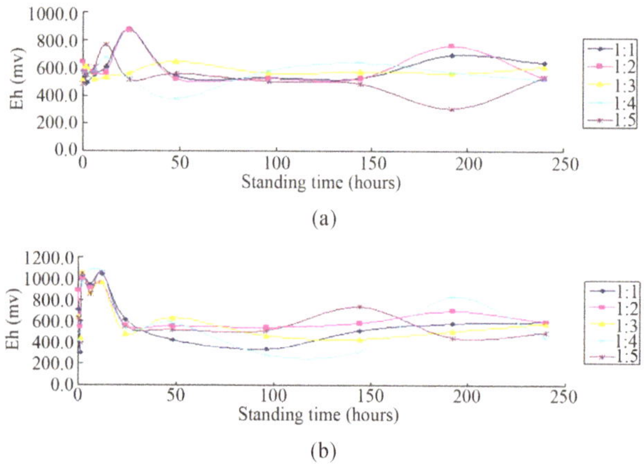

Figure 3. The variational curves of Eh values in Layer A_d (a) and Layer B_2 (b) measured by the first FJA-5 potentiometer. 1:1, 1:2, 1:3, 1:4 and 1:5 were the soil-water ratios.

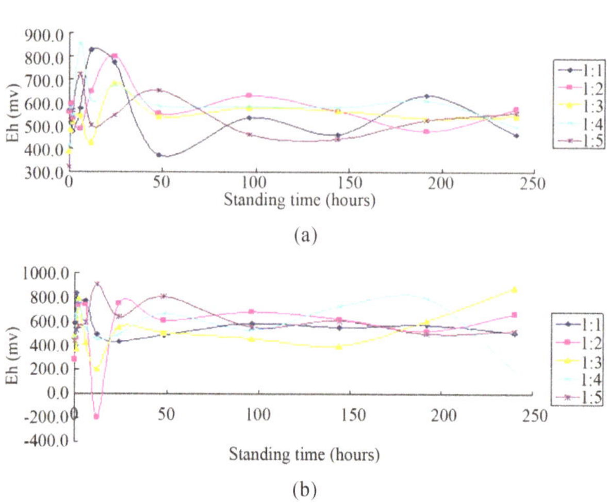

Figure 4. The variational curves of Eh values in Layer A_d (a) and Layer B_2 (b) measured by the second FJA-5 potentiometer. 1:1, 1:2, 1:3, 1:4 and 1:5 were the soil-water ratios.

ducing matters (*RED*), then the Eh value is higher. However, with soil-water ratio amplifying, the concentration of redox systems is decreasing little by little, the reducing matters' reduction capability is just improving and this moment the concentration of reducing matters (*RED*) will exceed the oxidizing matters (*OX*), as a result the Eh value is falling by degress. Furthermore soil is isolated from the atmosphere outside more and more seriously when standing time is infinitely prolonging, the reduction condition is more and more superior, and then the concentration of reducing matters (*RED*) is more and more greater. Thereupon the Eh value will be more and more lower.

From what had been analysed above, the approximate tendency of Eh values measured by the self-made potentiometer was coincident with the consequence deduced from Nernst Equation, while the total trend of Eh values detected by FJA-5 potentiometers was not. Willis (1932), Quispel (1947) and Bohn (1968) considered consensually that Eh was related to water levels, moisture content, soil structure and the amount of oxidational and reducible materials as well as it was only possible to detect oxidation-reduction conditions in soils when the measurements were carried out *in situ* after making Eh measurements on soil pastes or suspensions [18,29,30].

3.2. *In Situ* Analysis

Eh values measured *in situ* were showed by **Table 1**. In order to examine the precision and reappearance, coefficient of variation was employed. It was noticed that '0' (**Table 1**) should not be involved in averaging when calculating coefficient of variation.

It was clear in **Table 1** that coefficient of variation of Eh values got from the self-made potentiometer was much less than 1%, while that of values got from FJA-5 potentiometers was much bigger. Thus it was considered that the measurement results of the self-made potentiometer had relatively higher precision and better reappearance.

Soil redox potential is closely related to moisture, porosity, ventilated extent and content of organic matter [1,6,9], so them of soil *in situ* were revealed in **Table 2**. Seen from which, the contents of moisture and organic matter in Layer A_d were obviously higher than that of Layer B_2, as well as porosity and ventilated extent were smaller than that of Layer B_2, thereupon the Eh value of Layer A_d ought to be lower than Layer B_2. From what **Table 1** showed, the Eh values of Layer A_d measured by the self-made potentiometer were all lower than Layer B_2, while the results got from FJA-5 potentiometers were

Table 1. Soil Eh values measured *in situ*.

Soil Layer		Eh Values, mv			Coefficient of Variation, %
A_d	△	585.6	384.7	595.2	22.78
	○	386.1	385.1	386.1	0.15
	▽	520.6	459.7	0	8.59
B_2	△	323.7	285.5	337.4	8.52
	○	494.8	494.8	493.8	0.12
	▽	377.9	311.3	0	9.66

△, Eh values measured by the first FJA-5 potentiometer; ○, Eh values measured by the self-made potentiometer; ▽, Eh values measured by the second FJA-5 potentiometer.

Table 2. Soil moisture, porosity, ventilated extent and contents of organic matter.

Soil Layer	Moisture, %	Porosity, %	Ventilated Extent, %	Contents of Organic Matter, %
A_d	33.64	40.6	22.77	5.3556
B_2	12.35	49.72	28.65	1.3484

opposite. It has been proved that oxidation processes occupy primary superiority in formation of forest soil, however moisture and organic matter often accumulate in surface layer so that redox potential of surface soil will be lower than that of bottom soil [31]. This difference may vary from several or dozens of millivolts to hundreds of millivolts. Usually there is a buffer slightly influenced by moisture and organic matter below Layer A_1, such as Layer A_d [31]. In summary, Eh values measured by the self-made potentiometer could objectively reflect the oxidation-reduction conditions and formation characteristics of forest soil in field.

The significance of difference between Eh values measured by the self-made potentiometer and by FJA-5 potentiometers was tested. Similarly "0" (**Table 1**) should not be involved in averaging and the test results were displayed in **Table 3**. It was found in **Table 3** that the interclass differences (△ & ▽) were not significant ($P > 0.05$) regardless of Layer A_d or B_2, while in interblock differences, except that the difference between Eh values of Layer A_d measured by the self-made potentiometer and by the first FJA-5 potentiometer (○ & △) was not significant ($P > 0.05$), the others were very significant ($P < 0.01$). This test manifested that the measurement results of the self-made potentiometer had higher precision and accuracy than FJA-5 potentiometers. The measurement results obtained from FJA-5 potentiometers were hardly able to reflect the oxidation-reducetion conditions and formation characteristics of soil in field.

To move forward a single step, significance of differences between Eh values measured *in situ* and in laboratory by the self-made potentiometer were tested. The result demonstrated that there was a significant ($P < 0.05$) or very significant difference ($P < 0.01$) between Eh values measured *in situ* and at each soil-water ratio and standing time in Layer A_d (**Appendix 1**), and that for Layer B_2, only exsited very significant difference ($P < 0.01$) (**Appendix 2**). So it is safely concluded that if forest soils are brought back to laboratory and made into suspensions for redox potential, there will be a great difference compared with the measurement *in situ* and the Eh values measured in suspensions cannot completely reflect the actual oxidation-reduction conditions of forest soil.

Table 3. Significance of differences of Eh values.

Soil Layer	Source of Variation		F	P-value	F crit
A_d	Interblock	○ & △	3.9554	0.1176	7.7086
		○ & ▽	35.5390**	0.0039	7.7086
	Interclass	△ & ▽	0.1959	0.6809	7.7086
B_2	Interblock	○ & △	136.7974**	0.0003	7.7086
		○ & ▽	55.9376**	0.0017	7.7086
	Interclass	△ & ▽	1.4266	0.2983	7.7086

△, Eh values measured by the first FJA-5 potentiometer; ○, Eh values measured by the self-made potentiometer; ▽, Eh values measured by the second FJA-5 potentiometer; **Significance of difference is at the $P < 0.01$ level.

4. Conclusions

For resreaches on redox process of unhydromorphic soils such as forest soil, if soil samples are made into suspensions, their Eh values will change with changed soil-water ratio and standing time and it is difficult to determine appropriate soil-water ratio and standing time. The Eh values measured at varied soil-water ratio and standing time have significant or very significant differences with that measured *in situ*. So Eh measurement *in situ* should be gradually promoted.

The methods and technologies about measurement of unhydromorphic soil Eh *in situ* still need to face a lot of problems and challenges in study on soil electrochemistry. Through preliminary analysis on Eh values measured *in situ* and in laboratory, it can be definitely deemed that the measurement results of the self-made potentiometer have relatively higher precision and accuracy and that the self-made potentiometer can be applied to measuring forest soil Eh *in situ*.

5. Acknowledgements

The authors thank Prof. Yongxin Zhan and Prof. Liang Zhao (Department of Physics, College of Life and Basic Sciences, Sichuan Agricultural University) for guiding make of redox potentiometer.

REFERENCES

[1] Q. Y. Huang, "Soil Science," Agricultural Press in China, Beijing, 2006.

[2] W. E. Connell and W. H. Patrick, "Sulfate Reduction in Soil: Effects of Redox Potential and pH," *Science*, Vol. 59, No. 3810, 1968, pp. 86-87.

[3] S. Gotoh and W. H. Patrick, "Transformation of Manganese in a Waterlogged Soil as Affected by Redox Potential and pH," *Soil Science Society of America Journal*, Vol. 36, No. 5, 1972, pp. 738-742.

[4] S. Gotoh and W. H. Patrick, "Transformation of Iron in a Waterlogged Soil as Influenced by Redox Potential and pH," *Soil Science Society of America Journal*, Vol. 38, No. 1, 1974, pp. 66-71.

[5] A. X. Hou, G. X. Chen and Z. P. Wang, "Methane and Nitrous Oxide Emissions from a Rice Field in Relation to Soil Redox and Microbiological Processes," *Soil Science Society of America Journal*, Vol. 64, No. 6, 2000, pp.

2180-2186.

[6] L. N. Mandal, "Levels of Iron and Manganese in Soil Solution and the Growth of Rice in Water-Logged Soils in Relation to the Oxygen Status of Soil Solution," *Soil Science*, Vol. 94, No. 6, 1962, pp. 387-391.

[7] W. H. Patrick, "Nitrate Reduction Rates in a Submerged Soil as Affected by Redox Potential," *Transactions 7th Inernational Congress of Soil Science*, Vol. 2, 1960, pp. 494-500.

[8] W. H. Patrick and A. Jugsujinda, "Sequential Reduction and Oxidation of Inorganic Nitrogen, Manganese and Iron in Flooded Soil," *Soil Science Society of America Journal*, Vol. 56, No. 4, 1992, pp. 331-332.

[9] K. R. Reddy and W. H. Patrick, "Effect of Alternate Aerobic and Anaerobic Conditions on Redox Potential, Organic Matter Decomposition and Nitrogen Loss in a Flooded Soil," *Soil Biology & Biochemistry*, Vol. 7, No. 2, 1975, pp. 87-94.

[10] A. Swarup, "Influence of Organic Matter and Flooding on the Chemical and Electrochemical Properties of Sodic Soil and Rice Growth," *Plant and Soil*, Vol. 106, No. 1, 1988, pp. 135-141.

[11] Z. P. Wang, R. D. De Laune and W. H. Patrick, "Soil Redox and pH Effects on Methane Production in a Flooded Rice Soil," *Soil Science Society of America Journal*, Vol. 57, No. 2, 1993, pp. 382-385.

[12] T. R. Yu, "The Electrochemical Properties and Methodology of Soil," Science Press, Beijing, 1976.

[13] C. Y. Huang, "Instructors on Soil Experiment and Practice," Agricultural Press in China, Beijing, 1992.

[14] C. J. Li, S. S. Li, S. Chen, *et al.*, "Research on Eh-pH of Granite in Jiaocheng City of Shanxi Province," *Hydrogeology and Engineering Geology*, No. 5, 1998, pp. 43-45.

[15] Z. Li, Y. Wang and J. Kong, "A Study on Measurement of Soil Redox Potential," *Environmental Science and Management*, Vol. 33, No. 10, 2008, pp. 172-174.

[16] K. Marie, P. H. Masscheleyn and C. W. Lindau, "Production of Dinitrogen and Nitrous Oxide in Soil Suspensions as Affected by Redox Potential," *Water, Air and Soil Pollution*, Vol. 61, No. 1-2, 1992, pp. 37-45.

[17] W. H. Patrick, B. G. Williams and J. T. Moraghan, "A Simple System for Controlling Redox Potential and pH in Soil Suspensions," *Soil Science Society of America Journal*, Vol. 37, No. 2, 1973, pp. 331-332.

[18] A. Quispel, "Measurement of the Oxidation—Reduction Potentials of Normal and Inundated Soils," *Soil Science*, Vol. 63, No. 4, 1947, pp. 265-275.

[19] L. Z. Tang, K. Haibara, H. Toda, *et al.*, "Transformation of Fe^{2+}, Eh and pH Values of Forest Soil in Wet Land," *Acta Ecologica Sinica*, Vol. 25, No. 1, 2005, pp. 103-107.

[20] G. H. Willis, R. C. Wander and L. M. Southwick, "Degradation of Trifluralin in Soil Suspensions as Related to Redox Potential," *Journal of Environmental Quality*, Vol. 3, No. 3, 1974, pp. 262-265.

[21] K. Yu, Z. Wang and A. Vermoesen, "Nitrous Oxide and Methane Emissions from Different Soil Suspensions: Effect of Soil Redox Status," *Biology and Fertility of Soils*, Vol. 34, No. 1, 2001, pp. 25-30.

[22] C. Carrie, J. Wafer, R. Barrett, *et al.*, "Construction of Platinum-Tipped Redox Probes for Determining Soil Redox Potential," *Journal of Environmental Quality*, Vol. 33, No. 6, 2004, pp. 2375-2379.

[23] B. Eric-Van, B. Suzanne and T. Georges, "Continuous Multiple Measurement of Soil Redox Potential Using Platinum Microelectrode," *Soil Science Society of America Journal*, Vol. 66, No. 6, 2002, pp. 1813-1820.

[24] S. Fiedler, M. J. Vepraskas and J. L. Richardson, "Soil Redox Potential: Importance, Field Measurements and Observations," Academic Press, Ireland, 2007.

[25] W. H. Patrick, R. P. Gambrell and S. P. Faulkner, "Redox Measurements of Soils," In: D. L. Sparks and A. L. Page, Eds., *Methods of Soil Analysis. Part* 3. *Chemical Methods, Soil Science Society of America*, WI, Madison, 1996, pp. 1358-1390.

[26] S. Du and X. Z. Gao, "Regulations on Technology of Soil Analysis," Agricultural Press in China, Beijing, 2006.

[27] W. L. Si, "Analysis and Discussion on Frequently Asked Questions in Soil Eh Measurement," *Sinkiang Science and Technology of Oil*, Vol. 7, No. 2, 1997, pp. 68-69.

[28] S. D. Zhao, "General Chemistry," Agricultural Press in China, Beijing, 2007.

[29] L. G. Willis, "Oxidation—Reduction Potentials and the Hydrogen Ion Concentration of a Soil," *Spanish Journal of Agricultural Research*, Vol. 45, No. 1, 1932, pp. 571-575.

[30] H. L. Bohn, "Electromotive Forces of Inert Electrodes in Soil Suspensions," *Soil Science Society of America Process*, Vol. 32, No. 2, 1968, pp. 211-215.

[31] T. R. Yu and Z. C. Chen, "Chemical Processes in Soil Formation," Science Press, Beijing, 1990.

Appendix 1

Significance of differences between Eh values measured *in situ* and in laboratory (Layer A_d).

Soil-water Ratio	Standing Time, hours	Source of Variation		F	*P*-value	F crit
		Suspension Eh	*In situ* Eh			
		360.9	386.1			
	0	360.9	385.1	2708.48**	8.1589E-07	21.1977
		361.9	386.1			
		363.1	386.1			
	1	362.1	385.1	2380.50**	1.0558E-06	21.1977
		363.1	386.1			
		363	386.1			
	2	361	385.1	1045.46**	5.4547E-06	21.1977
		361	386.1			
		357.1	386.1			
	6	357.1	385.1	7396.00**	1.0959E-07	21.1977
		357.1	386.1			
		342	386.1			
	12	342	385.1	17239.69**	2.018E-08	21.1977
		342	386.1			
		343	386.1			
1:1	24	344	385.1	8102.65**	9.1315E-08	21.1977
		343	386.1			
		320.9	386.1			
	48	320.9	385.1	37869.16**	4.1832E-09	21.1977
		320.9	386.1			
		286.5	386.1			
	96	286.5	385.1	88684.84**	7.6282E-10	21.1977
		286.5	386.1			
		262	386.1			
	144	262	385.1	137863.69**	3.1567E-10	21.1977
		262	386.1			
		251.3	386.1			
	192	251.3	385.1	162731.56**	2.2656E-10	21.1977
		251.3	386.1			
		250	386.1			
	240	250	385.1	82539.85**	8.8062E-10	21.1977
		251	386.1			

Continued

		407.2	386.1			
	0	407.2	385.1	4134.49**	3.50435E-07	21.1977
		407.2	386.1			
		402	386.1			
	1	402	385.1	2371.69**	1.06369E-06	21.1977
		402	386.1			
		384.7	386.1			
	2	384.7	385.1	10.24*	0.03290081	21.1977
		384.7	386.1			
		388	386.1			
	6	389	385.1	37.84**	0.00354206	21.1977
		389	386.1			
		371.7	386.1			
	12	371.7	385.1	1780.84**	1.88485E-06	21.1977
		371.7	386.1			
		360	386.1			
1:2	24	361	385.1	2910.85**	7.06511E-07	21.1977
		360	386.1			
		338.1	386.1			
	48	338.1	385.1	10082.00**	5.8989E-08	21.1977
		339.1	386.1			
		298	386.1			
	96	299	385.1	34138.85**	5.14717E-09	21.1977
		299	386.1			
		247.2	386.1			
	144	247.2	385.1	172806.49**	2.00916E-10	21.1977
		247.2	386.1			
		238	386.1			
	192	238	385.1	196514.89**	1.55362E-10	21.1977
		238	386.1			
		231.3	386.1			
	240	231.3	385.1	214739.56**	1.30111E-10	21.1977
		231.3	386.1			

Continued

		346.1	386.1			
	0	346.1	385.1	14161.00**	2.9906E-08	21.1977
		346.1	386.1			
		344	386.1			
	1	345	385.1	7601.45**	1.03748E-07	21.1977
		345	386.1			
		346.2	386.1			
	2	346.2	385.1	7164.05**	1.16797E-07	21.1977
		345.2	386.1			
		323.3	386.1			
	6	323.3	385.1	35118.76**	4.86397E-09	21.1977
		323.3	386.1			
		298	386.1			
	12	298	385.1	34400.65**	5.06914E-09	21.1977
		299	386.1			
		287.3	386.1			
1:3	24	287.3	385.1	87261.16**	7.87909E-10	21.1977
		287.3	386.1			
		288	386.1			
	48	287	385.1	43601.05**	3.15567E-09	21.1977
		287	386.1			
		252.6	386.1			
	96	252.6	385.1	159600.25**	2.35541E-10	21.1977
		252.6	386.1			
		249	386.1			
	144	250	385.1	83354.45**	8.63493E-10	21.1977
		250	386.1			
		238.4	386.1			
	192	238.4	385.1	195452.41**	1.57056E-10	21.1977
		238.4	386.1			
		197	386.1			
	240	197	385.1	320695.69**	5.83386E-11	21.1977
		197	386.1			

Continued

		359.6	386.1			
	0	359.6	385.1	6162.25**	1.57835E-07	21.1977
		359.6	386.1			
		359.6	386.1			
	1	359.6	385.1	6162.25**	1.57835E-07	21.1977
		359.6	386.1			
		360	386.1			
	2	360	385.1	3065.45**	6.37119E-07	21.1977
		359	386.1			
		351.8	386.1			
	6	351.8	385.1	10383.61**	5.56129E-08	21.1977
		351.8	386.1			
		327	386.1			
	12	327	385.1	31081.69**	6.20939E-09	21.1977
		327	386.1			
		326	386.1			
1:4	24	327	385.1	15895.45**	2.37369E-08	21.1977
		326	386.1			
		291.8	386.1			
	48	291.8	385.1	79467.61**	9.50024E-10	21.1977
		291.8	386.1			
		262	386.1			
	96	261	385.1	69676.45**	1.23577E-09	21.1977
		261	386.1			
		188.7	386.1			
	144	188.7	385.1	349517.44**	4.9114E-11	21.1977
		188.7	386.1			
		186	386.1			
	192	187	385.1	178383.65**	1.88549E-10	21.1977
		187	386.1			
		182.1	386.1			
	240	183.1	385.1	186050.00**	1.73331E-10	21.1977
		182.1	386.1			

Continued

		335.3	386.1			
	0	335.3	385.1	22921.96**	1.14162E-08	21.1977
		335.3	386.1			
		337.2	386.1			
	1	337.2	385.1	10760.45**	5.17871E-08	21.1977
		336.2	386.1			
		335	386.1			
	2	335	385.1	23195.29**	1.11488E-08	21.1977
		335	386.1			
		327.5	386.1			
	6	327.5	385.1	30555.04**	6.42526E-09	21.1977
		327.5	386.1			
		302.4	386.1			
	12	302.4	385.1	62550.01**	1.53338E-09	21.1977
		302.4	386.1			
		283.1	386.1			
1:5	24	284.1	385.1	46818.00**	2.73693E-09	21.1977
		284.1	386.1			
		290	386.1			
	48	291	385.1	40698.05**	3.62187E-09	21.1977
		291	386.1			
		244	386.1			
	96	244	385.1	90865.85**	7.26638E-10	21.1977
		243	386.1			
		201.5	386.1			
	144	201.5	385.1	305587.84**	6.42495E-11	21.1977
		201.5	386.1			
		201	386.1			
	192	200	385.1	154179.05**	2.52396E-10	21.1977
		201	386.1			
		195.4	386.1			
	240	195.4	385.1	326155.21**	5.64019E-11	21.1977
		195.4	386.1			

**Significance of difference is at the $P < 0.01$ level; *Significance of difference is at the $P < 0.05$ level.

Appendix 2

Significance of differences between Eh values measured *in situ* and in laboratory (Layer B_2).

Soil-water Ratio	Standing Time, hours	Source of Variation		F	*P*-value	F crit
		Suspension Eh	*In situ* Eh			
		412.1	494.8			
	0	411.1	494.8	30776.81**	6.333E-09	21.1977
		412.1	493.8			
		410	494.8			
	1	410	494.8	64211.56**	1.4551E-09	21.1977
		410	493.8			
		409.5	494.8			
	2	410.5	494.8	32232.61**	5.7739E-09	21.1977
		409.5	493.8			
		385.5	494.8			
	6	385.5	494.8	106863.61**	5.2537E-10	21.1977
		385.5	493.8			
		375	494.8			
	12	375	494.8	128450.56**	3.6363E-10	21.1977
		375	493.8			
		337.2	494.8			
1:1	24	338.2	494.8	105093.27**	5.4322E-10	21.1977
		337.1	493.8			
		346.6	494.8			
	48	347.6	494.8	97947.38**	6.2537E-10	21.1977
		346.6	493.8			
		316.4	494.8			
	96	317.4	494.8	142151.12**	2.969E-10	21.1977
		316.4	493.8			
		282	494.8			
	144	281	494.8	204416.18**	1.436E-10	21.1977
		281	493.8			
		268.4	494.8			
	192	268.4	494.8	459955.24**	2.836E-11	21.1977
		268.4	493.8			
		261	494.8			
	240	262	494.8	244580.18**	1.003E-10	21.1977
		261	493.8			

Continued

		376.3	494.8			
	0	376.3	494.8	125670.25**	3.799E-10	21.1977
		376.3	493.8			
		377	494.8			
	1	378	494.8	61390.08**	1.592E-09	21.1977
		378	493.8			
		374.7	494.8			
	2	373.7	494.8	65268.85**	1.408E-09	21.1977
		373.7	493.8			
		352.6	494.8			
	6	353.6	494.8	90142.58**	7.383E-10	21.1977
		352.6	493.8			
		349	494.8			
	12	350	494.8	94786.58**	6.678E-10	21.1977
		349	493.8			
		312.2	494.8			
1:2	24	312.1	494.8	296246.53**	6.837E-11	21.1977
		312.1	493.8			
		306.3	494.8			
	48	306.3	494.8	318660.25**	5.909E-11	21.1977
		306.3	493.8			
		278	494.8			
	96	278	494.8	421720.36**	3.374E-11	21.1977
		278	493.8			
		278.1	494.8			
	144	277.2	494.8	221885.34**	1.219E-10	21.1977
		277.1	493.8			
		261	494.8			
	192	262	494.8	244580.18**	1.003E-10	21.1977
		261	493.8			
		258	494.8			
	240	259	494.8	250915.28**	9.53E-11	21.1977
		258	493.8			

Continued

		346.8	494.8			
	0	346.7	494.8	194393.67**	1.588E-10	21.1977
		346.8	493.8			
		346.9	494.8			
	1	346.9	494.8	193955.21**	1.595E-10	21.1977
		347	493.8			
		345.3	494.8			
	2	345.3	494.8	200256.25**	1.496E-10	21.1977
		345.3	493.8			
		342.1	494.8			
	6	342.1	494.8	208940.41**	1.374E-10	21.1977
		342.1	493.8			
		317.3	494.8			
	12	317.3	494.8	279590.06**	7.675E-11	21.1977
		317.4	493.8			
		324.1	494.8			
1:3	24	324.1	494.8	261223.21**	8.793E-11	21.1977
		324.1	493.8			
		320.1	494.8			
	48	320	494.8	271131.57**	8.162E-11	21.1977
		320	493.8			
		281.5	494.8			
	96	281.5	494.8	404025.19**	3.676E-11	21.1977
		281.6	493.8			
		279.3	494.8			
	144	279.3	494.8	207690.13**	1.391E-10	21.1977
		280.3	493.8			
		243	494.8			
	192	243	494.8	569119.36**	1.852E-11	21.1977
		243	493.8			
		239.2	494.8			
	240	239.2	494.8	586449.64**	1.745E-11	21.1977
		239.2	493.8			

Continued

		338.1	494.8			
	0	338.1	494.8	220054.81**	1.239E-10	21.1977
		338.1	493.8			
		338	494.8			
	1	338	494.8	110638.08**	4.901E-10	21.1977
		337	493.8			
		340.3	494.8			
	2	340.3	494.8	213906.25**	1.311E-10	21.1977
		340.3	493.8			
		329.5	494.8			
	6	329.5	494.8	244926.01**	1E-10	21.1977
		329.5	493.8			
		305.1	494.8			
	12	305.2	494.8	319317.24**	5.884E-11	21.1977
		305.2	493.8			
		274.4	494.8			
1:4	24	274.4	494.8	319317.24**	5.884E-11	21.1977
		274.4	493.8			
		276.6	494.8			
	48	276.7	494.8	422704.51**	3.358E-11	21.1977
		276.7	493.8			
		247.2	494.8			
	96	247.2	494.8	550267.24**	1.982E-11	21.1977
		247.2	493.8			
		245.1	494.8			
	144	245.1	494.8	559653.61**	1.916E-11	21.1977
		245.1	493.8			
		216.3	494.8			
	192	216.3	494.8	696390.25**	1.237E-11	21.1977
		216.3	493.8			
		215.2	494.8			
	240	215.2	494.8	701908.84**	1.218E-11	21.1977
		215.2	493.8			

Continued

		336.7	494.8			
	0	336.7	494.8	224012.89**	1.196E-10	21.1977
		336.7	493.8			
		336.4	494.8			
	1	336.5	494.8	222451.49**	1.212E-10	21.1977
		336.5	493.8			
		338.1	494.8			
	2	338.1	494.8	220054.81**	1.239E-10	21.1977
		338.1	493.8			
		319.9	494.8			
	6	319.9	494.8	274261.69**	7.976E-11	21.1977
		319.9	493.8			
		307.7	494.8			
	12	307.7	494.8	313936.09**	6.088E-11	21.1977
		307.7	493.8			
		280.3	494.8			
1:5	24	280.3	494.8	205761.13**	1.417E-10	21.1977
		281.3	493.8			
		279.5	494.8			
	48	279.5	494.8	415896.01**	3.469E-11	21.1977
		279.5	493.8			
		238.8	494.8			
	96	238.8	494.8	588289.00**	1.734E-11	21.1977
		238.8	493.8			
		240.2	494.8			
	144	240.2	494.8	581863.84**	1.772E-11	21.1977
		240.2	493.8			
		207	494.8			
	192	207	494.8	743733.76**	1.085E-11	21.1977
		207	493.8			
		206.4	494.8			
	240	206.4	494.8	746841.64**	1.076E-11	21.1977
		206.4	493.8			

**Significance of difference is at the $P < 0.01$ level; *Significance of difference is at the $P < 0.05$ level.

Permissions

The contributors of this book come from diverse backgrounds, making this book a truly international effort. This book will bring forth new frontiers with its revolutionizing research information and detailed analysis of the nascent developments around the world.

We would like to thank all the contributing authors for lending their expertise to make the book truly unique. They have played a crucial role in the development of this book. Without their invaluable contributions this book wouldn't have been possible. They have made vital efforts to compile up to date information on the varied aspects of this subject to make this book a valuable addition to the collection of many professionals and students.

This book was conceptualized with the vision of imparting up-to-date information and advanced data in this field. To ensure the same, a matchless editorial board was set up. Every individual on the board went through rigorous rounds of assessment to prove their worth. After which they invested a large part of their time researching and compiling the most relevant data for our readers. Conferences and sessions were held from time to time between the editorial board and the contributing authors to present the data in the most comprehensible form. The editorial team has worked tirelessly to provide valuable and valid information to help people across the globe.

Every chapter published in this book has been scrutinized by our experts. Their significance has been extensively debated. The topics covered herein carry significant findings which will fuel the growth of the discipline. They may even be implemented as practical applications or may be referred to as a beginning point for another development. Chapters in this book were first published by Scientific Research Publishing Inc.; hereby published with permission under the Creative Commons Attribution License or equivalent.

The editorial board has been involved in producing this book since its inception. They have spent rigorous hours researching and exploring the diverse topics which have resulted in the successful publishing of this book. They have passed on their knowledge of decades through this book. To expedite this challenging task, the publisher supported the team at every step. A small team of assistant editors was also appointed to further simplify the editing procedure and attain best results for the readers.

Our editorial team has been hand-picked from every corner of the world. Their multi-ethnicity adds dynamic inputs to the discussions which result in innovative outcomes. These outcomes are then further discussed with the researchers and contributors who give their valuable feedback and opinion regarding the same. The feedback is then collaborated with the researches and they are edited in a comprehensive manner to aid the understanding of the subject.

Apart from the editorial board, the designing team has also invested a significant amount of their time in understanding the subject and creating the most relevant covers. They scrutinized every image to scout for the most suitable representation of the subject and create an appropriate cover for the book.

The publishing team has been involved in this book since its early stages. They were actively engaged in every process, be it collecting the data, connecting with the contributors or procuring relevant information. The team has been an ardent support to the editorial, designing and production team. Their endless efforts to recruit the best for this project, has resulted in the accomplishment of this book. They are a veteran in the field of academics and their pool of knowledge is as vast as their experience in printing. Their expertise and guidance has proved useful at every step. Their uncompromising quality standards have made this book an exceptional effort. Their encouragement from time to time has been an inspiration for everyone.

The publisher and the editorial board hope that this book will prove to be a valuable piece of knowledge for researchers, students, practitioners and scholars across the globe.

List of Contributors

Li Lv, Heng Xi and Guozhu Han
Department of Pharmacology, College of Pharmacy, Dalian Medical University, Dalian, China

Margarita Kruchinina, Mikhail Voevoda, Svetlana Kurilovich and Abdrey Gromov
Institute of Internal Medicine, SB RAMS, Novosibirsk, Russia

Sergey Peltek
Institute of Cytology and Genetics, SB RAS, Novosibirsk, Russia

Vladimir Kruchinin, Sergey Rykhlitsky and Vladimir Volodin
Rzhanov Institute of Semiconductor Physics, SB RAS, Novosibirsk, Russia

Vladimir Generalov
State Research Center of Virology and Biotechnology, Ministry of Public Health and Social Development of the Russian Federation, Koltsovo, Russia

Valery M. Eskov, Alexander A. Khadartsev, Valery V. Eskov and Olga E. Filatova
Surgut State University, Surgut, Russia

Sergei A. Aseyev
Institute of Spectroscopy of the Russian Academy of Sciences, Troitsk, Russia

Peter M. Weber
Department of Chemistry, Brown University, Providence, USA

Anatoli A. Ischenko
Moscow M. V. Lomonosov State University of Fine Chemical Technologies, Moscow, Russia

Andre Ricard and Jean-Philippe Sarrette
Paul Sabatier University, Toulouse, France

María de los Desamparados Escribano Otero and Mohamed Nour Kayali Sayadi
Mass Spectrometry Unit, Faculty of Chemistry, Complutense University of Madrid, Madrid, Spain

Luis María Polo Díez
Department of Analytical Chemistry, Faculty of Chemistry, Complutense University of Madrid, Madrid, Spain

Sameer Amereih
Palestine Technical University-Kadoorie (PTUK), Tulkarm, Palestine

Thomas Meisel and Wolfhard Wegscheider
General and Analytical Chemistry Department, University of Leoben, Leoben, Austria

Zaher Barghouthi
Institute of Chemistry, Karl-Franzens University, Graz, Austria

John Jairo Ríos Acevedo and Luis María Polo Díez
Department of Analytical Chemistry, Faculty of Chemistry, Complutense University of Madrid, Madrid, Spain

Mohamed Nour Kayali Sayadi
Mass Spectrometry Unit, Faculty of Chemistry, Complutense University of Madrid, Madrid, Spain

Reshu Chaudhary and Monika Datta
Department of Chemistry, University of Delhi, New Delhi, India

Edelvio de Barros Gomes
Laboratory of Flavor and Chromatographic Analysis, Federal University of Sergipe, São Cristóvão, Brazil
Instituto Federal de Educação, Ciência e Tecnologia da Bahia, Canela, Brazil

Suyare Araújo Ramalho, Nayjara Carvalho Gualberto, Rita de Cássia Mendonça de Miranda and Narendra Narain
Laboratory of Flavor and Chromatographic Analysis, Federal University of Sergipe, São Cristóvão, Brazil

Nisha Nigam
Department of Chemistry, R. D. National College, Bandra, Mumbai, India

Víctor García Gaytán, María de las Nieves Rodríguez Mendoza and Libia I. Trejo-Téllez
Postgrado en Edafología, Colegio de Postgraduados, Texcoco, México

Marcos Soto Hernández
Postgrado en Botánica, Colegio de Postgraduados, Texcoco, México

Martha E. Pedraza Santos
Facultad de Agrobiología "Presidente Juárez", Uruapan, México

Guadalupe Valdovinos Ponce
Postgrado en Fitosanidad, Colegio de Postgraduados, Texcoco, México

Dirk Lassner, Christine Sabine Siegismund, Jenny Stehr, Maria Rohde and Ulrich Michael Gross
Institute of Cardiac Diagnostics and Therapy, Berlin, Germany

Felicitas Escher and Uwe Kühl
Institute of Cardiac Diagnostics and Therapy, Berlin, Germany
Department of Cardiology and Pneumonology, Charité-University Hospital Berlin, Campus Benjamin Franklin, Berlin, Germany

Carsten Tschöpe and Heinz-Peter Schultheiss
Department of Cardiology and Pneumonology, Charité-University Hospital Berlin, Campus Benjamin Franklin, Berlin, Germany

Lylia Hamidatou, Hocine Slamene, Tarik Akhal, Samir Begaa, Mohammed Messaoudi and Zouranen Boussaad
Neutron Activation Analysis Department, Nuclear Research Centre of Birine, Djelfa, Algeria

Marco Iammarino and Aurelia Di Taranto
Istituto Zooprofilattico Sperimentale della Puglia e della Basilicata, Foggia, Italy

Alfons Penzkofer
Faculty of Physics, University of Regensburg, Regensburg, Germany

Daigo Iwahata, Kazuki Nakamura, Rie Yamada, Hiroshi Miyano and Naoyuki Yamada
Institute for Innovation, Ajinomoto Co., Inc., Kawasaki, Japan

Md. Nazmul Hasan and M. Akhtaruzzaman
Institute of Nutrition and Food Science, University of Dhaka, Dhaka, Bangladesh

Md. Zakir Sultan
Centre for Advanced Research in Sciences, University of Dhaka, Dhaka, Bangladesh

Hongwei Xie, Jianhua Zhang, Jinchuan Chen, Faqiang Zhang, Linbo Li, Jianming Qi and Yanyun Chu
Institute of Nuclear Physics and Chemistry, China Academy of Engineering Physic, Mianyang, China

Puneet Bajaj
Department of Medicine, Division of Pulmonary, Allergy, and Critical Care Medicine, Section of Allergy and Immunology, Milton S. Hershey Medical Center, The Pennsylvania State University, Hershey, USA

Faoud T. Ishmael
Department of Medicine, Division of Pulmonary, Allergy, and Critical Care Medicine, Section of Allergy and Immunology, Milton S. Hershey Medical Center, The Pennsylvania State University, Hershey, USA
Department of Biochemistry and Molecular Biology, College of Medicine, The Pennsylvania State University, Hershey, USA

Geoffrey K. Wu
Office of Generic Drugs, Food and Drug Administration, Rockville, USA

Abhay Gupta, Mansoor A. Khan and Patrick J. Faustino
Division of Product Quality Research, Office of Pharmaceutical Science, Food and Drug Administration, Silver Spring, USA

Ming Fa, Xuejun Fan, Zhiqiang An and Ningyan Zhang
Texas Therapeutics Institute, Houston, USA

Kathleen Hoch and William P. Dubinsky
Proteomics Center, Brown Foundation Institute of Molecular Medicine, The University of Texas Health Science Center at Houston, Houston, USA

Mu Liu
College of Forestry, Sichuan Agricultural University, Ya'an, China
Promotion Center of Forestry Technology, Forestry Bureau of Fushun County, Fushun, China

Chengde Luo and Bin Dai
College of Forestry, Sichuan Agricultural University, Ya'an, China

Xiaoyun Liu and Gang Wan
Promotion Center of Forestry Technology, Forestry Bureau of Fushun County, Fushun, China